AF361728

Building Energy Audits-Diagnosis and Retrofitting

Building Energy Audits-Diagnosis and Retrofitting

Editor

Constantinos A. Balaras

MDPI • Basel • Beijing • Wuhan • Barcelona • Belgrade • Manchester • Tokyo • Cluj • Tianjin

Editor
Constantinos A. Balaras
National Observatory of Athens
(NOA)
Greece

Editorial Office
MDPI
St. Alban-Anlage 66
4052 Basel, Switzerland

This is a reprint of articles from the Special Issue published online in the open access journal *Energies* (ISSN 1996-1073) (available at: https://www.mdpi.com/journal/energies/special_issues/Building_Energy_Audits-Diagnosis_Retrofitting).

For citation purposes, cite each article independently as indicated on the article page online and as indicated below:

LastName, A.A.; LastName, B.B.; LastName, C.C. Article Title. *Journal Name* **Year**, *Volume Number*, Page Range.

ISBN 978-3-03943-829-7 (Hbk)
ISBN 978-3-03943-830-3 (PDF)

Contents

About the Editor

Constantinos A. Balaras is a mechanical engineer, Research Director at the Institute for Environmental Research and Sustainable Development, in the National Observatory of Athens, Greece, a public research organization. He received a Ph.D. & a Masters in Mechanical Engineering from Georgia Tech., and a B.S.M.E. from Michigan Tech. He has participated in over 50 R&D and demonstration projects financed by the European Commission, national ministries and organizations, and the private sector. He has co-authored more than 240 papers in international journals, chapter contributions in books, and proceedings of international conferences. His research interests include high-performance buildings and sustainable cities, energy audits-diagnosis and renovations, energy baselines and benchmarks, environmental impact of buildings, embodied energy, indoor environmental quality, thermal simulations, and solar cooling. He is an ASME Fellow, ASHRAE Fellow, EUR ING, and a member of the Hellenic Technical Chamber as a professional engineer.

Preface to "Building Energy Audits-Diagnosis and Retrofitting"

Building energy audits are used to systematically collect and analyze relevant data for the energy use profile of a building or group of buildings, to identify, quantify, prioritize, or rank cost-effective energy conservation and efficiency measures. They are also employed for the sustainability assessment of buildings, neighborhoods, cities, and regions. A variety of methods and tools are available that can be used for building energy audits, surveys, diagnosis, inspections, and assessments. Depending on the level of detail and overall approach, they are complemented with data collected from non-destructive testing, measuring, and monitoring. The collected information can then be used with calibrated tools to accurately assess energy efficiency and conservation measures. Actual energy consumption can also be monitored to quantify and verify energy saving and use different approaches to close the gap with calculations. Data from energy audit case studies and lessons learned from the field, along with data from depositories of energy performance certificates can be used to derive benchmarks for the energy use intensity and the embodied energy for different building typologies. This accumulated knowledge can then be exploited to assess mid- and long-term renovation of building stocks. Progressively more attention is focused on larger scale monitoring and assessing sustainable development of the built environment at an urban scale.

This book is a collection of 12 papers that cover a variety of aspects. The first two papers focus on historic buildings and their unique challenges and opportunities. They may be excluded from minimum energy codes like EPBD in Europe, but preserving their heritage and optimizing their use while improving indoor working and living conditions are priorities in several countries. Cabeza-Prieto et al. investigate the thermal behavior of an early 20th-century building by performing in situ measurements of the external wall thermal conductance that reveal significant impacts as a result of moisture, and then use these insights to improve the accuracy of thermal simulations. Andreotti et al. audit a historic palace and perform a hygrothermal assessment of an internally insulated masonry wall with in-situ monitoring that complements simulations to analyze different means for improved thermal performance.

Non-residential buildings account for about a quarter of the total EU building stock and include various building types and different building sizes and energy characteristics. At the same time, there is limited available information on their construction characteristics, installed systems for different services and energy use for the non-residential building sector, and the various types and branches of activity. A total of six papers address non-residential buildings and also focus on some of the most common building typologies that include offices and schools. Droutsa et al. explore 30,000 energy performance certificates from energy audits of whole non-residential buildings in Greece to derive energy baselines for 30 different building uses and their main services. Cho et al. propose a method to identify the major variables that contribute to electric energy use in non-residential buildings using clustering in machine learning and demonstrate its application in 11 buildings in eight different regions of South Korea. Doulos et al. focus on office buildings and present a post-occupancy evaluation of occupant satisfaction and acceptance in relation to daylight utilization equipped with automatic controls and supporting in-situ measurements in three offices. Englund et al. perform an energy audit and in-situ measurements in a 1960s secondary school building in Sweden in order to validate a simulation model and assess various energy conservation measures. Mohelnikova et al. consider 18 representative schools around the Czech Republic that have been audited to analyze their building envelope characteristics and energy consumption, while in-situ thermal and daylight measurements and simulations of indoor conditions are analyzed from one building. Zyczynska et al. monitor the actual energy use before and after comprehensive thermo-modernization in nine Polish educational buildings to quantify actual energy saving and compare with predictions.

Residential buildings are responsible for over a quarter of the EU's total final energy use. As we progress beyond nearly zero energy buildings, EPBD provisions encourage the integration of other energy-related aspects, such as the embodied energy of the materials used during the life cycle of the buildings and demand response to provide residential consumers with control signals and/or financial incentives to adjust their consumption at strategic times. Dascalaki et al. derive key metrics and baselines for the embodied energy intensity in representative Hellenic houses, exploiting data from short energy audits in local manufacturing facilities to complement a lifecycle inventory database and LCA calculations. Park investigates how a human comfort-based control approach can be used with demand response programs for home energy management to promote household participation. Several case studies validate the proposed approach and the results document significant energy saving during the demand response period and improve occupant comfort.

Advancing from individual buildings to groups of buildings, neighborhoods, and cities, the European Clean Energy Package recognizes energy communities as a way to organize collective energy actions around open, democratic participation and governance and the provision of benefits for the members or the local community. Furthermore, buildings and the built environment in cities are seen as both a source of, and solution to, today's economic, environmental, and social challenges. However, the audit process to collect data and rate their sustainability levels is a demanding process given the complexity of the issues involved. Along these lines, two concluding papers address energy communities and urban sustainability audits and ratings. Torabi Moghadam et al. present a method for implementing consumer stock ownership plans in renewable energies sources to identify potential buildings, perform analysis and involve target groups, and present a case study with Italian sites. Balaras et al. present a new multicriteria method and tools for assessing the main sustainability issues of the built environment using a manageable number of key performance indicators, and demonstrated in nine pilots performed in six Mediterranean countries.

Constantinos A. Balaras
Editor

Article

Moisture Influence on the Thermal Operation of the Late 19th Century Brick Facade, in a Historic Building in the City of Zamora

Alejandro Cabeza-Prieto [1], María Soledad Camino-Olea [1,*], María Ascensión Rodríguez-Esteban [2], Alfredo Llorente-Álvarez [1] and María Paz Sáez Pérez [3]

[1] E.T.S. de Arquitectura, Universidad de Valladolid, avda Salamanca, 18, 47014 Valladolid, Spain; alejandro.cabeza@uva.es (A.C.-P.); llorente@arq.uva.es (A.L.-A.)
[2] Campus Viriato, Universidad de Salamanca, avda Cardenal Cisneros, 34, 49001 Zamora, Spain; mare@usal.es
[3] Campus Fuentenueva, Departamento de Construcciones Arquitectónicas, Universidad de Granada, calle Severo Ochoa, s/n; 18071 Granada, Spain; mpsaez@ugr.es
* Correspondence: mcamino@arq.uva.es

Received: 21 January 2020; Accepted: 7 March 2020; Published: 11 March 2020

Abstract: To improve the energy performance of restored cultural heritage buildings, it is necessary to know the real values of thermal conductivity of its envelope, mainly of the facades, and to study an intervention strategy that does not interfere with the preservation of their cultural and architectural values. The brick walls with which a large number of these buildings were constructed, usually absorb water, leading to their deterioration, whereas the heat transmission through them is much higher (than when they are dry). This aspect is often not taken into account when making interventions to improve the energy efficiency of these buildings, which makes them ineffective. This article presents the results of an investigation that analyzes thermal behavior buildings of the early 20th century in the city of Zamora, Spain. It has been concluded that avoiding moisture in brick walls not only prevents its deterioration but represents a significant energy saving, especially in buildings that have porous brick masonry walls and with significant thicknesses.

Keywords: brick 1; moisture 2; heat flow 3; energetic rehabilitation 4; non-destructive test 5

---

1. Introduction

There is an important number of buildings built in the last centuries, distributed all over the world, which due to their architectural value are worthy of special protection during the actions that could be carried out in them: restoration, rehabilitation, and even in works of conservation. Many of the Spanish cities are a characteristic example of this fact, since a high percentage of them have historical centers of special relevance, with a great wealth of architectural heritage.

In order to protect this heritage, public administration have been passing laws, regulations and special plans. The main goal is to regulate the actions that can be done in these heritage buildings and to avoid modifications or unfortunate changes that could deface their original configuration.

The research focuses on centennial buildings, which do not usually comply with current regulations regarding their thermal behavior. These standards limit energy consumption, as published in this century in the different European Directives [1]. This is a relevant issue since these buildings are the images of these cities, and in many cases, identity symbols, such as it happens to Zamora and many other small inner cities, in the Autonomous Community of Castilla y León (Spain).

Among the different typologies of cultural heritage, this research focuses on buildings with pressed brick facades, where ornamentation is based on the combination of multiple geometric designs in panels, openings, imposts, and cornices, as differentiating elements. However, this is not only in cultural heritage buildings, but also in those where an intervention to thermally insulate the exterior is not possible, in order to improve the thermal efficiency of the envelope [2,3]

When calculations and estimates of energy demand are made due to losses through this type of facade, it is usual to work with the theoretical values contained in the regulations or auxiliary documents, without making specific checks that corroborate its application. Brick is a porous material that can absorb a significant amount of water: from rain, from the ground or from air humidity, and this humidity can cause thermal characteristics to vary considerably, showing a large difference in the dry state to the wet [4–10]. For this reason, it is necessary to perform an analysis that allows knowing the influence of moisture on the thermal behavior of the walls [11].

This study presents the results of the research that has been carried out to evaluate the difference of the thermal behavior of these facades [12], from dry to saturated state. A thermal flow test was realized that determines the real thermal behavior [4] in a representative facade of this typology, concerning a residential building in the city of Zamora (Spain), built in 1894. Of which there is documentation of the original project. This building is called "Matilde Mechán's house", designed by the architect Segundo Viloria [13], has three floors, and is located in the historic center near the Plaza Mayor de Zamora. This facade has been selected because bricks similar to those used in its construction have been located, which come from the same tilery. This allows testing to determine the characteristics of the materials, without extracting samples from the facade, such as: with the water absorption, density and porosity [14], related to its hygrothermal functioning. With the information obtained in the previous tests, simulations can be carried out by means of which the thermal behavior of the facade with very different moisture contents can be analyzed. Information is needed to better define the actions aimed at the energy rehabilitation of these buildings.

2. Methods

To get to know the behavior of the facades, several actions have been carried out: characterize the materials, analyze the application of the regulations to the values of thermal conductivity obtained according to the water content, perform a thermal flow test "in situ" [4] on the facade, and to subsequently carry out the simulations with the values obtained in these tests. The first simulation aims to verify the similarity between the results obtained in the thermal flow test "in situ" and those shown in the simulation. Subsequently, other series of simulations of the operation of this facade are carried out, considering different moisture contents and assuming that energy rehabilitation would be carried out by attaching a leaf of insulating material through the interior of the facade.

2.1. Characterization of Materials

The two types of bricks that, in general, were used in the construction of the facades at that time have been analyzed: pressed bricks and ordinary bricks [15]. Several bricks from demolitions of buildings of the same era and nearby buildings were located: pressed brick of $261 \times 127 \times 53$ mm. and ordinary brick of $266 \times 126 \times 46$ mm. Four bricks of each type were chosen that were cut in half and ground to make their faces perfectly smooth and parallel. In total, for the tests, eight specimens were used. The morphology of the specimens was determined by the requirements of the test machine that analyzes the thermal conductivity value and the dimensions of the bricks. The two types of brick had different manufacturing processes, the pressed brick was made by pressing the clay between two molds, and the ordinary brick is manufactured by extrusion [16]. Eight mortar specimens were also made with sand and lime in a 1/3 ratio to perform the same tests as with brick specimens, of $158 \times 89 \times 40$ cm. The specimens were left in the laboratory environment at 20 °C and 50% to 55% humidity for 28 days before testing.

The bricks were manufactured in the Tejera de San Antonio, the first industrial tilery of Zamora (late 19th century). It was located near the clay deposit (El Perdigón, Zamora, Spain) and had a great production, so it supplied bricks to all the buildings in the capital, during the late 19th and early 20th centuries [16]. For this reason, it has been possible to find some pieces to carry out the tests. To test the characteristics of the mortar, eight specimens were manufactured with sand from the area and lime in a ratio of three to one.

The 24 specimens were tested to obtain the value of λ, thermal conductivity, for which a quick thermal conductivity meter (QTM 710/700 model, from KEM, KYOTO ELECTRONICS) was used; the the laboratory temperature was 22 °C ± 1 °C and had a relative humidity of 50% ± 5%. The specimens were tested in various moisture states: dry, semi-saturated and saturated, by immersion in cold water. The procedure of European Standard EN 772-21) [17] has been followed to determine the water content.

Other tests were also performed, regarding bulk density [18] and porosity by mercury intrusion porosimetry test, according to ASTM D4404-18 [19]. Through the same test, the average dimension of the pores size was calculated, based on the hypothesis that it could be a characteristic of the materials that could influence thermal conductivity.

In addition, cold water absorption (European Standard EN 772-21) [17] has been verified, calculating the water content in m^3/m^3 instead of percentage by weight, as indicated in the standard, because it has considered that, using these units, the value is more easily comparable in materials that have different densities.

Subsequently, the thermal conductivity coefficient values obtained, in the wet state, were compared with those obtained by applying the formula of EN ISO 10456 [20], which indicates that the conversion of thermal values from one set of conditions to another set of conditions is performed according to the following expression:

$$\lambda_2 = \lambda_1 \, F_m \, F_T \, F_a \tag{1}$$

where:

λ_n thermal conductivity of the material conditions n, W(m.K);
F_m moisture conversion factor;
F_T temperature conversión factor;
F_a ageing conversión factor.

It should be noted that the tests have been carried out on the specimens under the same temperature conditions, so the temperature conversion factor is 1. The ageing conversion factor is not known, so the value 1 will also be used. The moisture conversion factor F_m is calculated, in turn, by the expression:

$$F_m = e^{\,f_\psi \,\times(\Psi_2 - \Psi_1)} \tag{2}$$

where:

f_ψ design moisture coefficient % by volumen;
ψ_{design} design water content % by volumen (m^3/m^3).

Therefore, in the case of the study, the relationship between the coefficients of thermal conductivity of the specimens of the same type of brick, but with different water content, can be compared using the formula:

$$\lambda_2 = \lambda_1 \, e^{f_\psi x(\Psi_2 - \Psi_1)} \tag{3}$$

In Table 4 of the standard EN ISO 10456 [18], it is obtained that the value of the moisture coefficient for the baked clay $f_\Psi = 10$, with a density between 1000 y 2400 kg/m^3, and for a mortar with a density between 250 and 2000 kg/m^3, its value would be $f_\Psi = 4$, valid for a moisture content between 0 and 0.25 m^3/m^3.

To obtain the temperature conversion coefficient for different temperatures, using the figure in table A.111 of the same standard, for burnt clay and mortar of all densities, the value would be $f_T = 0.001$ 1/K. This is equivalent to that, for a temperature difference of 20 °C, the conversion factor would be $F_T = 1.020$.

Once the thermal conductivity values of the component materials have been obtained, the masonry conductivity of a $\lambda_{design,\,mas}$ masonry, more depending on the values of its components [21], in this case the brick $\lambda_{design,\,unit}$ and the mortar $\lambda_{design,\,mor}$, taking into account the percentage of the area in the elevation, is obtained by the following formula:

$$\lambda_{desing,\,mas} = a_{unit} \times \lambda_{design,unit} + a_{mor} \times \lambda_{design,mor} \tag{4}$$

If the formula is applied to the two types of brick wall from which the facade is formed, the thermal conductivity value is obtained for the two leafs that make up the facade. This is the result of calculating the percentage of raised area brick and mortar, being the one of 95% and 5% pressed brick and the ordinary brick 92% and 8%. This is possible since these facades are formed by blight leafs, one with pressed bricks and another with ordinary bricks, locked by keys of the same pieces. The pressed brick leaf is executed with 3-mm joints and that of ordinary brick with 8-mm joints [16].

Masonry specimens were also made to test the water content that this type of masonry can have in a dry and saturated state and the moisture that can be absorbed from the environment by the procedure followed for the materials, European Standard EN 772-21 [17]. The ordinary brick specimens formed by eight bricks were placed in four rows of 270×265 mm base.

2.2. "in situ" Thermal Flow Test

The facade wall on which the "in situ" test was carried out [4,22,23] has not been subject to interventions and is kept in very good condition after more than 120 years of life. It is composed of two brick walls tied with rigging of Spanish blights, using the brick pressed outside, and the ordinary brick inside, as already mentioned. In the report of the original project of 1894, it is specified that, on the first floor, the wall thickness is 60 cm, very approximate value to the measurement made "in situ", in which 58 cm have been obtained.

The building was selected by: (1) Being inhabited, so that there is a constant indoor temperature; (2) Having the brick masonry facade, without any other material; (3) Not having undergone restoration or rehabilitation, which may have modified the original composition of the brick wall; (4) Being in an environment with extreme temperatures, below 0 °C in winter, to work in the most unfavorable conditions, and with following permission to place the instruments to do the essay. The test is carried out in the blind area of the facade of which there is greater surface area and is not carried out in singular areas or thermal bridges because the methodology used is better adapted [24].

The in situ test on this facade wall was carried out for 13 days, according to the methodology of the International Standard ISO 9869-1 [25], specifically between 13 and 25 March 2019. It is of a north-facing facade with a slight deviation to the east. This orientation was chosen with the intention of preventing the direct incidence of the sun from having a significant influence and so that it could cause alterations of the flow and surface temperatures (sun-air temperature). Of the 13 days of testing, 11 have been selected discarding the first and the last, because they are not full days and because of the small interferences that could exist during the assembly and disassembly of the measuring equipment.

A novelty was introduced with respect to the test standard and it is that two thermal flow plates were placed, one inside, to measure the flow through the facade from the inside, and another outside, to know the flow in the face outside to better calibrate the simulation (Figure 1a). With those thermal flow plates values are captured at different times of the day, which are very different, since there are important changes in temperatures outside. In addition to the plates, four probes were placed, two inside and two outside the wall, to measure air temperatures and surface temperatures.

The location of the thermal flow plates in the wall is determined by two conditions, on the one hand, allowing the cables to connect both plates and the probes on both sides of the facade with the data logger, which collects the data. On the other, away from the thermal bridges, which, as you can see, were captured by images made with a thermal imager (Figure 1b).

The equipment used in carrying out the test are listed below (Figure 2):

- Heat flow meter AMR model FQAD19T of Ahlborn (250 mm $\times$ 250 mm $\times$ 1.5 mm) made of epoxy resin (Figure 2a) (accuracy 0.02% of the measured value) suitable for flat plaster finish, which was placed inside, and a heat flow meter AMR model FQAD18TSI of Ahlborn (120 mm × 120 mm × 3 mm) made of silicon (Figure 2b), which adapts well to the most irregular surface of the brick facade (accuracy 0.02% of the measured value of the measured value).

- Four thermocouples (Figure 2b) to measure the surface temperature: indoor and outdoor, and the temperature: outdoor and indoor (accuracy ± 0.05 °C ± 0.05% of the measured value).
- For data storage of heat fluxes and surface temperatures, two Data Logger units model Almemo 2590 of the Ahlborn trademark (Figure 2d) (accuracy 0.03%) have been used.
- FLIR ThermaCAM B29 brand thermal imager, with a thermal sensitivity of 0.1 °C, temperature measurement range from −20 °C to + 100 °C, spectrum range of 7.5 to 13 μm, and emissivity value of the brick 0.9.

(a) (b)

Figure 1. (a) Placement of the thermal flow plate and probes on the exterior face of the facade; (b) Thermographic image of the facade.

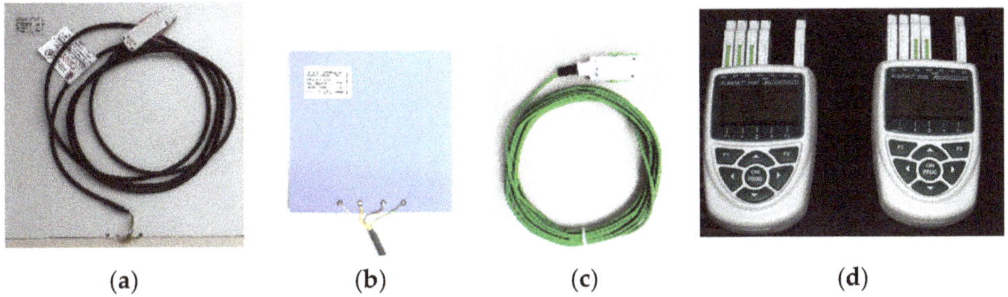

(a) (b) (c) (d)

Figure 2. (a) Thermal flow plate placed inside; (b) thermal flow plate placed outside; (c) thermocouple; (d) data logger.

With the surface temperature data and the value of the flow through the specimen, the thermal conductivity value can be calculated according to the procedure established in ISO 9869-1 [25–29] using the formula:

$$\Lambda = \frac{\sum_{j=1}^{n} q_j}{\sum_{j=1}^{n} \left(T_{sij} - T_{sej}\right)} \tag{5}$$

where:

Λ thermal conductance, en W/(m^2.K)

q density of heat flow rate = ϕ/A, en W/m^2;

T_{si} interior surface temperature, en °C;

T_{se} exterior surface temperature, en °C.

2.3. Energy Simulations Based on the Data Obtained in the Flow Test

With the data obtained in the "in situ" test, it is intended to validate the energy simulation tool to analyze, through simulations, situations in which the facade presents different water contents. For the simulation, a climate file is generated from the data collected by the outdoor air temperature probes. To establish the indoor temperature, an indoor HVAC (Heat Ventilation Air Condicioned) system is simulated that maintains a simulation surface temperature, practically equal to the surface temperature obtained by the probe during the "in situ" test. This is achieved by conditioning the operating temperature inside the space in the simulation at a ratio of 0.70 radiant. A wall similar in size to that of the "in situ" test is simulated, which is supposed to be the closing of a building that has a cubic shape, where the rest of the elements of the envelope are adiabatic. For the characteristics of the materials of which the wall to be simulated is composed, the values of the tests carried out on the materials are used taking into account the following simplifications: The wall is formed by brick leaf, as already described above, and the interior has a water content equal to that of the simulated wall in the laboratory, under similar conditions of water content to the air during the "in situ" test, and the outer leaf has a water content that is obtained from the value of the thermal conductance of the thermal flux test and of the values of the tests on the materials. That is, the water content of the outer leaf has been calculated starting from the rest of the values obtained in the tests. This simulation was carried out with the Energy plus version 8.3 program [30]. Subsequently, the results obtained have been compared with those released in situ. It is possible to know the degree of reliability of the simulation.

Once the simulation has been adjusted to the in situ test, and using the thermal conductivity values according to the water content obtained in the material characterization tests, it has been possible to perform other simulations that calculate the thermal flow of the facade when the rain has dampened by water or by which it rises by capillarity from the ground. The data obtained with these simulations are compared with those obtained in the actual test, and the differences that exist in the thermal flux transmission are analyzed:

- The first simulation has been carried out for an alleged case of rainwater that moistens the facade. According to document DB HS1 of the Technical Building Code (Spain) [31], a wall of the thickness of the brick stretcher is sufficient to prevent the passage of rainwater into the interior; for this reason, it has been simulated that only the leaf is moistened on the exterior and is done so gradually: 1/3 of the thickness is totally wetted 241 l/m^3 and has a λ = 1.96 W/(mK), another third of the facade is wetted at 66% 160 l/m^3 with λ = 1. 52 W/(m.K), and the remaining third is moistened to 33%, 80 l/m^3 with λ = 1.08 W/(m.K).

- The second simulation was carried out assuming that it is a boundary zone where the water rises by capillarity and it has been assumed that the two brick leafs were similarly moistened. For a water content of 0.015 m^3/m^3 (lthe facade is practically dry), λ $_{pressed\ brick}$ = 0.73 W/(m.K), λ $_{ordinary\ brick}$ = 0.74 W/(m.K), and λ $_{mortar}$ = 0.73 W/(m.K). For a water content of 0.077 m^3/m^3, λ $_{pressed\ brick}$ = 1.07 W/(m.K), λ $_{ordinary\ brick}$ = 1.07 W/(m.K), and λ $_{mortar}$ = 1.11 W/(m.K). For a water content of 0.125 m^3/m^3, λ $_{pressed\ brick}$ = 1.33 W/(m.K), λ $_{ordinary\ brick}$ = 1.31 W/(m.K), and λ $_{mortar}$ = 1.40 W/(m.K). For a water content of 0.165 m^3/m^3, λ $_{pressed\ brick}$ = 1.54 W/(m.K), λ $_{ordinary\ brick}$ = 1.52 W/(m.K), and λ $_{mortar}$ = 1.65 W/(m.K). For a water content of 0.210 m^3/m^3, λ $_{pressed\ brick}$ = 1.79 W/(m.K), λ $_{ordinary\ brick}$ = 1.79 W/(m.K), and λ $_{mortar}$ = 1.795 W/(m.K), and for a water content of 0.241 m^3/m^3, the values previously calculated. Then, other simulations have been carried out to relate the water content of this facade with the thermal flux that would pass through it, the value of the thermal conductance and the thickness of a leaf of insulating material that would be necessary, located inside, to maintain the dry values: flow and thermal conductance of the facade, depending on the water content.

3. Results

3.1. Materials Characterization

The value of the thermal conductivity of the specimens, calculated with the formulas of the trend lines, (Figure 3) are saturated more than three times that of the dried specimens [4]: for the pressed brick specimen $\lambda_{dry} = 0.65 W/(m.K)$ and λ_{241} l/m^3 = 1.96 W/(m.K), while for the ordinary brick specimen λ_{dry} = 0.67 W/(m.K) and λ_{243} l/m^3 = 1.93 W/(m.K), and for the mortar λ_{dry} = 0.64 W/(m.K) and λ_{231} l/m^3 = 2.05 W/(m.K). These values show the difference in thermal transmission between a dry and a saturated facade, especially in the type of facade being studied that has an important thickness.

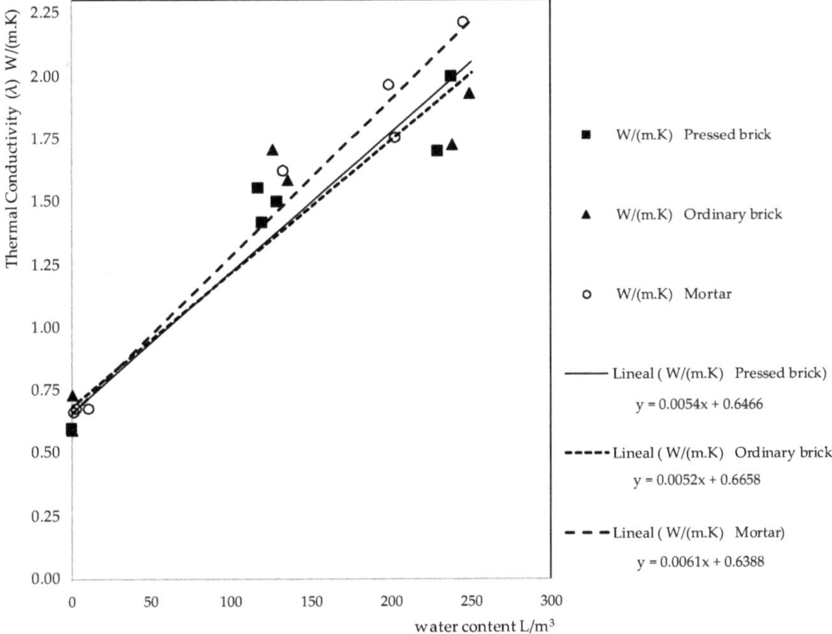

Figure 3. Thermal conductivity as function of water content of specimens.

In the tests of the materials, it can be seen that the two types of bricks that have been tested have similar values, probably because they are two solid bricks manufactured by the same ceramic in the same period of time. They were chosen by a high water absorption so the difference between the conductivity values between dry and wet brick would also be high (Figure 3). Having a high water absorption, the porosity is also high and the density is relatively low for solid bricks. Table 1 shows the density, porosity and average pore size results of the porosimetry test and the results of the water absorption test of the three materials.

In the absorption test of the ordinary brick and mortar specimen, values of 200 l/m^3 of difference were obtained between the dried specimen, after being taken out of the oven, and the saturated specimen. Once the sample was taken out of the oven for 2 weeks in the laboratory environment, similar to the interior of the house where the test was conducted, the test tube had absorbed 4 l/m^3.

Table 1. Material test values.

Material	Dimensionsmm	Apparent Density kg/m^3	Porosity %	Average Pore Diameter (μm)	Water Absorption m^3/m^3
pressed brick	$127 \times 97 \times 37$	1885	24.05	0.44	0.241
ordinary Brick	$113 \times 84 \times 30$	1877	24.32	5.64	0.243
mortar	$158 \times 89 \times 40$	1825	28.04	1.04	0.231

If the formulas of EN ISO 10456 [20] are applied for the conversion of thermal values from one set of conditions to another set of conditions, with different water content, by the formula (3) $\lambda_2 = \lambda_1 \, e^{f_\Psi x(\Psi_2 - \Psi_1)}$ based on the thermal conductivity values of the dry state materials obtained in the tests with those obtained using the coefficients of the standard, the following thermal conductivity values are obtained for saturated materials: for the pressed brick specimen λ_{241} l/m^3 = 7.23 W/(m.K), while for the ordinary brick specimen λ_{243} l/m^3 = 7.61 W/(m.K), and for the mortar λ_{231} l/m^3 = 1.61 W/(m.K). It can be seen that in a saturated state, the values markedly differ from those obtained in the tests.

In order to analyze more graphically what this increase in the value of thermal conductivity means, the thickness of a leaf of insulating material that would be necessary to be attached to the facade, on the inside, has been calculated to avoid losses due to the dampening of the facade, for an insulator whose characteristics are listed in Table 2.

Table 2. Characteristics of the insulating material used.

Material	Steam Resistivity (MNs/g)	Density kg/m^3	Specific Heat (J/kgK)	Termal Conductivity (W/mK)	Thermal Resistance (mk/W)
XPS-CO$_2$ Blowing	600	35	1400	0.034	24.41

The result of the calculations has been transferred to Figure 4, where the water content of the facade has been represented on the ordinate axis, the value of the thermal conductance of the facade enclosure studied is on the primary abscissa axis, and thickness of the insulating leaf necessary to maintain thermal insulation when the facade is wetted is on the secondary abscissa axis. To analyze this result, it should be taken into account that the thermal conductance of this facade is the same as a leaf of 17-mm insulating material.

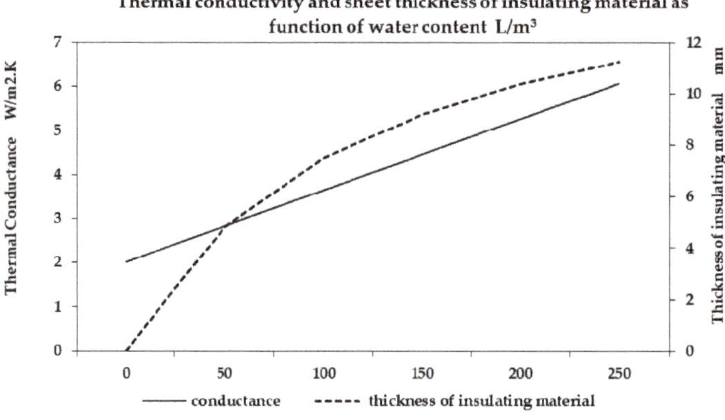

Figure 4. thermal conductance as a function of water content and leaf thickness of insulating material.

3.2. Thermal Flow Test

Figure 5 shows the data obtained after the "in situ" measurements of the facade. The thermal flux values measured by the plates: exterior and interior, interior and exterior surface temperatures, and interior and exterior temperatures. You can check the thermal wave offset between the external and internal flow characteristic of the enclosures with great thermal inertia. You can also see a minimum incidence of the sun on the facade at surface temperatures in the early hours of the day, (around 8:30 a.m.) when it is observed that the surface temperature increases slightly, maintaining the air temperature. This effect is of very little incidence in the thermal behavior of the enclosure, since it is a phenomenon of few minutes duration.

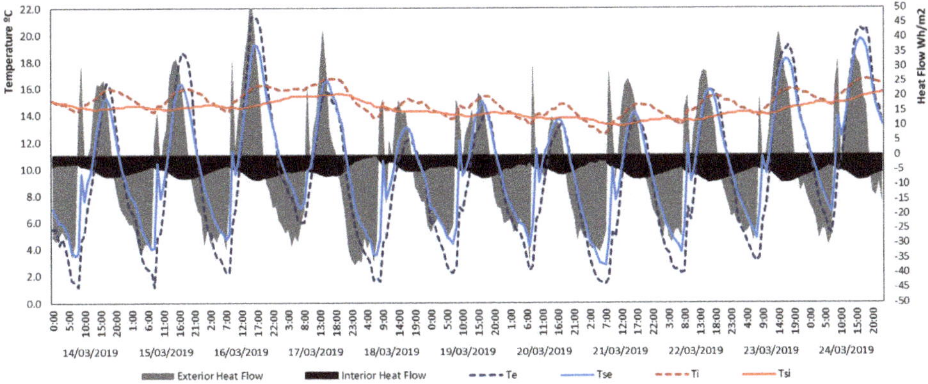

Figure 5. Results of the thermal flow test of the in situ facade.

Obtaining this data, allows us to know the real thermal behavior of the wall, if the wall accumulates thermal energy, and how it transmits it well inside or outside.

It also allows to evaluate the thermal demand due to the heat transmission through the external enclosures, data that are necessary to know for energy rehabilitation actions. With the values of the thermal flow of the plates located inside and outside, and the interior and exterior surface temperatures, applying the formula (5), the value of the thermal conductance of the facade can be estimated, as can be verified in the graph in Figure 6.

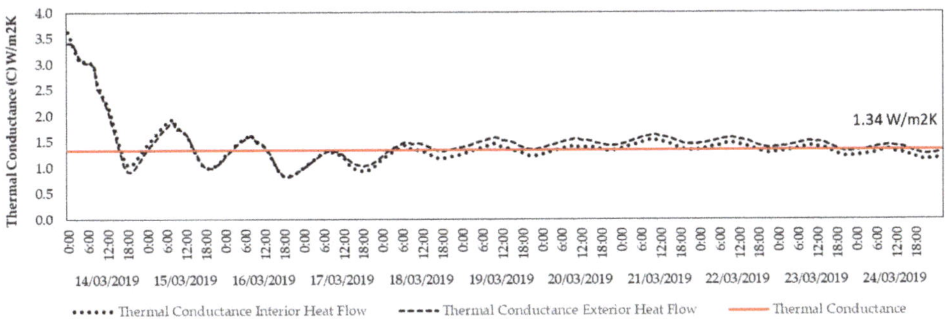

Figure 6. Thermal conductance W/m^2. K filed from the on-site test data with the formula (5).

For the simulations, this value of U has been used. The facade is formed by two leaves and the interior, as already indicated above, according to the CTE it would not be dampened with rainwater but with ambient humidity, which it has been calculated at $4 L/m^3$, so the value of the thermal conductivity of this leaf would be calculated with the formula (4) using the values of the tests performed on the materials of Figure 3. Once the conductance value of The facade has estimated the value of the conductivity of the outer leaf of pressed brick of 0.74 W/(m.K) and according to the results of the tests on the materials represented in Figure 2, the water content of this leaf is 15 L/m^3.

3.3. Simulations

For the simulations of the facades in the wet state, the thermal conductivity values that have resulted from the tests carried out in the present investigation have been used, according to the linear trend lines of Figure 3:

3.3.1. Simulations of the Behavior of the Facade in the Conditions of the Test "in situ"

The first simulation was carried out to verify if for the same conditions the results offered by the program are similar to those of the in situ test. To perform this simulation, it has been assumed that the water content of the wall is different in the pressed brick outer leaf. The outer leaf contains about 15 L/m^3 with a thermal conductivity λ = 0.74 W/(m.K), and the inner leaf is dry with some humidity due to the absorption of about 4 L/m^3 from the environment with a thermal conductivity λ = 0.66 W/(m.K). Figure 7 shows the simulation values.

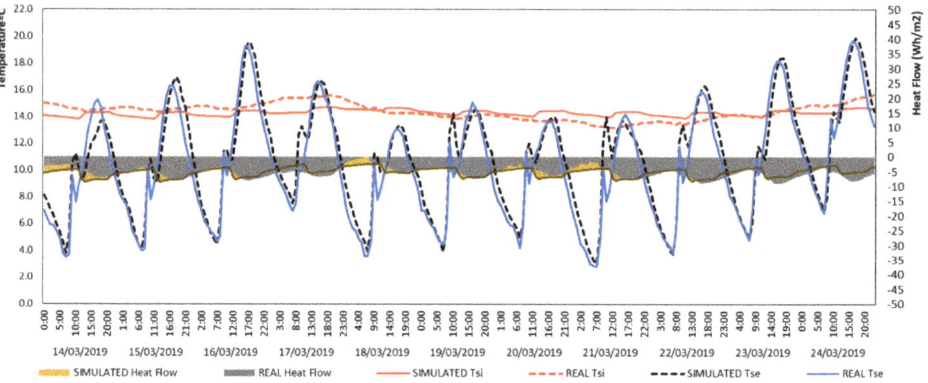

Figure 7. Simulated test results with the contour conditions of the in situ test.

3.3.2. Simulations of Facade Behavior in Other Humidity Conditions

Once the simulation has been validated by comparison with the results of the "in situ" test, other simulations have been carried out to analyze the thermal flow step when the facade is moistened with rainwater or with water rising from the ground. The comparative results of flow during the test and the flow in the two described situations of humidity by rainwater and humidity rising damp, have been transferred to Figure 8.

There is a difference in the flow between the simulation of the state of the facade when the test is performed and the simulation with the facade moistened by rainwater in the upper area. In the graph below, you can see the difference between the simulation of the in situ test and the simulation of the saturated facade due to the water that rises by capillarity.

As you can see, the flow through the facade is greater the more humid the facade, for the conditions of temperatures of the test in situ and during the test

In the case of the water-saturated facade, the increase is 75%, from 1430 to 2593 W/m^2; in the case of rainwater, the increase is approximately 10%, going to 1579 W/m^2.

The result of other simulations with the facade with different water content that allows to evaluate the difference in thermal flux, thermal conductivity, and the thickness of a leaf of insulating material that would be placed from the inside of facade to keep the thermal facade values dry, are represented in Figure 9.

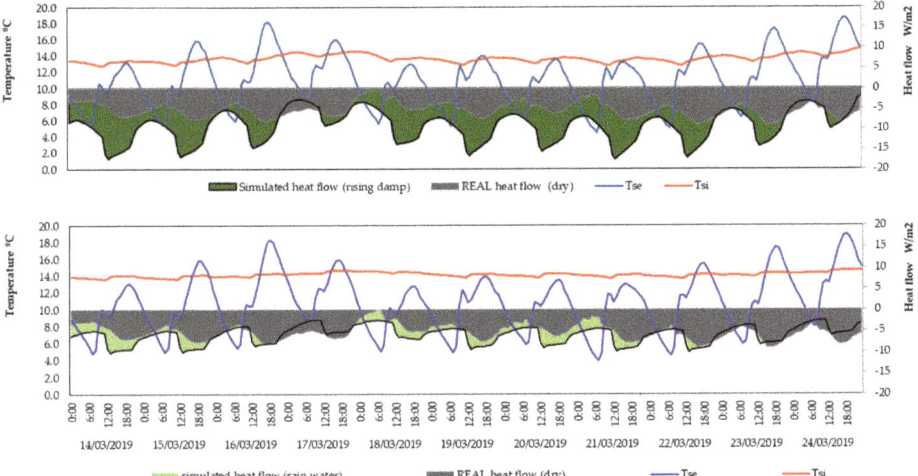

Figure 8. Top graph comparison between the thermal flow of the in situ test and the simulation in which the outer leaf is saturated with water. In the graphic above, the same type of comparison is made but in that case the wall is saturated with the water that ascends by capillary from the ground.

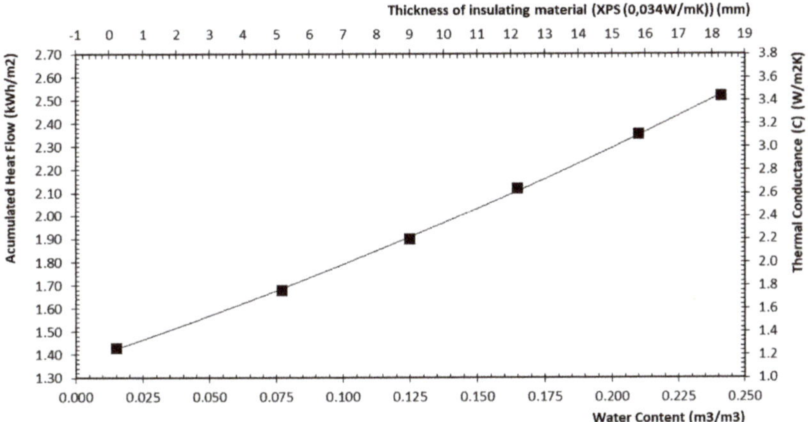

Figure 9. Comparative graph between the water content and the thickness of a leaf of insulating material to match the flow of a dry wall.

This graphic would be valid for buildings that had similar facades, under climatic conditions and internal temperature of the in situ test. As you can see in the graph, the heat flux that crosses the facade is almost double in humid conditions than in dry conditions, and the thickness of the leaf of insulating material would be 18 mm in order to equalize the greater heat loss due to the content of water from the facade.

4. Discussion

From the tests carried out on the materials, it can be verified that the coefficient of conductivity λ of the materials is higher in the wet state than in the dry state [32]. However, the results differ from those that could be obtained from the procedure of EN ISO 10456 [20], because the design moisture values coefficient % by volume are lower than the Standard EN 1745 [21,33].

In this case, for saturated bricks, the values are calculated much higher than those given by the test, while for mortar, the value resulting from the calculation with the standard is slightly lower than that of the test. In the results of the thermal conductivity tests, the two types of brick and mortar have similar dry and wet values; the values of apparent density and porosity are similar for both bricks, although for the mortar the density has a lower value and for the porosity has a higher value. The water absorption is also very similar for the two types of brick while the mortar absorbs a smaller amount of water than the bricks, although the porosity is higher, indicating that there is an important number of pores that are accessible for the water. The value of the average pore diameter is very different for the 0.44 pressed brick and for the 5.64 extruded brick, which seems to indicate that the density and porosity may be related to the thermal conductivity values, while the average pore diameter is not.

If the graph of the thermal flow test is analyzed in situ, it is observed that while the air and surface temperatures drop and rise gradually. Almost at the same time, the flow presents a lag with respect to the outside temperature. This phenomenon is due to the great thermal inertia of the facade factor that is important in historical buildings and is not usually taken into account in the calculations that are based on the thermal conductivity of the materials. Nor is the water content of the facades usually considered since the reference thermal conductivity values are in the dry state. Both in the flow test and in the tests carried out on the materials, it has been possible to verify the importance of these two factors to correctly evaluate the thermal behavior of the old, thick and massive walls, without air chambers.

If the simulations are analyzed, it can be verified that the higher the water content of this type of facade, the greater the value of the thermal flux that crosses them.

The tests established for buildings with modern facade systems are not useful for cultural heritage buildings. Without destructive tests, by means of the measurement of the thermal flow with plates of flow, one can have information to make simulations that allow to evaluate this thermal behavior.

5. Conclusions

This study concludes that the thermal conductance of the walls of facades composed of porous materials, such as brick, varies with the water content. The higher the water content, the greater the value of the thermal conductance. In the case of brick walls similar to the ones studied, the thermal conductance can be two to three times higher in a saturated wall than in a dry wall [29], because the water content they can absorb is high.

Estimates of energy loss through facade enclosures are made assuming materials are dry. This fact can lead to important mistakes when estimating energy consumption during interventions in buildings with enclosures similar to those studied in this work [5].

In relation to the suitability of the thermal flow test to know the moisture content of a masonry wall, it has been verified with the tests carried out that it is possible to estimate the water content of an enclosure. This can only be done if tests are carried out on test specimens of the constituent materials of the wall. This is very relevant information when you are going to intervene in a cultural heritage building. In the flow test performed, the results are similar if the flow plate is placed outside or inside.

Finally, it is important to highlight the value of the simulations, by offering with its application wide possibilities to energetically analyze the buildings with facades of important thicknesses and constituted by porous materials, before proposing an intervention. With these simulations the thermal inertia can be contemplated and the variation of the coefficients of thermal conductivity due to the water content of the materials.

Author Contributions: All authors have been involved in the preparation of the manuscript. All authors have read and agreed to the published version of the manuscript.

Funding: This research was funded by the Ministry of Economy and Competitiveness of the Spanish Government for the realization of the project "Proposal for evaluation of moisture that rises by capillarity in the brick walls heritage through non-destructive tests" BIA2015- 684449.

Acknowledgments: to Eduardo Torroja Institute of Construction Sciences of the Higher Council for Scientific Research (CSIC-IETcc) (Spain) for the tests carried out at said Institute whose results are incorporated into this article.

Conflicts of Interest: The authors declare no conflict of interest.

References

1. Directiva 2012/27/UE del Parlamento Europeo y del Consejo de 25 de octubre de 2012 relativa a la eficiencia energética, por la que se modifican las Directivas 2009/125/CE y 2010/30/UE, y por la que se derogan las Directivas 2004/8/CE y 2006/32/CE. Available online: https://eur-lex.europa.eu/legal-content/ES/TXT/PDF/?uri=CELEX:02012L0027-20180709&from=EN (accessed on 24 November 2019).
2. Calzolari, M. *Prestazione Energetica delle Architetture Storiche: Sfide e Soluzioni*; FrancoAngeli: Milan, Italy, 2016.
3. Belpoliti, V.; Bizzarri, G.; Boarin, P.; Calzolari, M.; Davoli, P. A parametric method to assess the energy performance of historical urban settlements. Evaluation of the current energy performance and simulation of retrofit strategies for an Italian case study. *J. Cult. Herit.* **2018**, *30*, 155–167. [CrossRef]
4. Litti, G.; Khoshdel, S.; Audenaert, A.; Braet, J. Hygrothermal performance evaluation of traditional brick masonry in historic buildings. *Energy Build.* **2015**, *105*, 393–411. [CrossRef]
5. Yu, S.; Cui, Y.; Shao, Y.; Han, F. Simulation Research on the Effect of Coupled Heat and Moisture Transfer on the Energy Consumption and Indoor Environment of Public Buildings. *Energies* **2019**, *12*, 141. [CrossRef]
6. Pavlík, Z.; Fiala, L.; Vejmelková, E.; Černý, R. Application of effective media theory for determination of thermal properties of hollow bricks as a function of moisture content. *Int. J. Thermophys.* **2013**, *34*, 894–908. [CrossRef]
7. Degiovanni, A.; Moyne, C. Conductivité thermique de matériaux poreux humides: Évaluation théorique et possibilité de mesure. *Int. J. Heat Mass Transf.* **1987**, *30*, 2225–2245. [CrossRef]
8. Azizi, S.; Moyne, C.; Degiovanni, A. Approche expérimentale et théorique de la conductivité thermique des milieux poreux humides—I. Expérimentation. *Int. J. Heat Mass Transf.* **1988**, *31*, 2305–2317. [CrossRef]
9. Bal, H.; Jannot, Y.; Gaye, S.; Demeurie, F. Measurement and modelisation of the thermal conductivity of a wet composite porous medium: Laterite based bricks with millet waste additive. *Constr. Build. Mater.* **2013**, *41*, 586–593. [CrossRef]
10. Vololonirina, O.; Coutand, M.; Perrin, B. Characterization of hygrothermal properties of wood-based products–Impact of moisture content and temperature. *Constr. Build. Mater.* **2014**, *63*, 223–233. [CrossRef]
11. Camino-Olea, M.S.; Cabeza-Prieto, A.; Llorente-Alvarez, A.; Sáez-Pérez, M.P.; Rodríguez-Esteban, M.A. Brick Walls of Buildings of the Historical Heritage. Comparative Analysis of the Thermal Conductivity in Dry and Saturated State. *IOP Conf. Ser. Mater. Sci. Eng.* **2019**, *471*, 082059. [CrossRef]
12. Pérez-Bella, J.M.; Dominguez-Hernandez, J.; Cano-Suñén, E.; del Coz-Diaz, J.J.; Rabanal, F.P.Á. A correction factor to approximate the design thermal conductivity of building materials. *Appl. Span. Facades. Energy Build.* **2015**, *88*, 153–164. [CrossRef]
13. Rodriguez-Esteban, M.A. La Arquitectura de Ladrillo y su Construcción en la Ciudad de Zamora (1888–1931). Ph.D. Thesis, Universidad de Valladolid, Valladolid, Spain, 2012.
14. Ten, J.G.; Orts, M.J.; Saburit, A.; Silva, G. Thermal conductivity of traditional ceramics. Part I: Influence of bulk density and firing temperature. *Ceram. Int.* **1951**, *1959*.
15. Rodríguez-Esteban, M.A. *La Arquitectura de Ladrillo y su Construcción en la Ciudad de Zamora (1888–1931)*; Instituto de Estudios Zamoranos Florián de Ocampo: Zamora, Spain, 2014; p. 179.
16. Rodríguez-Esteban, M.A.; Camino-Olea, M.S.; Sáez-Pérez, M.P. El ladrillo en la arquitectura ecléctica y modernista de la ciudad de Zamora: Análisis de los tipos, los aparejos y la ejecución de los muros. *Informes de la Construcción* **2014**, *66*, e035. Available online: http://informesdelaconstruccion.revistas.csic.es/index.php/informesdelaconstruccion/article/view/3488/3926 (accessed on 15 January 2020).

17. European Standard EN 772-21 Methods of test for masonry units. Part 21: Determination of water absorption of clay and calcium silicate masonry units by cold water absorption. 2011; AENOR.

18. Dondi, M.; Mazzanti, F.; Principi, P.; Raimondo, M.; Zanarini, G. Thermal conductivity of clay bricks. *J. Mater. Civ. Eng.* **2004**, *16*, 8–14. [CrossRef]

19. ASTM D4404-18. *Standard Test Method for Determination of Pore Volume and Pore Volume Distribution of Soil and Rock by Mercury Intrusion Porosimetry*; ASTM International: West Conshohocken, PA, USA, 2018.

20. European Standard EN ISO 10456. Building Materials and Products—Hygrothermal Properties—Tab-Ulated Design Values and Procedures for Determining Declared and DesignThermal Values. 2012.

21. European Standard EN 1745 Masonry and masonry products-Methods for determining thermal properties. 2013; AENOR.

22. Lucchi, E. Thermal transmittance of historical brick masonries: A comparison among standard data, analytical calculation procedures, and in situ heat flow meter measurements. *Energy Build.* **2017**, *134*, 171–184. [CrossRef]

23. Lucchi, E. Thermal transmittance of historical stone masonries: A comparison among standard, calculated and measured data. *Energy Build.* **2017**, *151*, 393–405. [CrossRef]

24. Baker, P. *U-Values and Traditional Buildings*; Historic Scotland Conservation Group: Glasgow, UK, 2011.

25. International Standard ISO 9869-1. Thermal Insulation—Building Elements –In-Situ Measurement of Thermal Resistance and Thermal Transmittance. Part 1. Heat Flow Meter Method. 2014.

26. Choi, D.S.; Ko, M.J. Analysis of Convergence Characteristics of Average Method Regulated by ISO 9869-1 for Evaluating In Situ Thermal Resistance and Thermal Transmittance of Opaque Exterior Walls. *Energies* **2019**, *12*, 1989. [CrossRef]

27. Choi, D.S.; Ko, M.J. Comparison of various analysis methods based on heat flowmeters and infrared thermography measurements for the evaluation of the in situ thermal transmittance of opaque exterior walls. *Energies* **2017**, *10*, 1019. [CrossRef]

28. Nardi, I.; Lucchi, E.; de Rubeis, T.; Ambrosini, D. Quantification of heat energy losses through the building envelope: A state-of-the-art analysis with critical and comprehensive review on infrared thermography. *Build. Environ.* **2018**, *146*, 190–205. [CrossRef]

29. Rotilio, M.; Cucchiella, F.; De Berardinis, P.; Stornelli, V. Thermal transmittance measurements of the historical masonries: Some case studies. *Energies* **2018**, *11*, 2987. [CrossRef]

30. EnergyPlus. U.S. Department of Energy (DOE). Available online: https://energyplus.net/ (accessed on 28 March 2016).

31. CTE DB-HS1. *Technical Building Code Basic Document-Basic Requirements Energy Saving*; Ministry of Development: Madrid, Spain, 2013; Available online: http://www.codigotecnico.org/index.php/menu-ahorro-energia (accessed on 15 October 2019).

32. Dell'Isola, M.; d'Ambrosio, F.R.; Giovinco, G.E.; Ianniello, E. Experimental analysis of thermal conductivity for building materials depending on moisture content. *Int. J. Thermophys.* **2013**, *33*, 1674–1685. [CrossRef]

33. Campanale, M.; Moro, L. Thermal conductivity of moist autoclaved aerated concrete: Experimental comparison between heat flow method (HFM) and transient plane source technique (TPS). *Transp. Porous Media* **2016**, *113*, 345–355. [CrossRef]

Article

Applied Research of the Hygrothermal Behaviour of an Internally Insulated Historic Wall without Vapour Barrier: In Situ Measurements and Dynamic Simulations

Mirco Andreotti [1], Dario Bottino-Leone [2], Marta Calzolari [3], Pietromaria Davoli [4], Luisa Dias Pereira [4,*], Elena Lucchi [2] and Alexandra Troi [2]

[1]   Istituto Nazionale di Fisica Nucleare, Sezione di Ferrara, 44122 Ferrara, Italy; mirco.andreotti@fe.infn.it
[2]   Eurac Research, Institute for Renewable Energy, 39100 Bolzano, Italy; dario.bottino@eurac.edu (D.B.-L.); elena.lucchi@eurac.edu (E.L.); alexandra.troi@eurac.edu (A.T.)
[3]   Department of Engineering and Architecture, University of Parma, 43124 Parma, Italy; marta.calzolari@unipr.it
[4]   Architettura Energia Research Centre, Department of Architecture, University of Ferrara, 44121 Ferrara, Italy; pietromaria.davoli@unife.it
*   Correspondence: dsplmr@unife.it; Tel.: +39-0532-293631

Received: 12 May 2020; Accepted: 17 June 2020; Published: 1 July 2020

Abstract: The hygrothermal behaviour of an internally insulated historic wall is still hard to predict, mainly because the physical characteristics of the materials composing the historic wall are unknown. In this study, the hygrothermal assessment of an internally thermal insulated masonry wall of an historic palace located in Ferrara, in Italy, is shown. In situ non-destructive monitoring method is combined with a hygrothermal simulation tool, aiming to better analyse and discuss future refurbishment scenarios. In this context, the original U-value of the wall (not refurbished) is decreased from 1.44 W/m^{2}K to 0.26 W/m^{2}K (10 cm stone wool). Under the site specific conditions of this wall, not reached by the sun or rain, it was verified that even in the absence of vapour barrier, no frost damage is likely to occur and the condensation risk is very limited. Authors proposed further discussion based on simulation. The results showed that the introduction of a second gypsum board to the studied technology compensated such absence, while the reduction of the insulation material thickness provides a reduction of RH peaks in the interstitial area by 1%; this second solution proved to be more efficient, providing a 3% RH reduction and the avoidance of further thermal losses.

Keywords: HeLLo; energy retrofit; non-destructive test; in situ; hygrothermal measurement; dynamic conditions; hygrothermal simulation; historic wall

---

1. Introduction

One of the most efficient ways to promote historic buildings (HB) sustainability is keeping them in use. Contributing to the life extent of HB, even those of heritage value, necessarily requires conservation improvements. In other words, safeguarding cultural heritage for future generations means trying to support the quality implementation of energy efficiency measures, needed to mitigate climate change [1] and to keep the buildings used. This means that, at this moment, the conservation aspect of HB can no longer be dissociated from its energy refurbishment. Nonetheless, the adaptation to such changes, e.g., adaptive re-use [2], retrofitting [3] and/or energy efficiency improvements actions [4], can also bring some risks. These risks are particularly likely to occur when dealing with thermal insulation of HB with patrimonial or heritage value. Here, adding interior thermal insulation layers to a façade is often the solely option, because the external insulation is frequently not suitable for preserving its aesthetical and cultural values. This action inevitably introduces changes to the hygrothermal behaviour (temperature and moisture conditions) of the historic walls, which can lead to interstitial condensation, frost damage or mould growth [5]. In other situations, the presence of water in different forms can also cause the reduction of the thermal performance of sub-components of HB's envelope, as pointed in [6]. Furthermore, they may lead to structural deterioration, saline efflorescence, or aesthetical decay.

Until a few years ago, most research on energy retrofit strategies and guidelines (materials, installation typologies [7,8]), focused on software simulations [9] and lab tests ([10,11]). Only recently have in situ measurements started to be performed. This delay is due to several aspects, for example: (i) long time and high costs of the research [12]; (ii) intrinsic uncertainties of implementation of the field campaigns; (iii) required invasive methods not compatible with HB protection guidelines; and (iv) field limitations because the results are often specific to each case-study. Despite these aspects, as pointed in [13] (p. 367) as future research directions, "in situ methods in historical buildings" are emerging and starting to take further steps towards the complex hygrothermal analysis of the internal insulation of HB as retrofit actions.

Many of these studies rose from the experience of European projects, such as: 3ENCULT [14], Co2olBricks [15], EFFESUS, [16]. The 3ENCULT project, which aimed "to bridge the gap between cultural heritage and climate protection, (…) communicating productively to find the right solution for a particular building" [17] (p. 10), contributed to the development of capillary active materials for the internal insulation. In [18], an output of the Co2olBricksproject, some of the problems concerning internal thermal insulation are addressed, alike moisture in brick wall construction. Authors tested five insulation materials and proposed an optimal selection, using the TOPSIS method with grey numbers. The EFFESUS project developed and tested two innovative insulation materials for the internal insulation of HB (blown-in aerogel blanket and insulating mortar), balancing their thermal performance and heritage conservation, in terms of reversibility, aesthetical impact, and material compatibility [19,20].

Several studies verified a significant knowledge gap between the measured thermal performance and the simulated behaviour, especially for historic walls ([21,22]), as in many situations, the wall composition is unknown [23], and the estimations are grounded on assumptions [24]. Hygrothermal simulation, more recently implemented, has also been contributing to bridge that gap [25]. Nonetheless, wrong estimations or excessive simplifications may have a severe impact on their thermal behaviour assessment, as well as on retrofit interventions [9]. Thus, field studies become mandatory to find the most suitable internal insulation solution, with regards to both the thermal and hygrothermal performance of HB. Many of these studies focus solely on the thermal performance of the walls, as the work developed by Bienvenido-Huertas et al. on the improvement of in-situ assessment of thermal transmittance (U-value) of historic walls [26], or the study of [27] on the energy efficiency proposal of a historical building in Naples, still neglect the undesirable effect that the presence of moisture in the wall structure might have. As pointed out in [28] (p. 117), "temperature and moisture conditions strongly influence thermal conductance of some materials", besides, it is also recognized that "condensed water increases the effective thermal conductivity of building materials" [29] (p. 1674).

In [30], Litti et al. gave a step forward proposing an indirect non-invasive monitoring procedure for the onsite evaluation of the thermal performance of traditional buildings' masonries, also quantifying the alteration due to the moisture distribution variation. More recently, the hygrothermal complexity of this matter has been fully embraced. In [31], authors studied the hygrothermal performance of four thermal insulation materials for an internally insulated brick wall in a cold climate. In [32], in its turn, in a deep study conducted in Denmark, authors presented long term in situ measurements of four cases (from 1877–1932) of internally insulated historic solid masonry walls, monitored at critical points, also creating numerical models which were validated against measurements. In this case, two different insulation systems with different insulation thicknesses were studied [32].

In this context, the paper presents part of the results of the HeLLo project—Heritage Energy Living Lab Onsite [33], which addresses one of the most relevant problems of HB energy refurbishment: the hygrothermal behaviour of internally insulated historic walls. In this project, to analyse this issue, in situ monitoring campaigns were performed. This study aims to assess thermal insulation technologies (including insulation material and installation system), among the most widespread in the market and used by professionals for new or existing buildings, to verify the hygrothermal behaviour when applied to HB. The final goal is to enhance the awareness of all the actors involved in the energy performance improvement of the HB envelope (designers, owners, heritage authority's members, and companies), for a conscious management of the entire process.

Within the paper, one of the studied thermal insulation technologies is deeply analysed. Initially, the case study is presented along with the methodology followed to develop the research (Section 2). In Section 3, the monitoring campaign is unveiled, along with the data acquisition system. Data collection was then used to validate 2D hygrothermal simulations performed in Delphin 6.0.20® (Section 4) [34], further discussed in Section 5. All these steps have thoroughly contributed to enrich the knowledge and safety levels of internal insulation of a historic brick wall, regarding the moisture performance, of which the main conclusions are shown in Section 6.

2. Case Study and Methods

2.1. Case Study Presentation

A real in situ laboratory was settled to assess and analyse the hygrothermal performance of different insulation thermal technologies applied to historic masonry walls [33]. The in situ experiment (as in [32]) was established in a historic palace located in Ferrara (Italy): Palazzo Tassoni Estense, a 15th century listed building part of an UNESCO site [35] (UNESCO—United Nations Educational, Scientific and Cultural Organization). This palace has been the subject of several studies, some of which have led to an architecture intervention of restoration [36].

The room selected to perform the study is located on the ground floor of a not-yet refurbished and naturally ventilated part of the palace (Figure 1). Due to the reminiscences of highly probable presence of two chimneys on two of the outer walls of this room (Figure 2), it was decided to undertake the study on the remaining wall, NE oriented, under a porch (Figure 3). "Though this situation does not correspond to a *'worst-case scenario'*" [37] (p. 2), authors recognize the limitation of their study. The results obtained are limited to this experiment condition: "rainwater might not reach directly the wall, but neither does the sun, i.e., both the capacities of wetting/absorbing and drying are limited" [37] (p. 2).

As this room is 700 m^3, aiming to minimize energy consumption and the impacts on the historic room, two in situ metering hot boxes adapted for HB were constructed, aiming to improve the overall experiment sustainability [38], i.e., two small rooms with controlled indoor hygrothermal conditions were created, inside a big, unoccupied and environmentally uncontrolled room.

Besides the two boxes construction, settled to assure a temperature difference (ΔT), between indoor and outdoor environment, the thermal insulation technologies were also installed according to the best practices; all of them were installed by technicians of the construction sector appointed directly by the companies (see also Section 2.2 for further details).

2.2. Technical Worktable and Selection of the Insulation Technologies

As part of the actions of the HeLLo project, a technical worktable with the national conservation authorities and material's companies was established [39], in order to assess the most suitable technological solutions to balance the needs of all the actors involved in the action: scientific aims, conservation aspects and building market's best practices. As previously outlined, the goal of the research is to test some insulation technologies, commonly widespread in the market for new or recent buildings, but suitable to be used in HB. All pros and cons for the energy retrofit have been considered and discussed. Three main criteria are considered, according to [40]:

- Conservation aspects, referred to the elements to preserve (i.e., decorations, finishing), to the aesthetical aspect of the finishing (e.g., proportions, materials, colours, textures), and to the reversibility (i.e., fastening system, and assembly and installation method);
- Energy efficiency referred to the final thickness, laying of the materials and U-value which justify the intervention—balance between thermal performance improvement and conservation aspects);
- Hygrothermal aspects, referred to the use of the vapour barrier. One of the innovative aspects of the HeLLo project is to verify the use of common insulation technologies without vapour barrier, differently from market suggestions for intervention in existing or new buildings, just to "stress" the performance of the tested stratigraphy and to keep the original vapour transport (i.e., low vapour resistance or vapour open materials), typical of HB materials [41]. This option enables summer drying potential of the historic envelope or the potential existing humidity in the wall (e.g., rising damp).

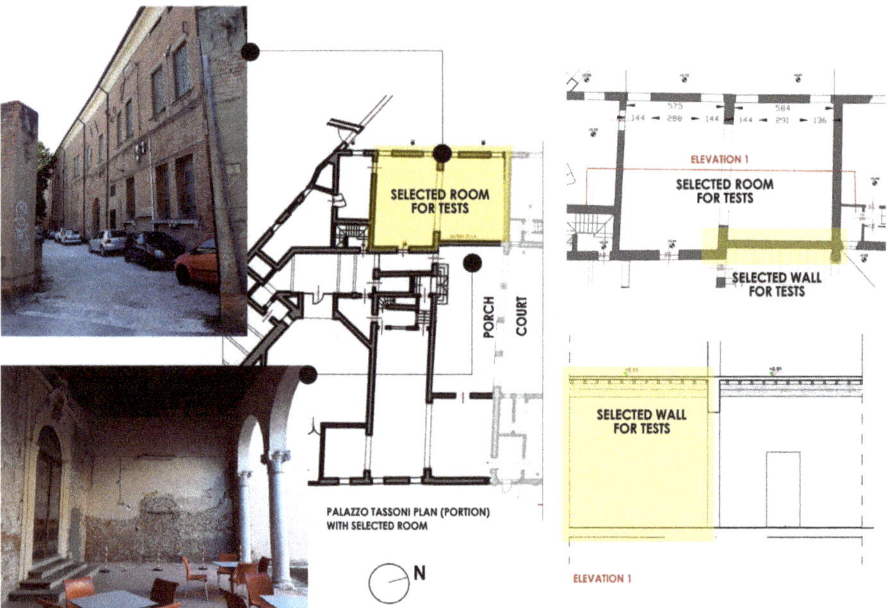

Figure 1. Palazzo Tassoni Estense: ground floor plan with the identification of the room where the experiment is carried out. Views of the external walls of the selected room: the one on the top left represents the front view from via Cammello, the one on the bottom left is a view of the wall under the porch (selected to be studied). The yellow shape underlines the selected room and the corresponding wall selected for the tests.

Figure 2. Internal views of the room. The picture 3 represents the wall selected for the tests.

Figure 3. External view of the studied wall (internal courtyard under the porch) [6].

The test aims to analyse the hygrothermal performance of three thermal insulation technologies. According to the discussion of the technical worktable, the following materials were chosen.

1. Stone wool boards: 40 + 60 mm thickness panels supported by their own steel frame and finished with gypsum boards (12.5 mm).
2. Cork boards: 50 mm thickness panels supported by their own timber structure, punctually fixed to the historic wall (very few anchor points) and finishing provided by gypsum fibre boards (12.5 mm);
3. Calcium silicate panels: 100 mm thickness panels glued, thanks to a mortar adhesive 8 mm thick, to the historic wall and given a 10 mm finishing mortar layer;

The paper presents the research activity related with the first described solution made by the use of the stone wool panels, widespread on the market even if herein tested without the vapour barrier commonly used when the stone wool panels are internally applied. Four different possible stratigraphies with stone wool were analysed during the technical worktable, with the aim to find the most suitable solution, both for the tests and to give interesting design directions for professionals that would like to use this material in future restoration and energy retrofit interventions. The analysed solution are: (i) use of a system composed by a gypsum board and/or gypsum fibre board jointed to a fibrous stone wool insulation panel, fastened to the wall by an adhesive mortar glue; (ii) stone wool insulation panel interposed between a metal frame (C-shape steel vertical profiles fixed to the wall thanks to U-shape steel adjustable brackets) with a gypsum board finish; (iii) stone wool insulation panel inserted between a metal frame (C-shape steel vertical profiles directly fixed to the wall) with a gypsum board finish; (iv) stone wool insulation panel interposed between a metal frame (C-shape vertical elements fixed to the floor) with a gypsum board finish. With respect to the previous solution, an additional insulation layer is inserted between the historic wall and the metal frame.

The first solution has been considered, by the heritage authorities, less suitable for the application on an historic wall, because of the necessity of the not reversible glue fastening system. For the same reason, the second solution has also been discarded due to the presence of the brackets, which requires an anchoring point piece by piece, affecting the conservation of the wall. The third solution is less intrusive for the historic walls, but the last option has been selected because it permits one to minimize the number of anchoring points between the wall and the dual system. Additionally, the additional insulation layer, compared to the third solution, allows solving the thermal bridge caused by the presence of the metal frame. All these aspects make this last option completely reversible and respectful for the historic wall.

Finally, the insulation stratigraphy selected by the technical worktable is composed by a first insulation panel 40 mm thick, a second insulation layer 60 mm thick, interposed between the metal frame, and the gypsum board finishing (12.5 mm).

Figure 4 shows the phase's sequence of the installation of the selected insulation system: in (a), the installation of the C metal structure outdistanced from the wall to minimize the number of anchor points on the historic wall; in (b), the insertion of the additional insulation layer between the metal frame and the wall; finally, in (c), the interposition of the second stone wool set of panels between the vertical steel elements, then covered by the gypsum board finishing.

Concerning the original historic wall, the historical analysis and the verification of the analogies with the literature allowed identifying the geometric and dimensional characteristics of the bricks: they are the "Bolognese" type (28 × 14 × 6 cm), as commonly used in contemporary architecture in the same geographical area. The non-invasive survey through the cracks currently present in the wall made possible the identification of the dimensions of the joints, which, even if very variable, have an average thickness of 2 cm. The overall thickness of the wall is 32 cm, including the internal and external plaster.

The calculated U-value of the not refurbished wall is 1.44 W/m^2K (estimated in steady state conditions), with the installed insulation system decreasing to 0.26 W/m^2K. The values used to calculate the U-value were the same of the hygrothermal simulation, and these are presented in Section 4.2.

Figure 4. Installation of the selected insulation system.

2.3. Monitoring System: New Metering Box and Hygrothermal Control Devices

As pointed out earlier, inside the 700 m^3 room, two smaller volumes (25 m^3/each) were constructed. In each volume (Figure 5), a standard indoor environment was set-up [indoor air temperature ((T_a) ≈ 20 °C and relative humidity (RH) ≈ 55%), according to the main international guidelines as EN ISO 7730 [42], EN ISO 13788 [43] or ISO 17772-1 [44]. Besides its special features (0.10 m high density stone wool insulation material, lined with a vapour barrier on the inside), each box was provided of a 2000 W heating convector (with 3 power levels), regulated by control system; two ultrasonic humidifiers (argo HYDRO digit, Argoclima S.p.A., Alfianello, Italy), 30 W each, self-regulated, which permitted controlling the indoor comfort parameters, Figure 6.

Figure 5. Internal view of the room, with the two boxes tied up to the studied wall.

Figure 6. Hygrothermal control devices.

The newly developed metering boxes (construction technology fully described in [38]) were built from a modular timber structure and provided with wheels, so that they can be (dis)assembled, repositioned and/or hereafter used in other spaces/experiments (Figure 5). Besides minimizing the impacts of the experiment on the room's surrounding walls, these also permitted improving the experiment energy efficiency. The individual box dimensions (2.50 × 2.50 × 4.01 m gross) allow vertical heat stratification and were determined to permit the study of up to two insulation systems in parallel.

3. Monitoring System and Period Campaign

To perform the in situ monitoring campaigns, authors used a non-commercial setup: they have developed a specifically built-up sensing technology method for the hygrothermal assessment of historic walls [45], later upgraded and tuned [38] to fit the requirements of the HeLLo project [33]. In brief, a low-cost and conservation compatible technology based on temperature (T) and relative humidity (RH) combined sensors (Telaire T9602, Amphenol Thermometrics, Inc. St. Marys, PA, USA) was used to measure the T-RH parameters of the environmental conditions (Figure 7a)—outdoor, inside the room and inside the box (herein considered the indoor environment), and in each point of the stratigraphy of the technology (Figure 7b). These sensors are connected and managed by a data acquisition system based on a master slave configuration [38]. Sensors have a typical temperature accuracy of ±0.5 °C, if T varies between 20–40 °C, or up to ±1 °C if T varies between 0–20 °C. Relative humidity instead has an accuracy of ±2%, when 20 ≤ RH (%) ≤ 80, decreasing up to ±4% if RH varies between 0–20% or 80–100% [45].

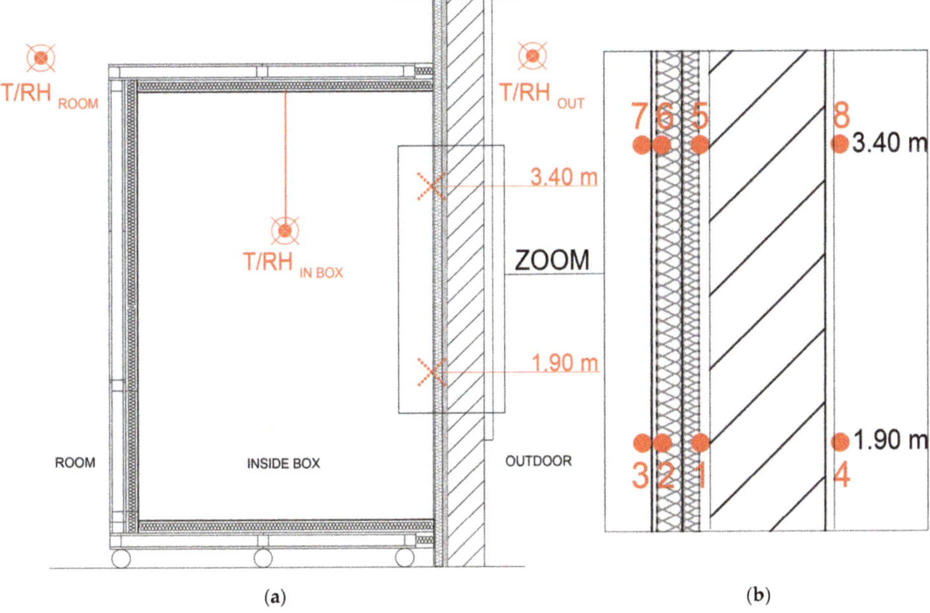

Figure 7. Scheme (vertical section) of the position of the monitoring sensors: (**a**) environmental conditions (room, inside box, outdoor); (**b**) wall stratigraphy.

Aiming to limit the influence of potential water capillarity absorption at the basis of the wall on the final results, sensors were strategically placed at a minimum height of 1.90 m from the floor. A second level of measuring points was added at 3.40 m (Figure 7). Although data were registered in different points of the stratigraphy, in this study, only the most probable condensation point—sensors 1 and 5, Figure 7b, is deeply analysed.

The monitoring period, between 11 December 2019 and 11 March 2020 is fully presented and described in Section 5.

Data were logged every 1 min. The results presented in Section 5 correspond to 10 min averages. The monitoring campaign, still on-going, begun during the 2019–2020 Ferrara's heating season (the hygrothermal control devices, Figure 6, were started on 23 October). In the current paper a long-term period of 3 months is shown, corresponding to the period between 11 December 2019 and 11 March 2020.

4. Hygrothermal Simulation Set-Up

4.1. Boundary Conditions

All the simulations are performed in accordance with EN 15026 [46], so they include the following physical phenomena: heat storage, heat conduction, moisture storage, liquid water convective transport and vapor diffusion. Simulation are performed using the software Delphin 6.0.20 [34], developed at the Technical University of Dresden (TUD) by the Building Climatology Department. The following section presents hygrothermal simulations, starting from the monitored internal and external climatic dataset (Section 4.1), to the geometric modeling and to the hygrothermal proprieties of the selected materials (Section 4.2).

The used climatic dataset was monitored in the framework of the mentioned research project HeLLo [33] and included hourly data of T and RH of the surrounding air, both for external and internal environment. Rain and solar action were not considered, as the studied wall is located under a porch of the courtyard of the building, so it is not exposed. Figure 8 shows the in situ monitored hourly data of T and RH for the internal and external climate, related to the period between 11 December 2019 and 11 March 2020, used also for the model validation.

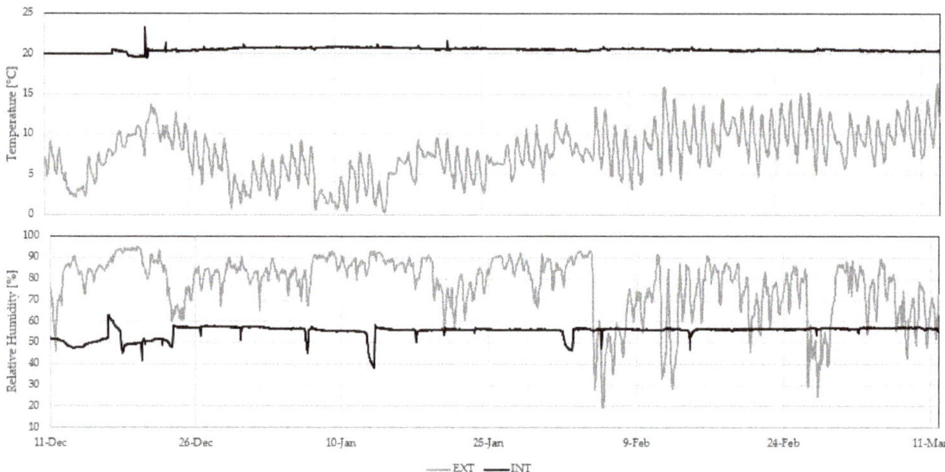

Figure 8. Hourly monitored data between 11 December 2019 and 11 March 2020, for external and internal climate: temperature (top panel) and relative humidity (bottom panel).

Anyway, to ensure a complete yearly dataset of the hygrothermal simulation, the residual months had to be integrated: regarding the external weather, an additional monitoring right in the adjoining garden was used, related to 2017 (provided by Prof. Michele Bottarelli, University of Ferrara). A preliminary analysis showed the compatibility of such data with the monitored dataset. Regarding interior climate, the residual period was calculated from the outdoor climate data, according to EN 15026 [46] and the WTA sheet 6.2 [47] adaptive climate model. Finally, all the main climatic coefficients and parameters, used for the simulation are presented in Table 1.

Table 1. Relevant parameters and coefficients used for the hygrothermal simulations, for external and internal boundary conditions.

	Quantity	Value	Unit
external	Heat transfer coefficient (exterior surface), $h_{c,ext}$	12	[W/m^{2}K]
	Vapour diffusion coefficient (exterior surface)	7.5×10^{-8}	[s/m]
internal	Temperature range for interior climate	20–25	[°C]
	Relative humidity range for interior climate (normal moisture load)	35–65	[%]
	Heat transfer coefficient (interior surface), $h_{c,int}$	8	[W/m^{2}K]
	Vapour diffusion coefficient (interior surface)	3×10^{-8}	[s/m]

4.2. Two-Dimensional Model

The hygrothermal model was designed on the basis of the horizontal section of the historic monitored wall (Figure 7). A two-dimensional (2D) hygrothermal model was constructed for the dynamic simulation from the horizontal section of the historic wall, as presented in Figure 9: in (a), the architectural survey is visible; while in (b), the 2D transport model used for dynamic simulations is found (the interfaces between the materials are indicated from A (external) to E (internal side)).

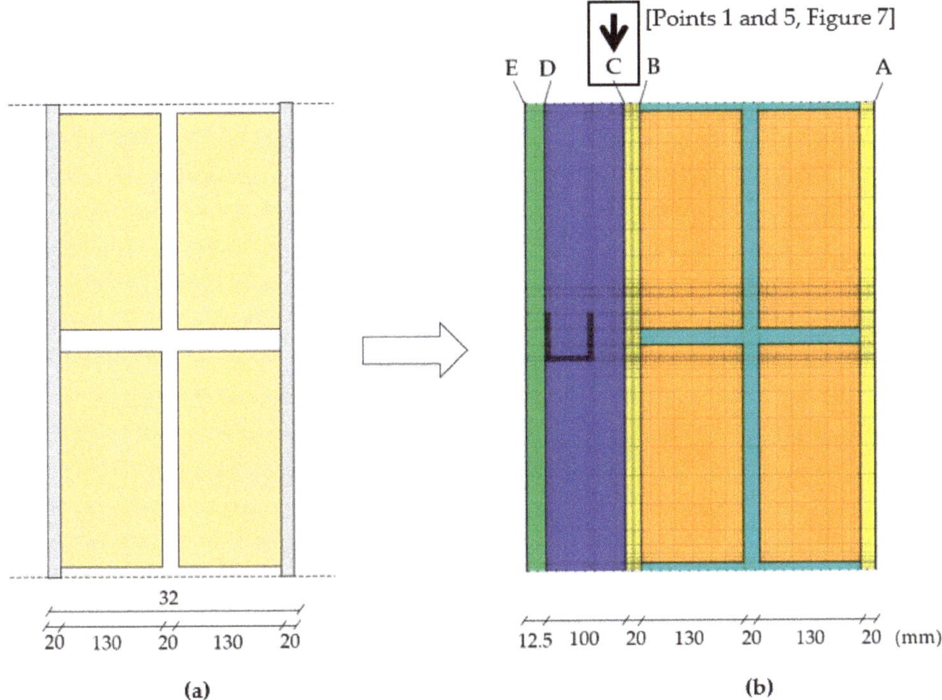

Figure 9. Two-dimensional model deducted for the dynamic simulation.

All the hygrothermal data regarding building materials are taken from Delphin's Material Database 6.0.20 [48] and were determined experimentally at the Climatology Laboratory at the University of Dresden. The building materials were selected from a specific historical cluster. In particular, brick and exterior plaster data refer to historic materials coming from a case study in Bologna and measured within the 3enCult European Project [14]. The hygrothermal characterization of the used stone wool, instead, comes from the material installed by the company involved in the HeLLo project. The characterization of the mentioned material for the dynamic calculation was performed in the laboratories of the Fraunhofer IBP (Institut of Building Physics) [49]; for this reason, data were appropriately formatted to fit the material characterization of the Delphin® database [48]. The main hygrothermal properties for all the used materials are presented in Table 2 and Figure 10. On the top of Table 2 (*), the materials that were characterized in the framework of the 3enCult Project are presented [14], while in Figure 10, the moisture storage function of the materials is shown: in (a), the moisture content is shown—$\theta_l(pc)$; in (b), the liquid water conductivity as a function of the capillary pressure $Kl(pc)$ is shown. In both, (a) and (b), the three vertical lines refer to the RH values of 50%, 95% and 99%.

Table 2. Main hygrothermal properties of the chosen materials.

Material	ID Delphin DB	ρ [kg/m^3]	C_p [J/KgK]	θ_{por} [m^3/m^3]	λ_{dry} [W/mK]	μ_{dry} [-]	A_w [kg/m^{2}s^{05}]
Lime Mortar	143	1570	1000	0.408	0.7	11.0	0.176
Historical Brick (*)	532	1759	1092	0.336	0.624	24.5	0.184
Lime plaster External (*)	520	1604	869	0.395	0.69	19	0.179
Lime plaster Internal	629	1498	802	0.435	0.412	9.3	0.018
Stone wool	-	70	1030	0.950	0.033	1	-
Steel	238	7800	470	-	47	-	-
Gypsum board	599	745	1826	0.719	0.177	11	0.179

Note: Density (ρ), Specific heat (Cp), porosity (θ_{por}), thermal conductivity (λ_{dry}), vapour resistance (μ_{dry}) and capillary absorption coefficient (Aw).

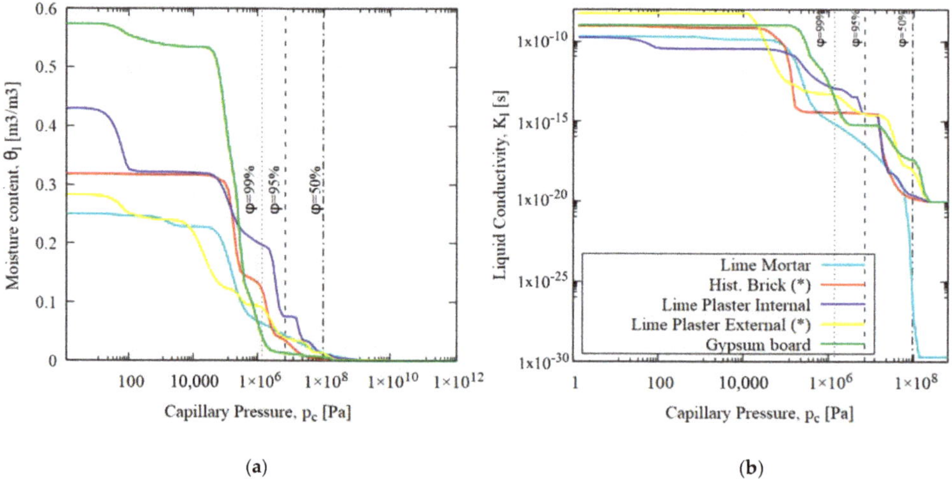

Figure 10. Moisture storage function of the materials used for the 2D simulation.

4.3. Selection of Outputs

The selected outputs to analyze the hygrothermal assessment were based on the recommendations of the WTA leaflet 6.5 [50] and on the possibility of comparing the simulated data with the monitored results. Although different points of the stratigraphy were simulated (Figure 9), here, only T and RH of the cold surface of the insulation, and most probable condensation point (Point C, Figure 9), were analysed, alike for the monitored data. In addition to the monitored period, RH was studied for the dry months (10 December–7 August).

The total moisture content of the wall was observed, to determine when the simulation reaches periodic behavior and is not influenced any more by the chosen initial conditions. Then, the evaluation period starts.

All the outputs were evaluated through a spatial average, to obtain a unique value for each time step.

5. Results

5.1. In Situ Monitoring

Figures 11 and 12 show the 10 min averaged values of measured T (°C) and RH (%) at the most probable condensation point (sensors 1 and 5, Figure 7b), i.e., the point between the insulation panels and the historic wall, at both heights 1.90 m and 3.40 m. As the measured parameters inside the room where the experiment is being developed follow the outdoor conditions, these would not add any specific value to the analysis; therefore, authors deliberately did not present them in these graphs.

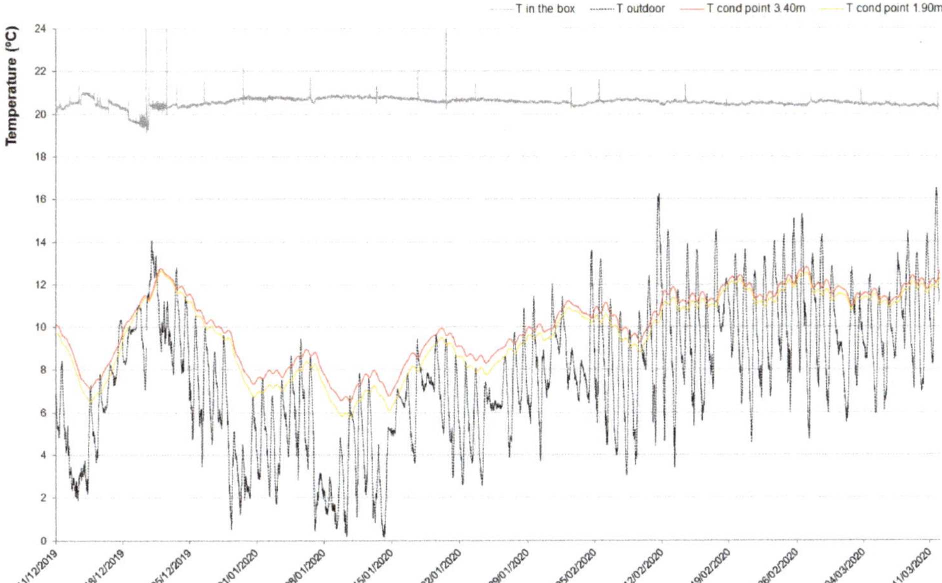

Figure 11. Graphical representation of the T values between 11 December 2019 and 11 March 2020.

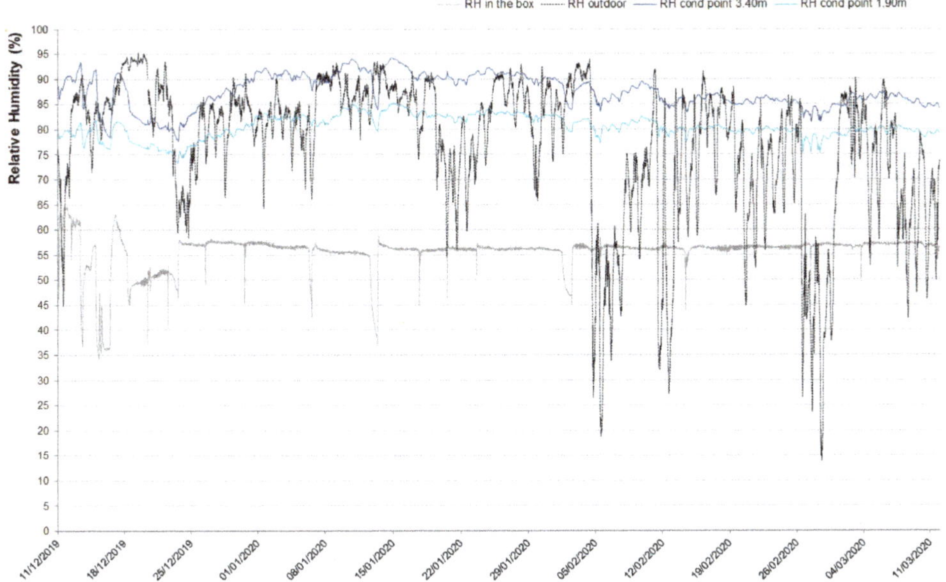

Figure 12. Graphical representation of the relative humidity (RH) values between 11 December 2019 and 11 March 2020.

As can be observed (Figure 11), T (°C), values measured on the inner side of the wall (just before the insulation, sensors 1 and 5, Figure 7b), clearly follow the outdoor trend. At the coolest moments, T was always above 5 °C, therefore the risk of frost damage did not occur. This data should be cautiously considered—in a colder winter, a different result could be expected. The maximum deviation between the line T 1.90 m and T 3.40 m is ΔT_{max} = 0.8 °C and has no significance, since the difference of values is within the accuracy of the T sensors.

The risk of interstitial condensation, even if limited, was present. According to the measured data authors would be tempted to state it did not happen, as RH maximum was 94.2%, measured at 3.90 m. Nonetheless, considering sensors accuracy, previously declared, it might have occurred. Through the observation of Figure 12, there seems to be a trend of a drying out process, as RH values were getting slightly lower in the beginning of March. Nonetheless, this hypothesis can only be discussed based on the simulation (Section 5.2), as for present monitored data do not allow one to confirm it. The maximum deviation between the line RH 1.90 m and RH 3.40 m is ΔRH_{max} = 12%. In this case, it is significant, since the difference of values is beyond the accuracy of the RH sensors. This may be caused by several factors depending on the historic wall: e.g., the non-homogeneity of materials, possible presence of cracks or the limited adherence of the insulation materials to the out-of-line wall.

Concerning the internal condition of the box, an observation can be outlined. Some insignificant peaks (in terms of time and amplitude) can be observed on the RH graph inside the box. These correspond to moments of maintenance procedures of the monitoring campaign (i.e., when the box was open and the conditions of the air inside the box naturally mixed with those of the room). These events do not influence the average internal conditions and the test.

5.2. Hygrothermal Simulation

The simulation reaches a periodic behavior in the second simulated year. The graph (Figure 13) shows the monitored period (11 December–11 March), plus the drying period for the investigated wall. Hourly averaged simulated values of T (°C) and RH (%) are shown at the most critical part of the section [51], point C (Figure 9).

As can be observed, at the coolest moments, T (°C) was always above 5 °C, therefore, the risk of frost damage was not evidenced.

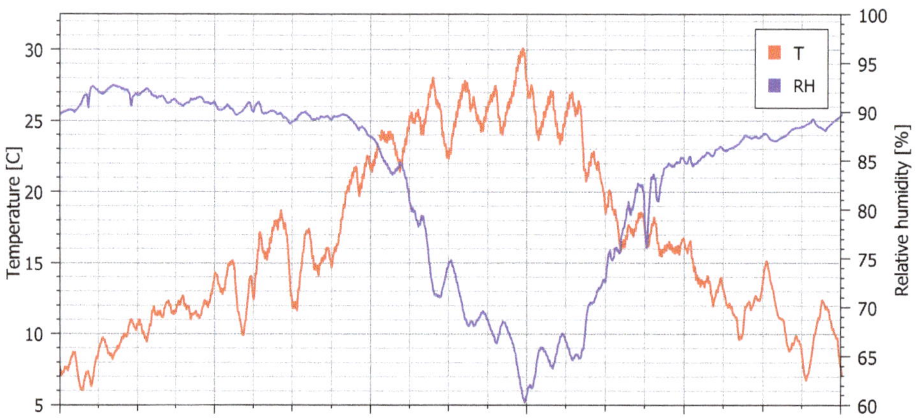

Figure 13. T (°C) and RH (%) profiles of the simulated wall in point C, during the 2nd year of simulation.

Concerning RH (%), although high values are exhibited during the studied wet period (11 December–11 March), it remains below 95%, the condensation risk threshold defined by the WTA Leaflet 6.5 [50], which indicates the absence of condensate. Moreover, Figure 13 illustrates a strong decrease of RH from the wet to dry period of the year, emphasizing a significant reduction of moisture content in summer. It reaches the driest value on 7 August. This fact is also visible in Figure 14, where a profile of RH during the most critical and the most favorable hours of the year are shown for the studied wall. In particular, in terms of moisture accumulation, it can be observed in Figure 14a the wall section during the wettest hour (on 26 January at 00:00) and in Figure 14b, the wall section during the driest hour (on 7 August at 12:00).

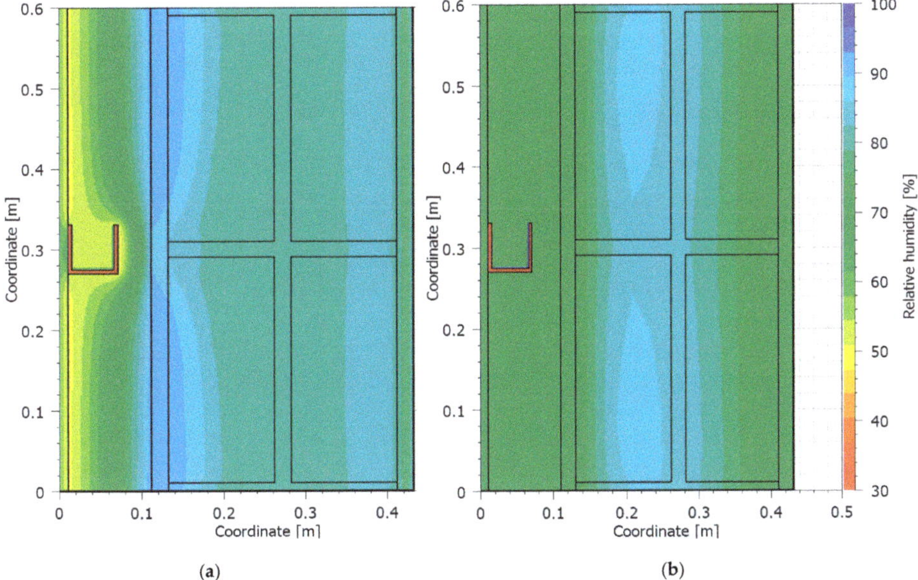

Figure 14. RH profile of the original simulated wall during the most critical (**a**) and the most favourable (**b**) hours of the year.

6. Discussion

6.1. Comparison of Monitored and Simulated Results

For the analysed 3-month period (11 December 2019 to 11 March 2020), several observations can be pointed out concerning both monitored and simulated data at point C (Figure 9) of the stratigraphy, namely:

- The average temperature values do practically coincide. Average T_{simul} = 9.9 °C, while average T_{monit} = [9.7–10.1] °C. Considering the accuracy of the sensors, it can be declared that averaged $T_{simul} = T_{monit}$;
- Both monitored and simulated values pointed to lowest T (°C) value > 5 °C, i.e., the risk of frost damage was not evidenced;
- Concerning RH, data must be observed more carefully, as monitored RH values at 3.40 m and 1.90 m differed more expressively than T. average monitored RH values varied between 80.6–87.6%, at 1.90 m and 3.40 m respectively, while average RH_{simul} = 90.8%. Bearing in mind the accuracy of the sensors, it can be stated that the average simulated values fit the average RH values measured at 3.40 m. The similarity of the simulated and monitored values is further reinforced by other statistical data: at 3.40 m, only 25% of data (Q3) was above 90.4%. The difference of monitored values between sensors at 1.90 m and 3.40 m, currently attributed to the possible presence of cracks, i.e., wall inhomogeneity, does not allow a fair comparison between RH at 1.90 m with the simulated data;
- These comments are also grounded on the synthesis presented on Table 3;
- Grounded on the aforementioned data, the authors considered that the simulation model was validated.

Table 3. Synthesis of the averaged hygrothermal data.

Title	Simulation	Monitoring at 3.40 m	Monitoring at 1.90 m
RH (%) [1]	90.8	87.6	80.6
T (°C) [2]	9.9	10.1	9.7

[1] At 80%, sensors accuracy ±2%; at 90%, ±3%. [2] At 10 °C, sensors accuracy ±0.5 °C.

6.2. Variation of the Simulation

6.2.1. New Models and Additional Outputs

Despite the absence of interstitial condensation suggested by the simulation for the given outdoor climate, but given the reasonable doubt pointed out by the monitored data, further simulation scenarios and outputs were explored.

The insulation system, described in Section 2.2, was simulated including different thickness of the stone wool (additionally to the installed 10 cm, also 6 cm and 8 cm) and the use of two gypsum boards as internal finishing (against the single layer installed). These simulations also allowed one to explore whether the biggest impact in moisture accumulation is credited to the ingress of moisture from the internal side, or to the reduction of the drying potential of the wall due to the interior insulation. Thus, considering the pros and cons, this evaluation could point to the most efficient retrofitting strategy, between reducing the thickness of insulation or improving the vapour resistance of the internal finishing.

Additionally, in order to quantify the performances of the insulation system and the cons of reducing its thickness in the first and second variant scenarios (6–8 cm), heat losses were calculated.

The heating period is defined as the part of the year in which the sinusoidal fit of the daily outdoor temperature is smaller than 10 °C [52]. For the selected dataset, this period corresponds to the days from 15 October to 15 April. Finally, a profile of the temperature in point C (Figure 9) was studied to compare different retrofit scenarios. A new moisture accumulation analysis was performed and the reaching periodic behaviour was also verified for the simulated variants.

6.2.2. Thermal Simulation

Figure 15 reports T (°C) in the cold side of insulation (point C, Figure 9), with a focus on the different simulation scenarios of 6–8 and 10 cm of stone wool (left panel). In order to quantify the expected thermal losses, heat flux during the simulated heating period is presented in the right panel.

The cumulative thermal losses for the studied heating period (15 October–15 April) are 96.27 kWh/m^2 for the uninsulated wall, 15.79 kWh/m^2 for the real case scenario (10 cm of stone wool), 19.74 kWh/m^2 for the 8 cm and 21.96 kWh/m^2 for 6 cm of stone wool as an insulation layer.

Figure 16 includes a thermal profile of the wall during the coldest day of the year (on 26 January at 00:00), for the most and least insulated scenarios, respectively 10 cm [Figure 16a] and the 6 cm (Figure 16b) of stone wool.

6.2.3. Moisture Accumulation Analysis

Figure 17 shows RH behind the insulation during the monitored period (11 December–11 March), plus the drying period (12 March–7 August) for the investigated wall. Moreover, scenarios with 6–8 cm of stone wool and 10 cm with two gypsum boards as the internal finishing layer are also presented, in order to compare the impact of different interventions. All the scenarios confirmed the initial simulation results: a strong decrease of RH from the wet to dry period of the year, emphasizing a significant reduction of moisture content in the summer.

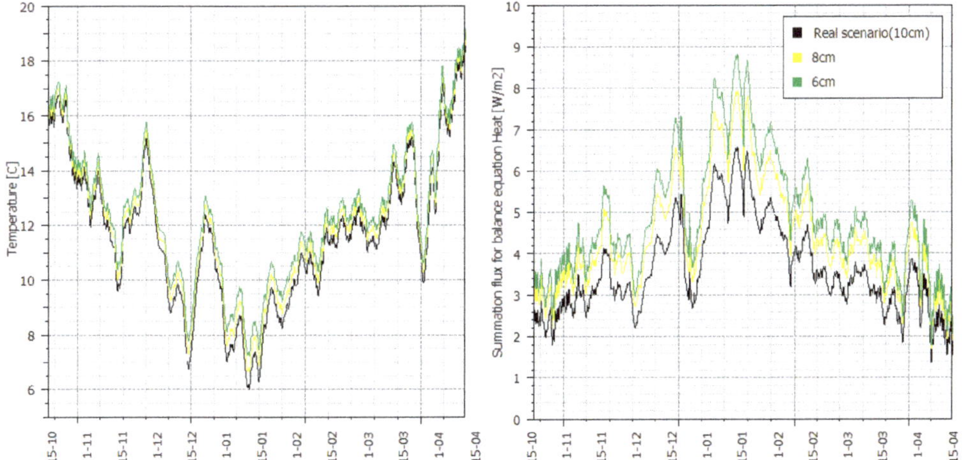

Figure 15. Left panel: hourly value of temperature in Point C (Figure 9). Right panel: hourly value of heat loss for the simulated walls, both during the heating period (15 October–15 April). Design scenarios of 6–8 cm of stone wool are included.

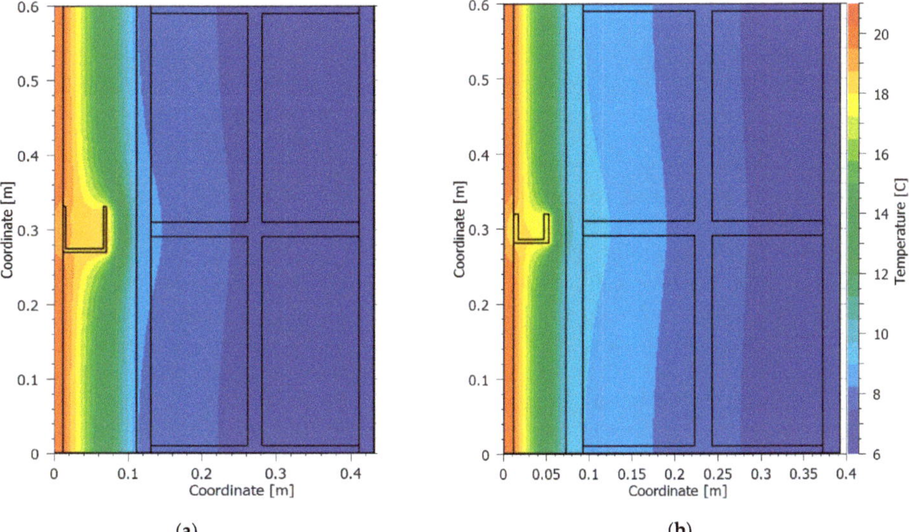

(**a**) (**b**)

Figure 16. Temperature profile of the simulated wall during the coldest hour for 10 cm and 6 cm insulation scenarios.

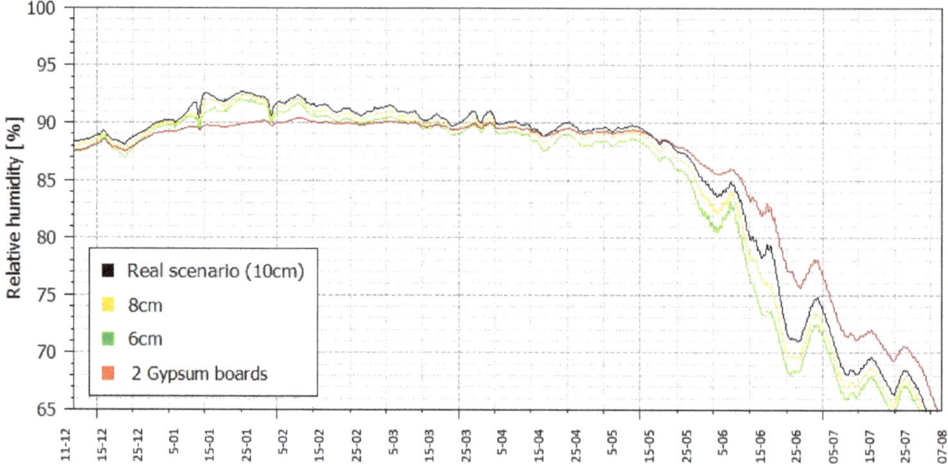

Figure 17. RH in point C (Figure 9) during the period 11 December–7 August, for the simulated wall. Design scenarios of 6–8 cm of stone wool and 2 gypsum boards as interior finishing are included.

6.2.4. Discussion of the Various Simulation Outputs

With respect to the real 10 cm insulated case, the thermal analysis of 6–8 cm scenarios (Section 6.2.2) has quantified the expected decrease of energy performance, due to the reduction of the insulation thickness, and the increase of T (°C) behind the insulation layer (point C, Figure 9), shown in Figure 15 (left panel) as an effect of the higher loss of heat reported and quantified in Figure 15 (right panel); also, in Figure 16 the thermal effect of insulation on the existing wall is visible: the use of 6 cm of stone wool as an insulation system (right panel) leaves the historic wall significantly warmer than using 10 cm (left panel).

Although, on the one hand, we can see that the use of thicker insulation systems corresponds to lower heat losses, we expect to see higher moisture accumulation in the wall, as a result of this decrease of interstitial temperature. This issue was presented in the moisture accumulation analysis (Section 6.2.3). Figure 17 presents RH at the interface between the insulation system and the existing wall, recognized as the most critical part of the section [51]. As a general comment, alike for the base case simulation, RH remained below 95% for all the cases.

The comparison with the first and the second scenarios of thickness reduction (6–8 cm) has shown a very small difference in RH, in a total range of 1%.

Contrariwise, in the third scenario, it resulted that improving the vapour resistance of the internal finishing with the use of the two gypsum boards, instead of one, may contribute to reduce moisture accumulation during the wetting period more efficiently than reducing the insulation thickness (Figure 17, first period).

Anyway, it is important to mention that, while the third scenario avoids reducing insulation thickness with consequent minor thermal losses, on the other hand, this design choice is responsible for a delay during the drying phase (Figure 17, second period). However, during the drying phase, moisture content significantly dries out from the interstitial area in all the studied cases. This phenomenon was also already visible in Figures 13 and 14 for the base case scenario.

7. Conclusions

The study presents an applied research of the hygrothermal behaviour of an internal insulated historic brick wall, assessed at the most critical point of the stratigraphy, both by in situ measurements and predicted performance through dynamic simulations.

The solution implemented for in situ monitoring balanced both thermal insulation improvement and the effects of the application over the historic wall: double stone wool boards (100 mm in total), supported by own steel frame, punctually fixed to the historic wall, finished with a single gypsum board.

The analysed monitored period lasted three months, during winter period 2019–2020. In situ data were also used for validating the 2D simulation model.

Under the current conditions—climate data and studied stratigraphy, the results of both in situ monitoring and simulation prediction, evidenced no risk of frost damage of the brick wall. With regards to the risk of interstitial condensation, the simulation showed no risk either. In situ data analysis suggested nonetheless a more cautious interpretation. Though the highest values remained below 95%, the condensation risk threshold, these data are dependent on sensors accuracy, which means that in fact the "true" RH could be over 95%. Furthermore, more severe climate could result in crossing the risk threshold. Nonetheless, it is worth highlighting that given the simulation results, during the drying phase moisture content significantly dries out from the interstitial area.

Authors also proposed the exploitation of three other simulation scenarios and outputs, namely: reducing the insulation thickness (6 and 8 cm) and decreasing the ingress of moisture into the wall by adding a second gypsum board layer to the initial 10 cm insulation technology (instead of the traditional vapour barrier).

It was found that reducing the thickness of the insulation material (by 2 cm or 4 cm) decreased the moisture content in the wall, but not very significantly. More meaningful, instead, was the result obtained for the same insulation technology (10 cm thickness) with two gypsum boards. In this case, the only verified drawback is the delay of the drying phase, absolutely compensated by the reduction of moisture accumulation during the wetting period.

The outcomes of this study are significant, not only to the scientific community, but mostly to practitioners, often missing guidance in energy refurbishment intervention of historic buildings. In other words, the thermal benefits of stone wool insulation with a certain thickness should not be compromised by the "fear" of moisture increase (once reducing it would not make much change). Instead, moisture accumulation might be improved—or better controlled—through the addition of a second gypsum board.

Author Contributions: Authors are listed in alphabetical order. Conceptualization, M.C., P.D., L.D.P. and E.L.; Methodology, M.C., L.D.P. and E.L.; Validation: M.A., D.B.-L., M.C. and L.D.P.; Formal analysis, M.A., D.B.-L. and L.D.P.; Investigation: M.A., M.C., P.D., L.D.P. and E.L.; Simulation: D.B.-L. and A.T.; Data curation: M.A.; Writing—original draft preparation, L.D.P; Writing—review and editing,: M.A., D.B.-L., M.C., P.D., L.D.P., E.L. and A.T.; Visualization: D.B.-L., M.C. and L.D.P.; Supervision P.D. and A.T.; Project administration: P.D. and L.D.P. All authors have read and agreed to the published version of the manuscript.

Funding: The results presented in this paper are part of the HeLLo project that has received funding from the European Union's Horizon 2020 research and innovation programme under the Marie Sklodowska-Curie grant agreement No. 796712.

Acknowledgments: The authors acknowledge ROCKWOOL® Italia S.p.A. for the material and the support to the project. They also acknowledge Giorgi Roberto and Lavorazione Legno for the execution of the metering box. Finally, authors thank the University of Ferrara for providing the access and usage of Palazzo Tassoni Estense as the case-study.

Conflicts of Interest: The authors declare no conflict of interest. The funders had no role in the design of the study; in the collection, analyses, or interpretation of data; in the writing of the manuscript, or in the decision to publish the results.

References and Note

1. European Commission, EU Research. *Cultural Heritage*; European Commission: Brussels, Belgium, 2013. [CrossRef]
2. Akande, O.K.; Odeleye, D.; Coday, A. Energy Efficiency For Sustainable Reuse Of Public Heritage Buildings: The Case For Research. *Int. J. Sustain. Dev. Plan.* **2014**, *9*, 237–250. [CrossRef]

3. European Commission. Commission Staff Working Document—European Framework for Action on Cultural Heritage, Brussels. 2018. Available online: https://ec.europa.eu/culture/library/commission-swd-european-framework-action-cultural-heritage_en (accessed on 6 January 2020).

4. Phoenix, T. Lessons learned: ASHRAE's approach in the refurbishment of historic and existing buildings. *Energy Build.* **2015**, *95*, 13–14. [CrossRef]

5. Vereecken, E.; Van Gelder, L.; Janssen, H.; Roels, S. Interior insulation for wall retrofitting—A probabilistic analysis of energy savings and hygrothermal risks. *Energy Build.* **2015**, *89*, 231–244. [CrossRef]

6. Calzolari, M.; Davoli, P.; Pereira, L.D. Analysis of the risks related to the energy retrofit interventions of historic buildings. *Recuper. Conserv.* **2019**, *154*, 88–95. (In Italian)

7. Galatioto, A.; Ciulla, G.; Ricciu, R. An overview of energy retrofit actions feasibility on Italian historical buildings. *Energy* **2017**, *137*, 991–1000. [CrossRef]

8. Güleroğlu, S.K.; Karagüler, M.E.; Kahraman, İ.; Umdu, E.S. Methodological approach for performance assessment of historical buildings based on seismic, energy and cost performance: A Mediterranean case. *J. Build. Eng.* **2020**, *31*, 101372. [CrossRef]

9. Akkurt, G.G.; Aste, N.; Borderon, J.; Buda, A.; Calzolari, M.; Chung, D.; Costanzo, V.; Del Pero, C.; Evola, G.; Huerto-Cardenas, H.E.; et al. Dynamic thermal and hygrometric simulation of historical buildings: Critical factors and possible solutions. *Renew. Sustain. Energy Rev.* **2020**, *118*. [CrossRef]

10. Asdrubali, F.; Baldinelli, G. Thermal transmittance measurements with the hot box method: Calibration, experimental procedures, and uncertainty analyses of three different approaches. *Energy Build.* **2011**, *43*, 1618–1626. [CrossRef]

11. Wakili, K.G.; Tanner, C. U-value of a dried wall made of perforated porous clay bricks Hot box measurement versus numerical analysis. *Energy Build.* **2003**, *35*, 675–680. [CrossRef]

12. Soares, N.; Martins, C.; Gonçalves, M.; Santos, P.; da Silva, L.S.; Costa, J.J. Laboratory and in-situ non-destructive methods to evaluate the thermal transmittance and behaviour of walls, windows, and construction elements with innovative materials: A review. *Energy Build.* **2019**, *182*, 88–110. [CrossRef]

13. Bienvenido-Huertas, D.; Moyano, J.; Marín, D.; Fresco-Contreras, R. Review of in situ methods for assessing the thermal transmittance of walls. *Renew. Sustain. Energy Rev.* **2019**, *102*, 356–371. [CrossRef]

14. 3ENCULT. 3ENCULT Efficient Energy for EU Cultural Heritage (n.d.). Available online: http://www.3encult.eu/en/project/welcome/default.html (accessed on 19 June 2016).

15. Co2olBricks. Co2olBricks—Climate Change, Cultural Heritage & Energy Efficient Monuments (2013). Available online: http://www.co2olbricks.eu/ (accessed on 28 August 2019).

16. EFFESUS. EFFESUS—Energy Efficiency for EU Historic Districts' Sustainability (2013). Available online: https://www.effesus.eu/ (accessed on 6 January 2020).

17. Troi, A.; Zeno, B. *Energy Efficiency Solutions for Historic Buildings: A Handbook*; Birkhäuser: Berlin, Germany; Basel, Switzerland, 2014. [CrossRef]

18. Zagorskas, J.; Kazimieras, E.; Turskis, Z.; Burinskien, M. Thermal insulation alternatives of historic brick buildings in Baltic Sea Region. *Energy Build.* **2014**, *78*, 35–42. [CrossRef]

19. Lucchi, E.; Becherini, F.; Concetta, M.; Tuccio, D.; Troi, A.; Frick, J.; Roberti, F.; Hermann, C.; Fairnington, I.; Mezzasalma, G.; et al. Thermal performance evaluation and comfort assessment of advanced aerogel as blown-in insulation for historic buildings. *Build. Environ.* **2017**, *122*, 258–268. [CrossRef]

20. Lucchi, E.; Roberti, F.; Alexandra, T. Definition of an experimental procedure with the hot box method for the thermal performance evaluation of inhomogeneous walls. *Energy Build.* **2018**, *179*, 99–111. [CrossRef]

21. Roque, E.; Vicente, R.; Almeida, R.M.S.F.; da Silva, J.M.; Ferreira, A.V. Thermal characterisation of traditional wall solution of built heritage using the simple hot box-heat flow meter method: In situ measurements and numerical simulation. *Appl. Therm. Eng.* **2020**, *169*, 114935. [CrossRef]

22. Adhikari, R.S.; Lucchi, E.; Pracchi, V. Energy modelling of historic buildings: Applicability, problems and compared results. In Proceedings of the 3rd European Workshop on Cultural Heritage and Preservation, Bolzano, Italy, 15–17 September 2013; pp. 119–125.

23. Calzolari, M. *Prestazione Energetica Delle Architetture Storiche: Sfide e Soluzioni. Analisi dei Metodi di Calcolo per la Definizione del Comportamento Energetico*; Franco Angelli: Milano, Italy, 2016; ISBN 9788891740885.

24. Belpoliti, V.; Bizzarri, G.; Boarin, P.; Calzolari, M.; Davoli, P. A parametric method to assess the energy performance of historical urban settlements. Evaluation of the current energy performance and simulation of retrofit strategies for an Italian case study. *J. Cult. Herit.* **2018**, *30*, 155–167. [CrossRef]

25. Webb, A.L. Energy retrofits in historic and traditional buildings: A review of problems and methods. *Renew. Sustain. Energy Rev.* **2017**, *77*, 748–759. [CrossRef]

26. Bienvenido-Huertas, D.; Rubio-Bellido, C.; Pérez-Ordóñez, J.L.; Oliveira, M.J. Automation and optimization of in-situ assessment of wall thermal transmittance using a Random Forest algorithm. *Build. Environ.* **2020**, *168*, 106479. [CrossRef]

27. Bellia, L.; Alfano, F.R.D.; Giordano, J.; Ianniello, E.; Riccio, G. Energy requalification of a historical building: A case study. *Energy Build.* **2015**, *95*, 184–189. [CrossRef]

28. Ficco, G.; Iannetta, F.; Ianniello, E.; Romana, F.; Dell, M. U-value in situ measurement for energy diagnosis of existing buildings. *Energy Build.* **2015**, *104*, 108–121. [CrossRef]

29. Dell'Isola, M.; D'Ambrosio Alfano, F.R.; Giovinco, G.; Ianniello, E. Experimental analysis of thermal conductivity for building materials depending on moisture content. *Int. J. Thermophys.* **2012**, *33*, 1674–1685. [CrossRef]

30. Litti, G.; Khoshdel, S.; Audenaert, A.; Braet, J. Hygrothermal performance evaluation of traditional brick masonry in historic buildings. *Energy Build.* **2015**, *105*, 393–411. [CrossRef]

31. Kloseiko, P.; Arumagi, E.; Kalamees, T. Hygrothermal performance of internally insulated brick wall in cold climate: A case study in a historical school building. *J. Build. Phys.* **2015**, *38*, 444–464. [CrossRef]

32. Hansen, T.K.; Bjarløv, S.P.; Peuhkuri, R.H.; Harrestrup, M. Long term in situ measurements of hygrothermal conditions at critical points in four cases of internally insulated historic solid masonry walls. *Energy Build.* **2018**, *172*, 235–248. [CrossRef]

33. HeLLo. EU H2020 MSCA-IF-ES HeLLo Project (2019). Available online: https://cordis.europa.eu/project/rcn/215475/factsheet/en (accessed on 7 April 2019).

34. Sontag, L.; Nicolai, A.; Vogelsang, S. *Validierung der Solverimplementierung des Hygrothermischen Simulationsprogramms Delphin*; Technical report; Technische Universität Dresden: Dresden, Germany, 2013.

35. UNESCO. Ferrara, City of the Renaissance, and its Po Delta (1995). Available online: http://whc.unesco.org/en/list/733 (accessed on 9 January 2019).

36. Davoli, P. Complexity, information surplus and interdisciplinarity management. The Rehabilitation of Tassoni Estense Palace in Ferrara. In *Conserving Architecture*; Jain, K., Ed.; AADI CENTRE: Ahmedabad, India, 2017; pp. 124–145, ISBN 978-81-908528-2-1.

37. Calzolari, M.; Davoli, P.; Pereira, L.D. From the dynamic simulations assessment of the hygrothermal behavior of internal insulation systems for historic buildings towards the HeLLo project. *Int. J. Environ. Sci. Dev.* **2020**, *11*, 278–285. [CrossRef]

38. Andreotti, M.; Calzolari, M.; Davoli, P.; Pereira, L.D.; Lucchi, E.; Malaguti, R. Design and construction of a new metering hot box for the in situ hygrothermal measurement in dynamic conditions of historic masonries. *Energies* **2020**, *13*, 2950. [CrossRef]

39. HeLLo. HeLLo Team Has Set-Up the Technical Worktables! 2019. Available online: https://hellomscaproject.eu/hello-team-has-set-up-the-technical-worktables/ (accessed on 15 January 2020).

40. STBA—Sustainable Traditional Buildings Alliance. Responsible Retrofit Guidance Wheel. Available online: http://responsible-retrofit.org/greenwheel/en (accessed on 6 January 2020).

41. English Heritage, Energy Efficiency and Historic Buildings: Application of part L of the Building Regulations to historic and traditionally constructed buildings. *Energy Effic. Hist. Build.* **2010**, 1–72.

42. International Organization for Standardization (ISO). *Ergonomics of the Thermal Environment—Analytical Determination and Interpretation of Thermal Comfort Using Calculation of the PMV and PPD Indices and Local Thermal Comfort Criteria*; ISO 7730:2005; ISO: Geneva, Switzerland, 2005.

43. Ente Italiano di Normazione (UNI). *Prestazione Igrotermica dei Componenti e Degli Elementi per Edilizia—Temperatura Superficiale Interna per Evitare L'umidita' Superficiale Critica e la Condensazione Interstiziale—Metodi di Calcolo*; UNI EN ISO 13788; UNI: Milano, Italy, 2013.

44. International Organization for Standardization (ISO). *Energy Performance of Buildings—Indoor Environmental Quality—Part 1: Indoor Environmental Input Parameters for the Design and Assessment of Energy Performance of Buildings*; ISO 17772-1:2017; ISO: Geneva, Switzerland, 2017.

45. Lucchi, E.; Dias Pereira, L.; Andreotti, M.; Malaguti, R.; Cennamo, D.; Calzolari, M.; Frighi, V. Development of a Compatible, Low Cost and High Accurate Conservation Remote Sensing Technology for the Hygrothermal Assessment of Historic Walls. *Electronics* **2019**, *8*, 643. [CrossRef]

46. Ente Italiano di Normazione (UNI). *Prestazione Termoigrometrica dei Componenti e Degli Elementi di Edificio—Valutazione del Trasferimento di Umidità Mediante una Simulazione Numerica*; UNI EN 15026; UNI: Milano, Italy, 2008.
47. *WTA 6.2 Leaflet—Simulation of Heat and Moisture Transfer*; WTA Publications: München, Germany, 2014; Volume 30.
48. TU Dresden, Institut fur Bauklimatik. Delphin 6.0.20. Material Database.
49. Fraunhofer IBP (Fraunhofer Institute for Building Physics). *WUFI Pro-6.2 [Computer Software]*; Fraunhofer IBP: Stuttgart, Germany, 2014.
50. *WTA 6.5 Leaflet—Interior Insulation*; WTA Publications: München, Germany, 2014.
51. Häupl, P.; Grunewald, J.; Fechner, H.; Stopp, H. Coupled heat air and moisture transfer in building structures. *Int. J. Heat Mass. Transf.* **1997**, *40*. [CrossRef]
52. Energieeinsparverordnung (EnEV). Verordnung über Energiesparenden Warmeschutz und Energiesparende Anlagentechnik bei Gebauden. Bundesgesetzblatt. 2007. Available online: https://www.enev-profi.de/wp-content/uploads/EnEV-2014-Lesefassung.pdf (accessed on 6 January 2020).

Article

Baselines for Energy Use and Carbon Emission Intensities in Hellenic Nonresidential Buildings

Kalliopi G. Droutsa [1,2,*], Constantinos A. Balaras [1], Spyridon Lykoudis [3], Simon Kontoyiannidis [1], Elena G. Dascalaki [1] and Athanassios A. Argiriou [2]

[1] Group Energy Conservation, Institute for Environmental Research & Sustainable Development, National Observatory of Athens, 118 10 Athens, Greece; costas@noa.gr (C.A.B.); skonto@noa.gr (S.K.); edask@noa.gr (E.G.D.)

[2] Laboratory of Atmospheric Physics, Department of Physics, School of Science, University of Patras, 265 04 Patras, Greece; athanarg@upatras.gr

[3] Enargia WG, Akrita 66, 24132 Kalamata, Greece; slykoud@gmail.com

* Correspondence: pdroutsa@noa.gr

Received: 27 February 2020; Accepted: 16 April 2020; Published: 23 April 2020

Abstract: This work exploits data from 30,000 energy performance certificates of whole nonresidential (NR) buildings in Greece. The available information is analyzed for 30 different NR building uses (e.g., hotels, schools, sports facilities, hospitals, retails, offices) and four main services (space heating, space cooling, domestic hot water and lighting). Data are screened in order to exclude outliers and checked for consistency with the Hellenic NR building stock. The average energy use and CO_2 emission intensities for all building uses are calculated, as well as the respective energy ratings in order to gain a better understanding of the NR sector. Finally, in an attempt to determine whether these values are representative for the various Hellenic NR building uses, their temporal evolution is investigated. The average primary energy use intensity is 448.0 kWh/m^2 for all NR buildings, while the CO_2 emissions reach 147.5 kgCO$_2$/m^2. The derived energy baselines reveal that indoor sports halls/swimming pools have the highest energy use, while private cram schools/conservatories have the lowest, due to their operational patterns. Generally, from the four services taken into account, lighting is the most energy consuming, followed by cooling, heating and finally domestic hot water. For a total of 11 building uses, more data from the certificates will be necessary for deriving representative baselines, but, when it comes to buildings categories, more data are required.

Keywords: nonresidential buildings; baselines; EUI; energy use intensities; carbon emission intensities; EPCs; energy performance certificates

---

1. Introduction

Built environment is a key target in European (EU) policies in order to achieve a sustainable and competitive low-carbon economy. EU buildings account for nearly 40% of energy consumption and 36% of CO_2 emissions [1], despite covering only 3.35% of the total land area [2]. About 35% of the building stock is old (over 50 years) and energy inefficient, while the annual renovation rate ranges between 0.4% and 1.2% depending on the country. Therefore, increased renovation of existing buildings may lead to significant savings both in total energy consumption and CO_2 emissions by about 5% [1].

Nonresidential (NR) buildings, accounting for about a quarter of the total EU building stock, comprise a very heterogeneous sector with various building types, different building sizes and energy characteristics. Limited information is available on construction characteristics, installed systems for different services and energy use for the entire NR sector as well as for the various types and branches of activity.

Over the past decades, the total final energy consumption in the EU Member States (EU-28) has remained practically stable, rising only from 1033.4 million tonnes of oil equivalent (Mtoe) in 1990 to 1060.0 Mtoe in 2017 [3]. However, the final energy consumption in NR buildings has boomed by 39.2%, from 110.7 Mtoe in 1990 to 154.0 Mtoe in 2017, keeping a relative stable share ranging from 11% to 15% of the total. The highest share in NR building's sector final energy consumption is electricity with 50.14%, followed by gas with 30.12% [4]. For 2050, the projection is that the NR buildings' sector will account for about 17% of the total final energy use [5].

1.1. Hellenic Nonresidential Buildings

Based on the most recent building census, the Hellenic Nonresidential Building Stock (NR BS) includes about 690,000 exclusive-use buildings (excluding churches, monasteries, industrial buildings and parking stations), representing 21% of the national building stock, with the majority being wholesale/retail trade and offices [6]. The estimated total floor area is about 73.52 Mm^2, with the majority being hotels and restaurants [3], while the total final energy in 2017 reached 16.05 Mtoe increased by 15.6% compared to 1990. NR buildings consumed 2.19 Mtoe in 2017, showing an increase by 236.4% since 1990 (Figure 1). Although until 2008 the NR sector's final consumption was steadily increasing by an average of 7% per year, it was one of the first sectors that was affected by the economic recession. This is reflected by the decrease of final energy consumption since 2009, with the exception of 2012, where a slight increase occurred due to an increase of electricity consumption. From 2015, there was again a clear upward trend. Detailed information for the different NR building types is limited to only to a few studies [7,8].

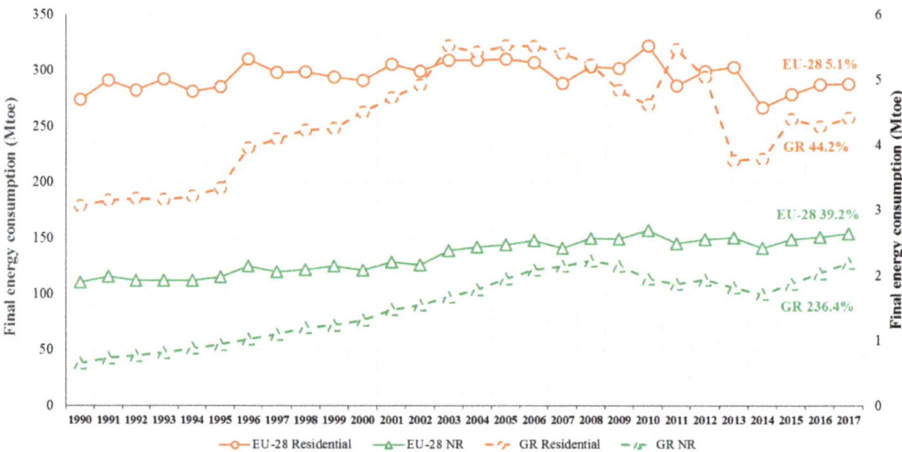

Figure 1. Time evolution of final energy consumption in residential (circles) and nonresidential (NR) (triangles) buildings for EU-28 (solid line, primary axis) and Greece (dashed line, secondary axis). The percentages refer to the corresponding increase of the final consumption in 2017 compared to 1990. (Data source [3]).

To boost energy performance of buildings, the EU has established an ambitious legislative framework that includes the Directive on Energy Efficiency (EED) recently revised by the Directive (EU) 2018/2002 amending Directive 2012/27/EU on energy efficiency, setting an EU target for energy consumption at less than 1273 Mtoe of primary energy and/or 956 Mtoe of final energy [9]. Another main instrument for addressing these challenges is the Energy Performance of Buildings Directive (EPBD) recently amended by the Directive 2018/844, promoting cost-effective building renovations, with the vision of a decarbonized building stock by 2050 [10]. Both directives are part of the new energy rulebook *Clean Energy for All Europeans* [11], according to which each Member State should draft an integrated 10-year national energy and climate plan (NECPs) for the time period 2021 to 2030.

Building energy certification plays a central role in the enhancement of the energy performance of buildings. The energy performance certificate (EPC) is a core instrument of the EPBD for existing and new buildings. The general idea behind the EPC is to influence the building market, by providing information on a building's energy use and typical energy costs and by making recommendations for cost-effective improvements [12]. Available information included in the EPCs should also be of great value for decision makers and energy planners at regional and national scales.

1.2. Literature Overview

While national statistical agencies provide detailed information on the residential building stock, there is a general lack of data on nonresidential buildings. Only a small number of studies have attempted to fill this gap by providing estimates on the total final and specific energy use. In most cases, these studies have focused on particular building services like offices, schools, health care or hotels. Some field work provide actual operational energy data from energy audits using utility bills or monitoring data, while others are theoretical or parametric investigations using building simulations. The challenge though remains when considering the diversities and complexities of large pool of buildings and building stocks.

As the EPC databases are progressively enriched with more data, studies have emerged in different countries that exploit this information to map the characteristics of different categories of buildings and gain a better insight on their energy performance. The majority of them focus on residential buildings. This is to be expected since most EPCs throughout Europe have been issued for dwellings or houses that are commonly rented or sold, and residential buildings constitute the majority of the existing building stock.

For example, in Italy, studies have been carried out by analyzing approximately 90,000 energy certificates of flats in order to establish a method for estimating primary energy for space heating and the main building characteristics (e.g., envelope U-values) that influence their energy performance [13]. According to another study in Italy, exploiting about 17,600 energy certificates for residential buildings, average primary energy for the representative old buildings (up to 1992), ranges between 392–178 kWh/m^2 for existing buildings, 321–158 kWh/m^2 for partially renovated buildings and 113–53 kWh/m^2 after major renovations [14]. In Spain, almost 130,000 energy performance certificates for existing residential buildings were examined [15]. Most were rated at a low energy class E (53.6%). On average, single-family houses use 248.0 kWh/m^2, which is higher than individual dwellings that average 183.2 kWh/m^2. In Switzerland, the thermal performance of the residential building stock was analyzed using about 10,400 cantonal EPCs. The results revealed that about 75% of the buildings' envelope are below new buildings' standards and about 50% of the surface area is still heated by inefficient oil-fired boilers, confirming the high potential for significant energy savings from thermal renovations of the Swiss residential building stock [16]. Finally, an initial study in Greece [17] analyzing early EPC data, concluded that the average primary energy use for residential buildings is 260 kWh/m^2, with thermal energy accounting for 72%. MFH have a lower energy consumption averaging 241 kWh/m^2 compared to 367 kWh/m^2 for SFH.

As the number of issued EPCs grows with time, so does the available number of certificates that have been issued for NR buildings. As a result, studies have progressively been performed exploiting this valuable resource of information. For example, an overview of the NR sector in different countries focusing on building characteristics, energy performance, efficiency measures and energy savings is presented in [18]. A more detailed focus is given to office buildings, which represent the most common building category in the database. The average consumption of existing offices before renovation is 203 kWh/m^2, while for new office buildings it is considerably lower at 122.9 kWh/m^2.

In Catalonia, Spain, researchers exploiting data from about 14,000 certificates found that offices have an average annual energy use of 207.4 kWh/m^2. Their energy ranking is mostly labeled in classes-C and –D (64%), while high energy ratings at the top ranked energy class A and –B are limited to ~8% of the database [19]. NR buildings in Spain, were found to have slightly better energy performance than residential buildings, with an average energy consumption of 317.8 kWh/m^2 [15].

In Sweden, the current energy consumption baseline for NR buildings is defined by exploiting data from about 186,000 EPCs issued for some common commercial building types [20]. Accordingly, the average energy use is 175 kWh/m^2 for hotels and restaurants, 151 kWh/m^2 for rental premises-offices, 169.2 kWh/m^2 for healthcare facilities, 171.8 kWh/m^2 for schools and 174 kWh/m^2 for sports facilities [20]. In another study, energy use in the public sector office stock in England and Wales is explored using a database of 2,600 certificates. The electrical and fossil-thermal energy use profiles for different office types range between 68–211 kWh/m^2 and 132–53 kWh/m^2 respectively [21]. The energy use in English schools was derived using data from 8500 EPCs of primary or secondary schools [22]. Electrical energy use ranged between 44–50 kWh/m^2, and thermal energy use between 138–140 kWh/m^2 [22]. Finally, from an earlier study of a limited number of certificates for NR buildings in Greece, the average primary energy use averaged 459 kWh/m^2, ranging from 211 kWh/m^2 for schools to 1023 kWh/m^2 for hospitals [17].

This work exploits for the first time the most comprehensive data included in the national EPC electronic repository (buildingcert) over a nine-year period (i.e., from January 2011 to the end of 2019). The main novelty and objective of the present work is to present a well-structured, modular methodology for the exploitation of data included in EPC databases, in order to fill the gap of knowledge on the performance of the existing NR building stock. The proposed methodology provides a complete approach, starting with the data quality control, the data clustering into building uses and building categories, and the definition of practical baselines on the energy use and CO$_2$ emissions for different clusters. The work demonstrates the application using the national EPC database in Greece and further investigates whether the derived baselines can be regarded as typical for the Hellenic building stock. The paper is structured as follows: Section 2 provides an overview of the national EPC database. Section 3 presents the method used in the analysis for deriving the baselines and for assessing the time evolution of the indicators. Section 4 applies the overall approach to the national EPC database and presents the results of the analysis. Finally, Section 5 discusses the main findings, outlines the limitations of the work and concludes by summarizing the main findings.

2. National EPC Database

The national regulation on the energy performance of buildings, KENAK, which was the EPBD transposition in Greece, was implemented in 2010 and updated in July 2017, imposing stricter building requirements. The regulation is supported by four Technical Guidelines that govern its practical implementation, first published in 2010 and revised in 2012 and in 2017. The normative calculations for estimating the building's energy demand are in accordance to the European standards (e.g., EN 13790) using the quasi-steady-state monthly method and are included in the official national calculation engine (TEE-KENAK) [23]. The national tool was initially developed in 2010 and was periodically updated to comply with the evolution of the national Technical Guidelines.

Based on the Technical Guidelines, the end uses taken into account for the assessment of the building's energy performance are heating, cooling, domestic hot water (DHW) and lighting. Mechanical ventilation is included in heating and cooling accordingly. Other office and miscellaneous plug loads (e.g., electronic or computing equipment) and process loads (e.g., elevators, escalators) are not considered.

According to the Hellenic labelling scheme, the energy rating of a building is based on the ratio of its calculated primary energy use to that of the corresponding reference building. As a result there are nine energy classes ranging from class G (lowest performance) up to class A+ (highest performance). The reference building is an exact copy of the audited building, complying with the minimum requirements for specific envelope and systems characteristics defined in the national regulation, e.g., U-values, system efficiencies etc. The reference building is ranked by definition at energy class B.

A building can be characterized as a nearly zero-energy building (nZEB) when it is ranked at least at energy class A for new constructions, and at energy class B+ for renovated buildings [24].

Based on the national Technical Guidelines, Greece is divided in four climate zones. The coverage was determined on the basis of the heating degree days (HDD), i.e., zone A in the south with mild conditions (averaging 859 HDD), to zone D in the north with the coldest conditions (averaging 2260 HDD).

In addition, the Technical Guidelines define a total of 60 different building types for the NR building sector, grouped into seven building categories (BC I-VII). In this work, the building types are further grouped into 30 building uses (BU 1-30) identified in Table 1, based on the criterion that each group includes similar building types sharing the same common assumptions and default values (i.e., building types hospital and clinic are grouped in one building use BU20—hospital/clinic). For example, BU20 refers to hospital/clinic with continuous annual operation (i.e., 24 hr/day, 7 days/week, 12 months/year); operative temperature at 22 °C for heating and 26 °C for cooling and operative relative humidity at 35% and 50%, respectively; fresh air requirements at 10.5 $m^3/hr/m^2$ heated floor area; DHW consumption, depending on the bed capacity, at 22.00–43.90 m^3/bed; annual artificial lighting hours at 7571 hr, internal heat gains from occupants at 27 W/m^2 heated floor area and from appliances at 7.5 W/m^2 heated floor area (internal heat gains from lighting are calculated based on the installed power) [25]. Note that, for some building uses (e.g., BU17—primary/secondary school), the DHW is considered negligible and is not taken into account in the calculations.

As of January 2011, issuing an EPC is compulsory for all buildings that are being sold and for whole buildings that are being rented. From January 2011 until December 2019, a total of about 1,820,000 EPCs have been issued, out of which 16.9% are for NR buildings. These include whole buildings, i.e., stand-alone buildings with the same use (e.g., office, school) and part buildings, i.e., building units (e.g., office spaces) in a multi-floor building. The exploited information from the EPCs include the calculated primary energy use per unit floor area (kWh/m^2) for the total consumption (EUIp) and for the different end uses, i.e., space heating (EUIp,H), space cooling (EUIp,C), domestic hot water (EUIp,DHW) and lighting (EUIp,L), as well as the total CO_2 emissions per unit floor area ($kgCO_2/m^2$). According to KENAK, the national primary energy conversion factors per energy carrier are 2.9 for electricity, 1.1 for heating oil and 1.05 for natural gas. Correspondingly, the carbon emission national conversion factors are 0.989 $kgCO_2/kWh$, 0.264 $kgCO_2/kWh$ and 0.196 $kgCO_2/kWh$.

Table 1. Energy use (kWh/m^2) and CO_2 emission (kg/m^2) intensities for the different building categories and building uses in the NR Dbase.

Building Category Building Use	# Bldgs/Heated Area (m^2)	EUIp,H Av [Min–Max]	EUIp,C Av [Min–Max]	EUIp,DHW Av [Min–Max]	EUIp,L Av [Min–Max]	EUIp Av [Min–Max]	CO_2 Emissions Av [Min–Max]
BCI. Temporal Residence	**6393/3,778,978**	**57.2**	**129.5**	**40.5**	**170.3**	**397.3**	**127.9**
BU1—Hotel (annual)	940/1,066,594	132.5 [0.1–586.2]	143.5 [5.5–451.8]	70.2 [0.0–287.9]	203.3 [2.9–689.0]	549.2 [102.0–1347.6]	172.9 [25.4–445.2]
BU2—Hotel (summer)	2270/1,844,069	7.4 [0.1–29.7]	120.0 [8.9–453.5]	24.2 [0.0–122.4]	143.9 [12.7–393.0]	259.4 [66.4–797.9]	97.0 [15.2–272.4]
BU3—Guest house (annual)	992/279,233	168.0 [0.5–665.4]	146.7 [1.7–503.3]	64.0 [0.0–288.1]	224.4 [12.2–583.6]	602.8 [67.2–1584.0]	191.1 [31.9–540.7]
BU4—Guest house (summer)	2073/524,446	8.7 [0.1–35.2]	123.3 [3.5–425.9]	22.0 [0.0–117.1]	159.7 [2.9–449.9]	313.4 [11.9–849.3]	103.3 [18.1–289.9]
BU5—Guest house (winter)	9/3882	258 [117.2–503.4]	6.0 [1.0–17.0]	22.9 [13.2–37.0]	146.2 [118.2–204.3]	433.6 [263.1–659.3]	128.5 [79.2–177.5]
BU6—Boarding school/Quarters/Dormitory	109/60,754	143.9 [0.5–763.7]	152.2 [8.1–403.7]	63.4 [0.0–329.9]	238.6 [80.4–475.8]	597.5 [155.8–1414.5]	194.6 [52.7–497.7]
BCII. Public assembly	**5955/1,573,991**	**206.6**	**257.4**	**101.2**	**130.9**	**695.8**	**220.2**
BU7—Restaurant	2597/439,908	235.5 [0.4–905.5]	272.0 [3.2–766.1]	165.8 [0.0–410.8]	140.2 [0.2–500.1]	813.4 [150.1–1945.7]	257.1 [51.0–660.9]
BU8—Pastry/Coffee shop	2153/252,214	353.5 [0.1–1405.1]	321.7 [5.7–938.2]	49.8 [0.0–81.5]	174.9 [0.3–564.5]	899.7 [100.8–2243.5]	277.6 [33.4–1294.5]
BU9—Night/Music hall	428/154,744	104.5 [0.2–371.5]	121.1 [8.8–408.4]	53.5 [0.0–86.0]	45.4 [5.8–208.7]	324.5 [74.6–707.3]	103.5 [24.9–239.4]
BU10—Theater/Cinema	50/29,467	103.8 [14.9–337.6]	206.5 [36.9–431.2]		74.3 [22.7–290.4]	384.6 [183.3–701.8]	126.8 [62.6–239.6]
BU11—Exhibition hall/Museum	120/135,487	79.0 [3.1–317.8]	166.0 [8.4–414.0]		68.1 [4.8–210.3]	312.5 [65.8–632.5]	102.6 [22.4–210.5]
BU12—Conference hall/Auditorium/Courthouse	16/23,703	71.3 [14.0–211.9]	135.2 [51.1–296.1]		105.1 [40.9–254.2]	311.6 [145.0–657.3]	104.0 [48.4–211.3]
BU13—Bank	69/43,748	66.7 [4.4–215.4]	77.7 [4.4–229.6]		114.4 [17.7–219.8]	257.8 [69.6–487.1]	85.5 [23.7–160.4]
BU14—Multi-purpose venue	278/160,915	107.1 [0.3–396.9]	138.0 [5.3–432.0]		83.1 [1.3–250.8]	328.2 [58.6–740.8]	106.9 [20.0–252.9]
BU15—Indoor Sports hall/Swimming Pool	244/333,805	241.4 [0.1–853.4]	384.0 [41.2–792.5]	196.0 [0.0–704.6]	182.6 [48.3–455.0]	1003.9 [162.4–1971.9]	318.6 [55.4–673.1]
BCIII. Educational	**1336/1,645,060**	**105.1**	**12.5**		**64.2**	**181.6**	**53.3**
BU16—Kindergarten	152/45,203	134.8 [12.2–427.6]	4.1 [0.2–13.9]		49.6 [15.7–99.4]	177.6 [41.5–491.1]	50.5 [14.0–159.8]
BU17—Primary/Secondary School	898/1,360,392	107.8 [1.8–421.7]	8.8 [0.0–38.6]		55.6 [14.1–118.3]	172.0 [41.7–647.4]	49.2 [13.9–150.9]
BU18—University/College/Lecture rooms	116/192,895	87.6 [5.5–316.8]	40.4 [3.2–111.7]		126.0 [17.8–308.8]	254.0 [51.3–567.2]	82.7 [17.5–192.3]
BU19—Private cram school/Conservatory	170/46,571	80.7 [1.4–326.6]	10.9 [0.0–45.4]		75.9 [11.8–150.9]	167.2 [26.2–396.1]	53.6 [8.9–128.1]
BCIV. Health and Social Welfare	**796/2,068,699**	**221.3**	**212.3**	**27.7**	**197.0**	**657.8**	**205.6**
BU20—Hospital/Clinic	161/1,644,060	234.1 [4.3–771.6]	232.5 [37.2–820.1]	28.7 [0.0–112.4]	205.4 [37.7–446.0]	700.3 [222.4–1707.0]	218.9 [69.9–555.0]
BU21—Health care/Rural outpatient clinic/Consultation room	230/162,686	154.8 [9.2–630.9]	121.5 [8.3–302.5]	13.6 [0.0–41.8]	157.5 [8.3–318.5]	446.7 [101.8–1009.4]	141.3 [34.8–328.9]
BU22—Foundling hospital/Nursing home/Asylum	119/141,938	255.0 [8.6–822.6]	208.4 [28.4–577.5]	37.6 [0.0–159.5]	256.9 [78.2–621.1]	757.1 [219.5–1620.3]	235.4 [66.5–530.8]
BU23—Nursery	286/120,016	95.7 [1.7–345.9]	63.7 [0.7–208.4]	20.6 [0.0–70.8]	63.8 [10.8–142.3]	243.5 [50.7–548.1]	75.9 [17.3–165.8]
BCV. Justice/public order/safety	**22/38,474**	**160.6**	**127.4**		**278.5**	**566.5**	**182.6**
BU24—Police station	22/38,474	160.6 [15.7–637.2]	127.4 [37.2–222.8]		278.5 [60.8–579.1]	566.5 [288.2–896.3]	182.6 [83.7–278.1]
BCVI. Commercial	**11,564/3,293,795**	**143.2**	**159.9**	**1.9**	**148.6**	**453.4**	**147.6**
BU25—Shopping mall/Large retail building	435/759,465	63.9 [0.9–312.0]	141.7 [3.1–504.8]		149.4 [2.6–389.4]	354.6 [54.6–822.6]	118.6 [26.7–280.8]
BU26—Small retail building/Drugstore	11,022/2,508,652	167.1 [0.1–729.6]	165.3 [0.1–588.4]		148.3 [0.2–342.3]	480.5 [63.5–1399.8]	155.5 [9.9–477.1]
BU27—Fitness center	53/22,397	138.0 [1.3–575.4]	173.7 [8.6–507.0]	268.2 [0.1–565.3]	164.1 [29.3–335.4]	744.1 [248.8–1325.9]	237.0 [79.2–452.6]
BU28—Barber shop/Hair salon	54/3280	239.6 [24.3–577.0]	156.4 [16.5–441.9]	66.4 [0.0–105.3]	146.3 [45.0–335.4]	608.8 [320.8–1058.1]	195.7 [78.9–361.2]
BCVII. Office/Library	**2728/2,308,452**	**100.7**	**121.2**			**354.3**	**115.4**
BU29—Office	2706/2,292,503	100.5 [0.4–483.4]	121.4 [0.8–438.5]		133.2 [5.2–299.7]	354.7 [29.7–952.8]	115.5 [16.7–325.3]
BU30—Library	18/15,949	130.5 [10.9–421.9]	99.7 [24.8–250.2]		81.4 [38.6–148.3]	303.5 [144.6–581.0]	94.9 [49.3–160.0]
NR Dbase	**28,970/14,707,447**	**128.0**	**147.2**	**25.5**	**147.5**	**448.0**	**142.9**

Empty cells refer to BUs that have no DHW in the calculations according to the national regulation.

3. Method

The following sections provide an overview of the approach used in this work to exploit the available information from the EPCs and derive practical baselines on the energy use and CO_2 emissions for NR buildings. The evolution of these baselines over time are also investigated in order to conclude whether they can be considered as representatives. The main steps of the method used in this work is illustrated in Figure 2 and elaborated in the following sections.

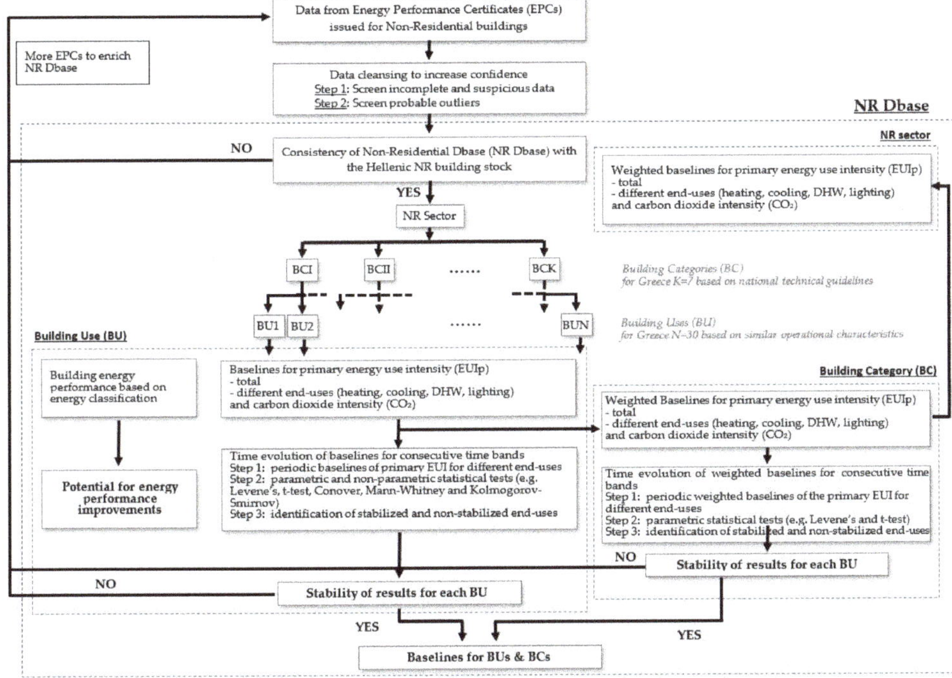

Figure 2. The main steps of the method.

3.1. Data Cleansing

Since EPC databases are one of the main sources of information about the energy use for various types of buildings, the quality of stored data is of vital importance [26]. By 2014, some kind of quality control processes of the EPCs were established in all EU member states [27]. In Greece, the first level data quality controls are performed by the responsible authority [28]. To further increase data confidence and ensure compliance, the available data were also screened before being used in the following analysis. The data cleansing process involved a series of checks in order to remove incomplete, erroneous, or otherwise suspicious data, thus ensuring that the data used are within reasonable limits and that derived baselines are reliable [29]. Accordingly, EPCs were excluded if they were:

- Not complying with the national regulation requirements;
- Issued by penalized energy inspectors;
- Issued in the initial phase of the national energy efficiency subsidy program;
- Incomplete or not finally submitted;
- Including negative values, heated floor area greater than or less than 20% of the total floor area, zero CO_2 emissions (when the energy carrier is not exclusively biomass), EUIp less than 5 kWh/m^2 (for buildings with no RES) and greater than 8000 kWh/m^2;

The resulting valid data for NR buildings reached about 235,000 EPCs or 76% of the initial database. As expected, most of the valid EPCs (86%) were issued for part buildings. However, since whole buildings are considered more representative of the various building uses, the following analysis is based on the available data for whole buildings. Accordingly, the valid data for NR whole buildings are about 31,700 EPCs.

Most data sets contain errors that can be identified as outliers having unusually large or small values. Such data may influence the statistical analysis and bias the results. A widely used method to screen data for outliers is the so-called Tukey boxplot method [30]. The method employs the interquartile range (IQR) as a measure of how spread-out the values are, being equal to the difference between 75th (Q3) and 25th (Q1) percentiles. Tukey defined Q1 − 1.5 × IQR and Q3 + 1.5 × IQR as "inner fences", Q1 − 3 × IQR and Q3 + 3 × IQR as "outer fences", the observations between an inner fence and the corresponding outer fence as "possible outliers", and anything beyond outer fences as "probable outliers".

In order to decide on the proper selection of the cut-off point one needs to know the shape of the empirical distribution of the data. Each end use of the different BUs was treated separately, since in the NR sector, BUs have very different energy profiles. Skewness and kurtosis are good indicators for nonsymmetrical distributions and the size and direction of the tails. Skewness is the degree of distortion from a symmetrical distribution. Skewness values between −0.5 and 0.5 indicate fairly symmetrical data, while skewness values less than −1 or greater than 1 correspond to skewed data. Kurtosis is actually the measure of outliers present in the distribution. The standard normal distribution has a kurtosis of three [31]. In more than 90% of the cases, the skewness of the end uses' empirical distribution differed significantly (at the 95% confidence level) from that corresponding to a normal distribution, reaching values higher than 10 in some cases. Similarly, more than 85% of the kurtosis values differed significantly (at the 95% confidence level) from those corresponding to a normal distribution, being, in some cases, higher than 100. The results indicated nonsymmetrical data distributions with long, heavy tails to the right (high positive skewness values).

The distance from normality was confirmed by a set of Kolmogorov–Smirnov tests against the data following the normal distribution [32]. Accordingly, as an additional quality control step, the valid EPCs for NR whole-buildings were further screened in order to identify and exclude outliers, using the aforementioned Tukey boxplot method. For more than half of the valid EPCs setting the cut-off point at the inner fence would label more than 15% (and up to 37%) of the cases as outliers, whereas for the outer fence the percentage of the data points considered outliers was more reasonable, less than 10% for about 75% of the cases (and up to 24%).

Applying the additional screening on the valid data for NR whole buildings (setting the cut-off point at the outer fence), the resulting database for NR whole buildings comprises about 28,800 EPCs, referred to as the NR Dbase in the following analysis.

3.2. NR Dbase

In order to justify the use of the resulting average values as baselines for the different BUs, the work assesses whether the NR Dbase could be considered representative of the Hellenic NR BS. The distribution of available data is defined for the number of buildings and size (i.e., heated floor area) and is then compared to the corresponding distribution of the NR BS from official national statistics [3,6]. In addition, the spatial distributions of the audited buildings according to climate zones, as well as to the four major socioeconomic regions (NUTS1 regions) are defined and compared to the corresponding distribution of the NR BS [6]. In case of similarity in profiles, the NR Dbase can be considered representative. Otherwise the results should be used with caution and the definition of baselines should be repeated at a later date when more data become available.

3.3. Primary Energy Use and CO_2 Intensities

The calculated primary energy use intensity is derived for the total consumption and for the four end uses taken into account for each BU, since these independent variables can give a better insight than the aggregated total. Similarly, the total CO_2 emissions per unit floor area are also derived for all BUs. The corresponding averages of the primary energy use and CO_2 intensities, for the different building categories, are calculated as the weighted average of all building uses included in the category, using the total heated area as weight. Finally, the average primary energy use and CO_2 intensities for the entire Hellenic NR sector are calculated as the weighted average of the corresponding values of the different building categories.

3.4. Building Energy Performance

In order to evaluate the energy performance of NR buildings, the distribution of the audited buildings into the nine energy classes of the national rating system is considered. In an attempt to compare the BUs as to their energy behavior, the energy classes are given a score from 1 (highest energy class A+) to 9 (lowest energy class G). The weighted average energy score for each BU (based on the number of buildings in each energy class) is then calculated, providing an indication about the average energy behavior of each BU.

Further analysis also investigates the distribution of energy use intensity per energy class for all the BUs, as well as for the entire NR sector. This can provide some insight on the gap of the audited buildings from the excellent energy performance.

3.5. Time Evolution of Intensities

Issued EPCs for whole NR buildings have increased from 184 by the end of 2011 to 28,790 by the end of 2019. The following analysis focuses on the differences between the periodic average EUIp values over time, in order to conclude whether they can be considered stabilized and therefore representative.

As a first step, the time series of the EPCs in NR Dbase is organized in the cumulative datasets of consecutive years. For the available national database, this refers to nine time bands, covering the entire period of implementation and available EPCs from 2011 until 2019. For example, the first time band includes the data from the EPCs issued up to 2011, the second time band includes all the data issued up to 2012, and so forth till the ninth (last) time band that includes the entire database up to 2019. The periodic averages of EUIp and CO_2 and the respective standard errors are then calculated for all 30 BUs for each time band.

As a second step, the degree of differentiation of the specific indicators (e.g., EUIp for the different end uses, CO_2 emissions) is investigated among pairs of consecutive time bands. For the available national database, this refers to eight pairs of the available data. For example, the first pair considers the data from the EPCs issued up to 2011 and those issued up to 2012. The second pair considers the data from the EPCs issued up to 2012 and those issued up to 2013 and so forth till the eighth pair with data from the EPCs issued up to 2018 and those issued up to 2019. Initially, two parametric tests are performed for each building use for the comparison of their (a) variances, and (b) averages.

Available statistical methods testing for the homogeneity of variance include Barlett's, Hartley's, Cochran's and Levene's tests [33]. From these, Levene's test is the most common assessment for homogeneity of variance, since it does not require normality of the underlying data. In this work, the vast majority of EPCs' probability distribution per BU is non-normal, as already discussed in Section 3.1. Accordingly, the Levene's test [34] is used to evaluate the possible changes in the variances of the indicators (e.g., total primary energy use intensity and for the different end uses) among pairs of consecutive time bands. The null hypothesis being tested is that the population variances of the two samples are equal, so in other words there is no change in their variance with the addition of new data.

One of the most common tests for equality of averages of two unrelated samples is the *t*-test [35]. In this work, the two independent samples' *t*-test was used for equal or unequal variances, depending on the Levene's test results, in order to check for possible changes in the average values among consecutive time bands. The null hypothesis is that the population average values of the two unrelated samples are equal.

Supplementary to the parametric tests, three nonparametric tests are also applied to compare the respective frequency distributions. The complementary nonparametric tests include the Conover test for equality of variances, the Mann–Whitney (M–W) test for equality of the averages, and the Kolmogorov–Smirnov (K–S) homogeneity test [32].

A crucial step in these parametric and nonparametric tests is evaluating how compatible the sample data is with the null hypothesis. This probability is called the p-value. The null hypothesis is rejected for small p-values (typically ≤0.05) as they indicate strong evidence against it. The null hypothesis is retained for large p-values (>0.05).

These two parametric and three nonparametric tests are applied for all BUs, for the specific indicators across the eight pairs of consecutive time bands. The results include the statistic values and the p-values of the respective tests. A dataset is characterized as stabilized (S) when there are no failures of the null hypotheses for any of the five tests, for all pairs of the time bands. If any two pairs of time bands fail in the variance tests, the dataset is characterized almost stabilized (AS), provided that the average tests and the K–S homogeneity test have no failures. In all other cases, the dataset is considered as nonstabilized (NS).

A similar approach is followed for testing the results for the different building categories (BCI–BCVII), applying the two parametric tests. Accordingly, the Levene's test and the appropriate two independent samples' *t*-test were used to test the weighted averages of EUIp for the different end uses and for the CO_2 emissions. The use of nonparametric tests is not suitable since the database is composed of weighted averages for the different indicators.

4. Results

4.1. NR Dbase

The profiles of the number of buildings and the total floor area for the NR BS and NR Dbase are illustrated in Figure 3. The buildings included in the NR Dbase comprise only 4.1% of the NR BS, but the total floor area the percentage reaches 23.5%. Figure 3b indicates that the breakdown of the data available in the NR Dbase into the various building uses is similar to that of the national building stock.

The spatial distribution of buildings according to climate zones as well as NUTS1 region for the NR BS and NR Dbase are illustrated in Figure 4. The available data in the NR Dbase are distributed throughout the country, resembling the distribution of the national building stock, according to NUTS1. On the other hand, it appears that the number of certificates issued in the two southern zones (A and B) is relatively higher than the northern/colder zones (C and D). These observations should be taken into account when considering national average values.

4.2. Primary Energy Use and CO_2 Intensities

The derived indicators of energy use and CO_2 emissions for the different building categories and building uses are summarized in Table 1. The lowest calculated EUIp corresponds to BU19—private cram school/conservatory averaging 167.2 kWh/m^2, since these buildings have limited operating and artificial lighting hours and small internal heat gains from occupants and appliances, while DHW is not considered. On the other hand, BU15—indoor sports hall/swimming pool buildings have the highest calculated EUIp, averaging 1003.9 kWh/m^2, due to their high fresh air requirements, high internal loads and high DHW demand. For information, the corresponding values for residential whole buildings range from 252.7 kWh/m^2 in multifamily houses (MFH) to 383.5 kWh/m^2 in single-family houses (SFH). The average CO_2 emissions for NR and residential buildings range from 49.2 kg/m^2 (BU17—primary/secondary school) to 318.6 kg/m^2 (BU15) and from 72.2 kg/m^2 in MFH to 99.5 kg/m^2 in SFH.

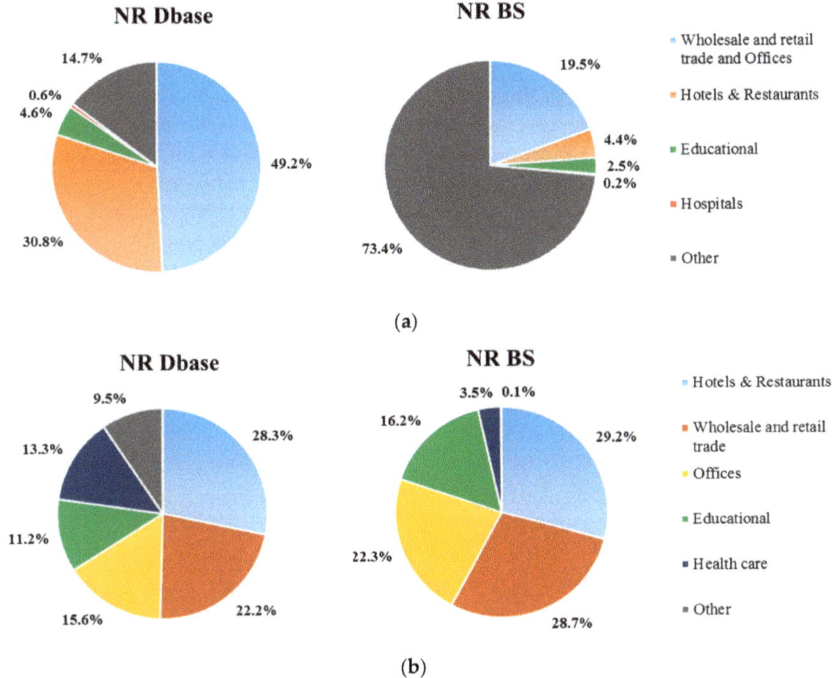

Figure 3. Distribution of (**a**) number of buildings and (**b**) total floor area of buildings in the nonresidential database from the certificates (NR Dbase) and the national building stock data (NR BS).

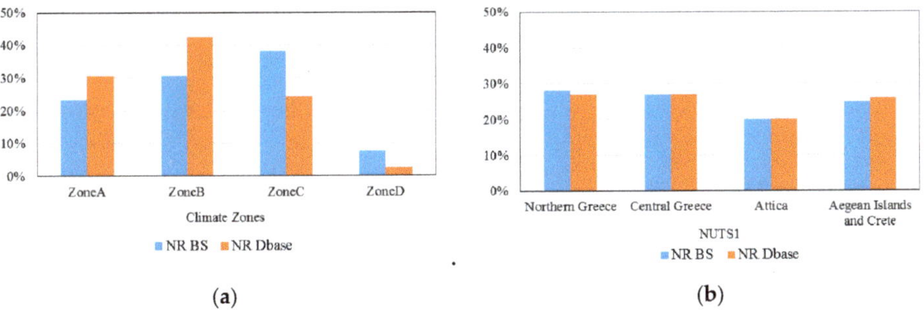

Figure 4. Spatial distribution of nonresidential buildings in the database from the certificates (NR Dbase) and the national building stock data (NR BS) (**a**) to climate zones and (**b**) to NUTS1 classification.

Analysis of the calculated primary energy per end use (Table 1) reveals that, in NR buildings, lighting is the most energy-consuming service in 40% of the BUs, followed by space heating (33%), space cooling (23%) and DHW (4%). Lighting is about half of the total primary energy for some building uses like BU2—hotel (summer) and BU18—university/college/lecture rooms, which may be attributed to the combination of high requirements for light levels and older lighting technology used. As expected, space heating is the most significant end use for school buildings, contributing up to 70% of the total in BU16—kindergarten and 63% in BU17—primary/secondary school, due to their unique operational characteristics (e.g., limited to the heating season and mainly morning hours). Space cooling reaches about half of the total energy use in BU10—theater/cinema and BU11—exhibition hall/museum, mainly due to their large volumes, high internal gains and fresh air requirements. The energy use for

DHW starts from zero for several BUs, which covers all their needs with solar collectors and reaches up to 36% for BU27—fitness center. Once again, keep in mind that for some BUs the national technical guidelines neglect the DHW demand.

The public assembly buildings (BCII) usually include building uses with many operating hours, large volumes, latent loads and internal heat gains and as a result, they exhibit the highest weighted average EUIp (695.8 kWh/m^2) and CO_2 emissions (220.2 kg/m^2). On the other hand, educational buildings (BCIII) have the lowest weighted average EUIp (181.6 kWh/m^2) and CO_2 emissions (53.3 kg/m^2), since they have limited operating hours and they don't have special energy requirements (Figure 5). As expected, DHW is not a significant end use for NR buildings. Space heating is the predominant end use for BCIII—educational buildings (58% of the total primary energy use), space cooling for BCII—public assembly buildings (37%), while lighting for BCV—justice/public order/safety and BCI—temporary residence buildings (49% and 43% respectively). For the remaining BCs, the contributions of the three end uses are more balanced.

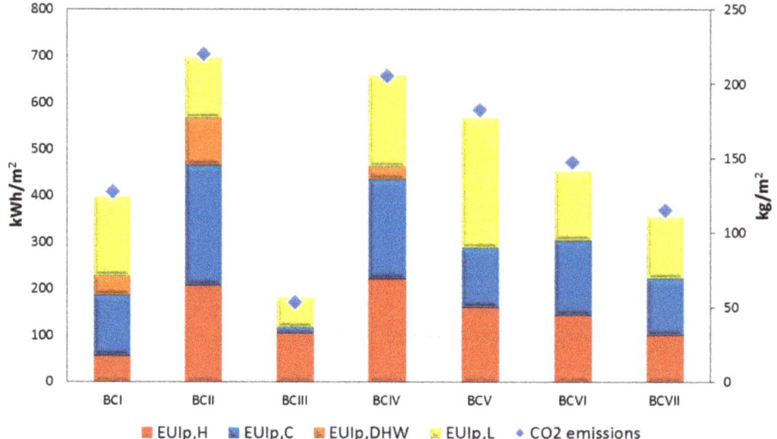

Figure 5. Weighted average primary energy use intensities (EUIp) per end use (columns, primary axis) and CO_2 emissions (diamonds, secondary axis) for the seven building categories (BCI-VII).

Looking at the Hellenic NR sector, the weighted average EUIp and CO_2 emissions are 448.0 kWh/m^2 and 142.9 kg/m^2; this is about 26% and 52% greater than the corresponding weighted averages for the residential sector. Lighting is the most energy-consuming service (33%), followed by cooling (32%), heating (29%) and DHW (6%).

4.3. Energy Performance

The data analysis confirmed the relatively low energy performance of existing NR buildings in Greece (Figure 6a). About 30% of the buildings are rated at class D; yet, this is better than the Hellenic residential buildings for which about 50% are rated at class G [36]. Similar results are reported in Spain [18] where NR buildings have a slightly better energy performance than residential buildings, with 26.4% of NR buildings rated at class D, followed by class E (22.8%) and class C (19.8%).

The weighted average energy score for each building use is presented in Figure 6b. Generally, all BUs have similar energy performance, with the average energy performance ranging between class C and class E. The building uses with the higher energy performance (best average energy scores) are BU2—hotel (summer), BU24—police station and BU13—bank. On the other hand, the worst energy performance (lowest scores) corresponds to BU17—primary/secondary school, BU28—barber shop/hair salon and BU26—small retail building/drugstore.

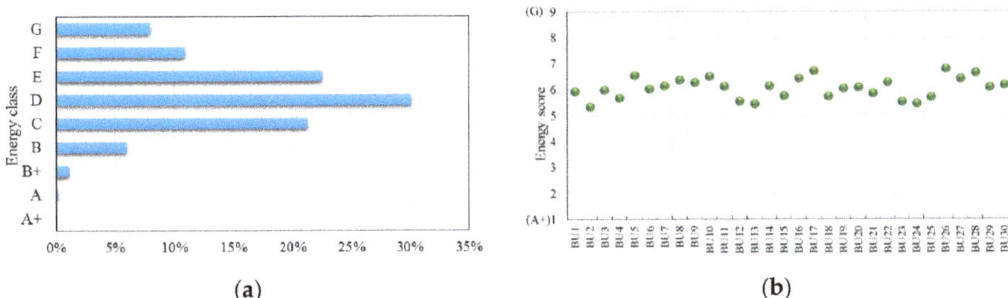

Figure 6. The distribution of NR buildings in energy classes (**a**) and the weighted energy score of NR buildings (**b**).

As expected, the average EUIp increases from the highest to the lowest energy class, ranging from 84.6 kWh/m^2 (class A+) to 760.3 kWh/m^2 (class G), as presented in Figure 7. Considering that 41% of the NR buildings are ranked in energy class E or lower, there is a significant potential for energy performance improvements by renovating the existing building stock. More insight is gained by referring to the distribution of average EUIp per energy class for the different building uses (Figure 8). The overall trend is similar, with the exception of some building uses (i.e., BU5, BU13, BU20 and BU24) for which the available sample is currently very small for detailed discretization. Overall, there is a very large range of EUIp between the highest and lowest energy class. The average decrease in EUIp between class G and class B (which corresponds to good energy behavior, according to KENAK) is about 65%. Comparatively, to reach the nZEB, the EUIp of the existing buildings will have to be improved by 76%, which demonstrates the challenges ahead for the national efforts to meet the ambitious energy targets in the coming decades, towards a decarbonized building stock by 2050.

4.4. Time Evolution of Intensities

The periodic baselines of the primary energy use and CO$_2$ intensities, as well as their respective standard error, according to the nine time bands, are summarized in Tables 2–7, for all building uses and building categories. Overall, there is no clear trend regarding the periodic averages amongst the different time bands. On the other hand, in most cases, standard errors are decreasing. This trend indicates that progressively the average values are becoming more accurate since there are smaller deviations in the available data.

A total of five statistical tests were applied to investigate the differentiation of the specific indicators across the eight pairs of the consecutive time bands, namely two parametric (Levene's and *t*-test) and three nonparametric (Conover, M–W and K–S) tests (see Section 3.5). The *t*-test and M–W test used to test the equality of averages, the Levene's and the Conover tests were for the equality of the variances, while the K–S test was used for the equality of the homogeneity.

As an example, the detailed results from all these tests (statistic values and p-values) as well as the retention or the rejection of the respective null hypothesis (equality of variances, averages and distributions) are summarized in Table 8 for one indicative building use (i.e., BU2—hotel (summer)). The failure codes (F-codes) summarize the rejections of the null hypothesis from the various tests, using a three-digit convention: the first digit refers to the results from the equality of averages tests; the second to the equality of the variances tests; the third to the results from the homogeneity test. For the first two digits, "0" represents retention of the null hypothesis in both parametric and nonparametric tests, while "1" signifies the rejection of the null hypothesis in at least one of the tests. For the third digit, "0" represents the retention of the null hypothesis in the K–S test, while "1" signifies its rejection. Accordingly, the nomenclature of "000" indicates that there is no failure in any of the tests, while "111" represents failure in all tests. As another example, for EUIp, L under the T2 time band "101" represents failure in the average tests (in this case both the *t*-test and the M–W test fail), no failure in any variance tests (Levene's and Conover test) and failure in the K–S test.

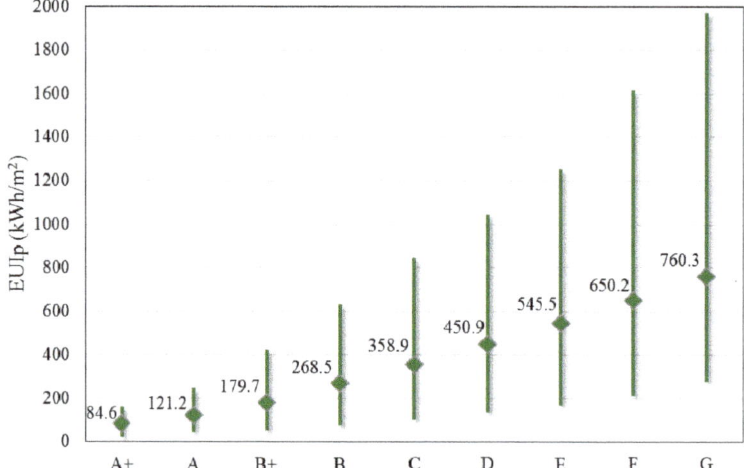

Figure 7. Calculated maximum, minimum and average (diamond) total primary energy use intensity for the different energy classes.

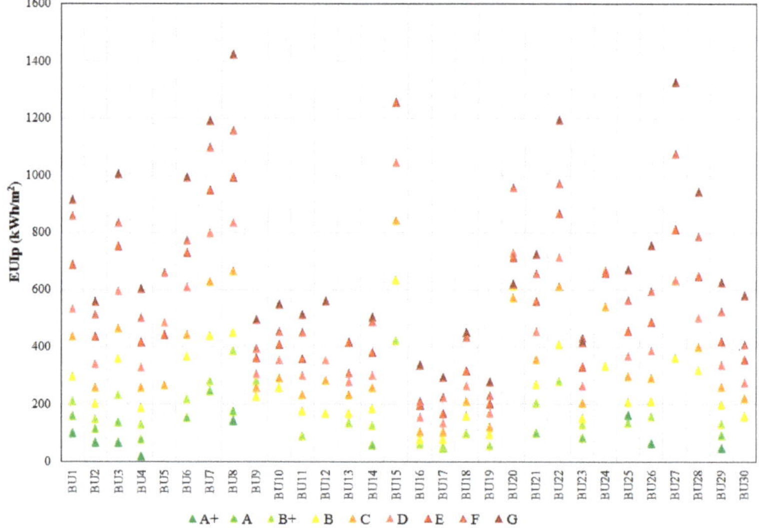

Figure 8. Calculated average primary energy use intensity per energy class (from A+ to G) for the various building uses (1 to 30, defined in Table 1).

Similar results have been produced for all BUs. The failures of the various null hypothesis for the different energy use and CO_2 emission intensities are summarized in Table 9, following the same three-digit nomenclature that was previously elaborated. To facilitate the interpretation of the results, failures are illustrated with shaded cells. Note that the datasets of the first two pairs of time bands (T1 and T2) that cover the early implementation years of the certification scheme, are not considered representative, due to the very small number of cases. Accordingly, these two pairs (T1 and T2) are not included in Table 9 and were not considered in the following analysis.

Table 2. Time evolution of number of EPCs and calculated average total primary energy use for the different building categories and building uses in the NR Dbase.

BC				Consecutive Time Bands					
	Upto2011	Upto2012	Upto2013	Upto2014	Upto2015	Upto2016	Upto2017	Upto2018	Upto2019
BU	Number of Buildings/EUIp ± Standard Error of the Mean (kWh/m^2)								
BCI	**57/399.4 ± 32.0**	**654/428.0 ± 8.0**	**1686/407.6 ± 4.8**	**2112/405.3 ± 4.2**	**2394/401.1 ± 3.9**	**3328/407.6 ± 3.4**	**4587/404.8 ± 2.9**	**5465/399.0 ± 2.6**	**6393/397.3 ± 2.4**
BU1	11/658.5 ± 68.3	98/605.1 ± 23.4	208/589.5 ± 15.1	268/581.9 ± 12.9	324/577.1 ± 11.7	484/565.7 ± 9.2	610/566.5 ± 8.4	758/551.9 ± 7.6	940/549.2 ± 6.9
BU2	40/265.7 ± 19.5	243/291.2 ± 7.8	651/294.3 ± 4.1	814/294.8 ± 3.6	943/291.1 ± 3.4	1181/292.2 ± 3.0	1686/295.0 ± 2.4	1947/294.4 ± 2.3	2270/295.4 ± 2.1
BU3	2/739.4 ± 193.6	118/608.5 ± 18.2	281/619.9 ± 11.6	343/619.6 ± 10.7	384/613.9 ± 10.1	559/603.4 ± 8.8	743/600.4 ± 7.7	874/601.6 ± 7.1	992/602.8 ± 6.7
BU4	3/315.8 ± 35.2	190/326.8 ± 7.9	535/327.5 ± 4.6	670/324.0 ± 4.1	724/321.0 ± 3.9	1075/314.7 ± 3.2	1468/314.0 ± 2.8	1789/315.1 ± 2.6	2073/313.4 ± 2.4
BU5			1/659.3	4/437.9 ± 100.8	4/437.9 ± 100.8	4/437.9 ± 100.8	4/437.9 ± 100.8	9/433.6 ± 46.6	9/433.6 ± 46.6
BU6	1/640.0	5/575.9 ± 93.3	10/593.5 ± 68.1	13/587.4 ± 58.2	15/557.5 ± 57.2	25/585.7 ± 53.9	76/563.3 ± 25.0	88/581.0 ± 23.2	109/597.5 ± 21.9
BCII	**50/863.8 ± 67.7**	**1053/741.3 ± 12.2**	**2073/710.4 ± 8.0**	**2473/703.0 ± 7.3**	**2743/692.0 ± 6.8**	**3726/672.8 ± 5.9**	**4581/662.2 ± 5.2**	**5249/695.1 ± 5.1**	**5955/695.8 ± 4.9**
BU7	21/969.1 ± 108.9	452/901.2 ± 16.5	920/844.6 ± 9.9	1095/835.2 ± 8.8	1195/826.6 ± 8.4	1602/819.0 ± 7.0	1996/807.1 ± 6.0	2283/811.8 ± 5.7	2597/813.4 ± 5.4
BU8	17/995.8 ± 144.5	407/919.5 ± 20.0	782/881.7 ± 12.4	914/874.4 ± 11.2	993/869.0 ± 10.6	1384/863.5 ± 8.6	1687/863.1 ± 7.6	1897/881.8 ± 7.4	2153/899.7 ± 7.2
BU9	5/286.6 ± 30.1	104/336.8 ± 11.1	208/324.0 ± 7.8	242/319.1 ± 7.0	252/318.6 ± 6.9	324/318.4 ± 5.8	369/317.9 ± 5.5	398/321.1 ± 5.3	428/324.5 ± 5.3
* BU10		8/387.4 ± 23.8	13/344.7 ± 21.6	19/389.1 ± 30.2	20/390.6 ± 28.7	30/366.0 ± 21.7	38/372.6 ± 19.8	44/375.8 ± 17.9	50/384.6 ± 16.8
* BU11	1/217.6	15/342.1 ± 26.8	28/314.5 ± 24.8	34/322.9 ± 23.9	45/323.8 ± 19.5	62/306.7 ± 16.4	80/310.5 ± 14.3	103/315.9 ± 12.6	120/312.5 ± 11.6
* BU12		1/355.1	4/287.9 ± 68.6	4/287.9 ± 68.6	6/286.4 ± 47.2	7/271.6 ± 42.5	11/273.0 ± 29.6	14/313.7 ± 37.7	16/311.6 ± 35.1
* BU13	3/10.3 ± 68.7	6/293.0 ± 33.9	10/289.6 ± 20.3	16/287.6 ± 17.1	24/287.6 ± 16.5	37/271.8 ± 13.6	55/262.3 ± 10.1	59/261.8 ± 9.6	69/257.8 ± 9.5
* BU14		26/298.6 ± 21.5	54/301.5 ± 14.0	77/297.9 ± 11.4	104/305.5 ± 10.3	150/308.5 ± 8.7	191/311.6 ± 7.9	234/318.4 ± 7.8	278/328.2 ± 7.6
BU15	3/1133.7 ± 68.8	34/1019.1 ± 48.4	54/979.8 ± 38.3	72/1001.2 ± 34.1	104/970.1 ± 27.3	130/977.7 ± 23.5	154/969.5 ± 22.2	217/1009.8 ± 21.4	244/1003.9 ± 20.6
BCIII	**15/224.2 ± 27.3**	**150/176.8 ± 5.5**	**247/179.3 ± 4.6**	**381/187.3 ± 3.8**	**657/174.8 ± 2.8**	**840/171.1 ± 2.5**	**948/172.1 ± 2.3**	**1183/179.7 ± 2.2**	**1336/181.6 ± 2.1**
* BU16	1/206.1	6/205.8 ± 34.6	13/167.5 ± 20.2	26/170.9 ± 15.3	59/166.1 ± 9.4	87/169.8 ± 9.0	99/165.9 ± 8.5	133/182.0 ± 8.2	152/177.6 ± 7.4
* BU17	13/186.0 ± 26.5	118/169.6 ± 5.7	154/169.2 ± 5.1	261/179.1 ± 4.5	491/169.1 ± 3.3	594/163.6 ± 2.9	650/164.3 ± 2.8	816/172.8 ± 2.7	898/172.0 ± 2.6
* BU18	1/290.8	10/252.5 ± 20.3	26/257.2 ± 21.6	31/251.0 ± 18.5	36/245.7 ± 16.9	59/253.6 ± 13.3	75/255.3 ± 11.4	91/254.1 ± 10.3	116/254.0 ± 8.7
* BU19		16/177.3 ± 15.4	54/158.9 ± 6.4	63/161.7 ± 6.6	71/162.3 ± 6.0	100/165.3 ± 5.6	124/162.9 ± 4.9	143/165.1 ± 4.5	170/167.2 ± 4.4
BCIV	**10/793.5 ± 147.7**	**60/555.7 ± 44.1**	**145/551.6 ± 19.9**	**211/572.7 ± 20.4**	**302/571.4 ± 14.5**	**412/607.3 ± 13.3**	**545/607.2 ± 11.4**	**683/633.8 ± 10.2**	**796/657.8 ± 9.5**
BU20	2/265.1 ± 19.1	3/471.7 ± 206.9	10/529.6 ± 63.7	25/592.0 ± 57.9	47/621.1 ± 40.5	65/650.0 ± 34.2	92/657.0 ± 28.0	129/673.4 ± 24.0	161/700.3 ± 21.1
BU21	1/245.6	17/460.1 ± 43.1	38/470.1 ± 29.0	57/464.3 ± 24.1	81/463.2 ± 17.9	112/456.8 ± 15.7	164/437.6 ± 11.9	203/448.5 ± 10.4	230/446.7 ± 10.1
BU22	4/993.8 ± 162.8	10/865.9 ± 90.1	31/815.8 ± 47.7	39/776.5 ± 41.6	48/747.1 ± 35.4	69/698.4 ± 33.5	86/721.6 ± 30.5	112/753.1 ± 26.1	119/757.1 ± 25.5
BU23	3/219.1 ± 23.7	30/248.3 ± 14.2	66/258.1 ± 10.8	90/257.7 ± 8.9	126/251.8 ± 8.2	166/247.2 ± 7.0	203/241.4 ± 6.5	239/245.5 ± 6.1	286/243.5 ± 5.6
BCV		**5/560.4 ± 90.4**	**8/578.4 ± 72.8**	**8/578.4 ± 72.8**	**9/575.2 ± 64.3**	**10/558.2 ± 60.0**	**15/595.5 ± 43.3**	**19/586.4 ± 35.7**	**22/566.5 ± 35.0**
* BU24		5/560.4 ± 90.4	8/578.4 ± 72.8	8/578.4 ± 72.8	9/575.2 ± 64.3	10/558.2 ± 60.0	15/595.5 ± 43.3	19/586.4 ± 35.7	22/566.5 ± 35.0
BCVI	**43/327.0 ± 20.0**	**1006/423.7 ± 4.6**	**2792/429.2 ± 2.6**	**3576/427.3 ± 2.4**	**4017/419.1 ± 2.3**	**6351/428.6 ± 1.9**	**8555/432.7 ± 1.7**	**9876/443.2 ± 1.6**	**11564/453.4 ± 1.6**
* BU25	9/250.9 ± 14.5	48/342.7 ± 15.1	120/363.9 ± 11	163/352.6 ± 10.4	191/344.7 ± 9.6	262/345.6 ± 8.1	354/348.6 ± 7.1	388/354.2 ± 7.0	435/354.6 ± 6.7
* BU26	34/378.8 ± 28.2	950/444.1 ± 5.1	2646/446.2 ± 2.9	3382/446 ± 2.6	3795/443.4 ± 2.4	6035/452.5 ± 2	8129/457.9 ± 1.8	9396/469.3 ± 1.8	11022/480.5 ± 1.8
BU27		5/568.4 ± 110.9	17/647.3 ± 45.5	20/656.4 ± 41.9	20/656.4 ± 41.9	29/666.5 ± 35.2	39/662.6 ± 28.9	46/711.0 ± 33.0	53/744.1 ± 34.3
BU28		3/511.4 ± 116.2	9/640 ± 63.1	11/624.5 ± 53.5	11/624.5 ± 53.5	25/593.6 ± 37.4	33/578.2 ± 32.8	46/601.8 ± 28.5	54/608.8 ± 27.0
BCVII	**9/293.5 ± 42.5**	**251/335.3 ± 7.7**	**577/341.7 ± 5.0**	**769/338.5 ± 4.1**	**974/334.0 ± 3.7**	**1601/345.6 ± 2.8**	**2000/346.7 ± 2.5**	**2314/350.1 ± 2.4**	**2724/354.3 ± 2.3**
* BU29	9/293.5 ± 42.9	249/335.2 ± 7.6	571/341.8 ± 5.3	762/338.5 ± 4.4	963/334.1 ± 3.9	1588/345.9 ± 3.2	1986/347.1 ± 2.9	2300/350.4 ± 2.8	2706/354.7 ± 2.6
* BU30		2/360.6 ± 87.8	6/334 ± 35.9	7/333.4 ± 30.3	11/322.3 ± 23.2	13/310.5 ± 22.8	14/299.5 ± 23.8	14/299.5 ± 23.8	18/303.5 ± 27.5

* BUs that have no DHW in the calculations according to the national regulation.

Table 3. Time evolution of calculated average primary energy use for space heating for the different building categories and building uses in the NR Dbase.

BC				Consecutive Time Bands					
BC	Upto2011	Upto2012	Upto2013	Upto2014	Upto2015	Upto2016	Upto2017	Upto2018	Upto2019
BU	EUIp,H ± Standard Error of the Mean (kWh/m^2)								
BCI	**98.5 ± 19.2**	**70.2 ± 3.8**	**59.4 ± 2.2**	**60.2 ± 1.9**	**58.8 ± 1.8**	**61.7 ± 1.5**	**59.8 ± 1.3**	**57.7 ± 1.2**	**57.2 ± 1.1**
BU1	228.5 ± 46.7	146.8 ± 12.2	143.2 ± 8.1	145.3 ± 6.8	140.7 ± 6.1	133.3 ± 4.7	136.5 ± 4.3	132.4 ± 3.8	132.5 ± 3.5
BU2	5.7 ± 0.6	8.1 ± 0.4	7.1 ± 0.2	7 ± 0.2	7 ± 0.2	7 ± 0.2	7.2 ± 0.1	7.2 ± 0.1	7.4 ± 0.1
BU3	126 ± 31.2	172.2 ± 10.8	177.4 ± 7.6	180.5 ± 7.2	177.6 ± 6.6	173.8 ± 5.6	170.2 ± 4.8	167.4 ± 4.4	168 ± 4.2
BU4	14.8 ± 6.6	9.5 ± 0.5	8.7 ± 0.3	8.4 ± 0.2	8.2 ± 0.2	8.2 ± 0.2	8.5 ± 0.2	8.5 ± 0.2	8.7 ± 0.1
BU5			503.4	284.1 ± 96.7	284.1 ± 96.7	284.1 ± 96.7	284.1 ± 96.7	258.5 ± 44.5	258.5 ± 44.5
BU6	376.7	187.7 ± 58.1	169.8 ± 33.7	159 ± 31.2	156.6 ± 28.9	195.8 ± 34.4	131 ± 15.8	136.8 ± 14.5	143.9 ± 13.4
BCII	**259.1 ± 36.9**	**230.0 ± 6.2**	**210.0 ± 3.9**	**205.9 ± 3.5**	**199.6 ± 3.2**	**195.3 ± 2.7**	**192.3 ± 2.4**	**203.5 ± 2.3**	**206.6 ± 2.2**
BU7	314.4 ± 57.1	273.3 ± 9.0	237.7 ± 5.7	233.6 ± 5.1	230.3 ± 4.8	227.5 ± 4.0	222.6 ± 3.5	229.8 ± 3.4	235.5 ± 3.3
BU8	463.9 ± 102.0	382.5 ± 15.1	347.1 ± 9.4	339.4 ± 8.5	336.5 ± 8.1	331.8 ± 6.6	330.2 ± 5.8	341.1 ± 5.7	353.5 ± 5.5
BU9	66.2 ± 30.9	106.1 ± 6.9	100.1 ± 4.9	98.3 ± 4.4	97.5 ± 4.3	98 ± 3.8	98.1 ± 3.6	101.4 ± 3.6	104.5 ± 3.6
BU10		124.5 ± 14.0	100.6 ± 13.4	107.5 ± 12.0	103.8 ± 12.0	94.1 ± 8.8	102.8 ± 10.6	102.8 ± 9.3	103.8 ± 8.3
BU11	31.5	93.1 ± 19.6	80 ± 11.6	83.6 ± 12.0	84.3 ± 9.4	79 ± 7.4	79.9 ± 6.8	79.7 ± 6.0	79.1 ± 5.3
BU12		70.9	49.3 ± 9.7	49.3 ± 9.7	57.2 ± 13.2	51 ± 12.7	58.8 ± 9.0	71.4 ± 13.2	71.3 ± 12.2
BU13	104.8 ± 34.3	87.2 ± 22.0	71.8 ± 14.9	75.4 ± 13.6	75.4 ± 10.9	71.9 ± 9.3	67.8 ± 7.0	68.1 ± 6.6	66.7 ± 6.0
BU14		81.7 ± 10.2	85.8 ± 7.7	84.5 ± 6.2	89.2 ± 5.6	92.6 ± 4.9	94.9 ± 4.6	100.8 ± 4.8	107.1 ± 4.8
BU15	138.1 ± 66.9	228.4 ± 25.3	208.2 ± 19.0	216.2 ± 16.6	212.5 ± 13.3	215.8 ± 12.2	216.8 ± 11.3	245 ± 11.3	241.4 ± 10.6
BCIII	**105.9 ± 20.2**	**101.3 ± 5.4**	**96.9 ± 4.0**	**109.0 ± 3.7**	**103.5 ± 2.6**	**98.6 ± 2.3**	**98.7 ± 2.1**	**106.4 ± 2.1**	**105.1 ± 1.9**
BU16	138.5	149.5 ± 36.7	117.8 ± 21.0	121.9 ± 15.3	113 ± 8.8	116 ± 8.4	112.3 ± 7.9	129.2 ± 8.0	123.8 ± 7.2
BU17	112.6 ± 24.9	104 ± 5.8	101.1 ± 5.1	115.4 ± 4.5	107 ± 3.3	100.6 ± 2.9	100.9 ± 2.9	108.8 ± 2.8	107.8 ± 2.6
BU18	93.5	72.1 ± 16.4	75.2 ± 9.8	71.8 ± 8.6	69 ± 7.6	78.3 ± 7.3	79 ± 6.4	84.2 ± 6.2	87.6 ± 5.5
BU19		85.3 ± 17.3	71.6 ± 6.5	74.7 ± 6.8	76.4 ± 6.2	79 ± 5.6	75.6 ± 4.9	77.3 ± 4.5	80.7 ± 4.4
BCIV	**449.7 ± 102.9**	**253.5 ± 31.0**	**183.6 ± 13.5**	**194.1 ± 10.5**	**203.9 ± 7.7**	**199.3 ± 6.5**	**186.9 ± 5.5**	**206.0 ± 5.0**	**221.3 ± 4.8**
BU20	133.7 ± 3.1	197.5 ± 63.9	150.3 ± 26.2	193.1 ± 27.9	219.9 ± 21.3	209.4 ± 16.8	196.5 ± 13.9	215.7 ± 12.5	234.1 ± 12.0
BU21	67.7	198.8 ± 40.2	191.7 ± 26.2	185.4 ± 20.1	177.1 ± 14.8	163.4 ± 12.7	141.3 ± 9.7	153 ± 8.8	154.8 ± 8.5
BU22	576.1 ± 133.0	421 ± 100.1	296.6 ± 41.5	281.4 ± 34.0	261.3 ± 28.6	233.4 ± 22.8	240.5 ± 20.7	254.4 ± 17.8	255 ± 17.5
BU23	67.2 ± 10.8	90.8 ± 11.1	98.6 ± 8.8	101.2 ± 7.1	103 ± 6.4	97.8 ± 5.4	94.3 ± 4.8	97.5 ± 4.5	95.7 ± 4.1
BCV		**234 ± 42.8**	**263.8 ± 62**	**263.8 ± 62.0**	**246.3 ± 57.4**	**228.2 ± 54.5**	**190.9 ± 40.5**	**177.1 ± 33.0**	**160.6 ± 29.9**
BU24		234 ± 42.8	263.8 ± 62	263.8 ± 62.0	246.3 ± 57.4	228.2 ± 54.5	190.9 ± 40.5	177.1 ± 33.0	160.6 ± 29.9
BCVI	**118.5 ± 12.4**	**135.7 ± 3.1**	**126.1 ± 1.8**	**123.5 ± 1.6**	**118.7 ± 1.4**	**124.9 ± 1.2**	**127.0 ± 1.0**	**135.1 ± 1.1**	**143.2 ± 1.1**
BU25	143.3 ± 22.6	92.8 ± 9.9	78.7 ± 5.9	68.1 ± 4.7	64.3 ± 4.1	63.9 ± 3.7	63 ± 3.2	63.8 ± 3.1	63.9 ± 2.9
BU26	101.6 ± 16.3	146.9 ± 4.0	140.1 ± 2.1	138.8 ± 1.9	137.8 ± 1.8	143.9 ± 1.5	148.1 ± 1.3	157.7 ± 1.3	167.1 ± 1.3
BU27		165.6 ± 37.7	127.9 ± 22	122.1 ± 19.2	122.1 ± 19.2	110.4 ± 16.9	101.1 ± 13.1	117.9 ± 13.7	138.0 ± 17.1
BU28		201.4 ± 63.1	250.6 ± 42.6	240.3 ± 37.9	240.3 ± 37.9	237.1 ± 27.5	220.7 ± 23.9	230.4 ± 20.3	239.6 ± 20.4
BCVII	**110.5 ± 39.8**	**86.1 ± 3.8**	**85.9 ± 2.5**	**84.9 ± 2.2**	**84.1 ± 2.0**	**90.8 ± 1.6**	**93.2 ± 1.5**	**96.0 ± 1.5**	**100.7 ± 1.4**
BU29	110.5 ± 39	85.7 ± 5.0	85.6 ± 3.4	84.6 ± 2.9	83.6 ± 2.6	90.6 ± 2.1	93 ± 1.9	95.9 ± 1.8	100.5 ± 1.8
BU30		178.2 ± 129.3	123.9 ± 41.2	120.3 ± 35.0	137 ± 27.2	125.5 ± 24.1	119.6 ± 23.1	119.6 ± 23.1	130.5 ± 26.2

Table 4. Time evolution of calculated average primary energy use for space cooling for the different building categories and building uses in the NR Dbase.

				Consecutive Time Bands					
BC	Upto2011	Upto2012	Upto2013	Upto2014	Upto2015	Upto2016	Upto2017	Upto2018	Upto2019
BU				EUIp,C $\pm$ Standard Error of the Mean (kWh/m^2)					
BCI	**111.0 ± 9.5**	**129.9 ± 2.6**	**124.4 ± 1.5**	**123.4 ± 1.4**	**122.7 ± 1.3**	**122.5 ± 1.1**	**123.4 ± 0.9**	**126.3 ± 0.9**	**129.5 ± 0.8**
BU1	183.9 ± 33.8	151 ± 7.4	141.8 ± 4.7	140.2 ± 4.3	142.1 ± 4.0	139.8 ± 3.2	138.9 ± 2.8	140.9 ± 2.6	143.5 ± 2.4
BU2	80.7 ± 10.4	114.7 ± 4.7	111.9 ± 2.6	111.3 ± 2.3	109.2 ± 2.1	109.2 ± 1.9	113 ± 1.6	115.9 ± 1.5	120 ± 1.4
BU3	91.9 ± 1.0	136 ± 7.5	137.6 ± 4.2	135.8 ± 3.8	139.4 ± 3.8	137.1 ± 3.1	137.9 ± 2.8	143.3 ± 2.7	146.7 ± 2.5
BU4	130.4 ± 25.0	124.4 ± 4.4	125.3 ± 2.7	122.8 ± 2.4	122 ± 2.3	117.7 ± 1.9	117.7 ± 1.6	121.9 ± 1.5	123.3 ± 1.4
BU5			2.8	2.1 ± 0.2	2.1 ± 0.2	2.1 ± 0.2	2.1 ± 0.2	6 ± 2.1	6 ± 2.1
BU6	136.8	148.2 ± 53.0	152.8 ± 26.2	178.9 ± 25.8	161.6 ± 25.2	156.2 ± 19.5	132.4 ± 7.9	142.9 ± 7.8	152.2 ± 7.1
BCII	**217.8 ± 23.7**	**263.7 ± 5.0**	**254.7 ± 3.3**	**251.6 ± 3.0**	**250.8 ± 2.8**	**245.1 ± 2.4**	**241.4 ± 2.1**	**255.1 ± 2.0**	**257.4 ± 1.9**
BU7	256.8 ± 30.0	295.3 ± 7.0	276.2 ± 4.3	272.2 ± 3.8	270.2 ± 3.6	266.9 ± 3	263.8 ± 2.6	269.1 ± 2.4	272 ± 2.3
BU8	344 ± 51.2	321.4 ± 8.2	309.8 ± 5.3	309.1 ± 4.8	307.6 ± 4.5	306.4 ± 3.7	307.5 ± 3.3	315.7 ± 3.2	321.7 ± 3.0
BU9	128 ± 16.7	125 ± 6.9	120.2 ± 4.7	118.7 ± 4.3	119 ± 4.2	119.3 ± 3.6	120 ± 3.4	120.3 ± 3.3	121.1 ± 3.2
BU10		213.1 ± 12.6	169.3 ± 20.0	190 ± 17.7	185.3 ± 17.5	192 ± 13.6	197.4 ± 12.3	197.9 ± 12.1	206.6 ± 12.1
BU11	101.5	192.5 ± 16.4	181.3 ± 19.3	180.6 ± 17.6	175.3 ± 14.9	161.6 ± 12.3	164 ± 10.8	166.1 ± 9.3	166 ± 8.5
BU12		163.2	152.9 ± 52.3	152.9 ± 52.3	148.4 ± 34.4	139.2 ± 30.5	132.5 ± 21.8	134.7 ± 17.9	135.2 ± 17.2
BU13	81.7 ± 13.9	84.5 ± 11.9	91.2 ± 8.8	85.2 ± 8.1	93.3 ± 9.4	83.9 ± 6.9	79 ± 5.1	79.9 ± 4.9	77.7 ± 4.6
BU14		135.5 ± 16.0	132.9 ± 9.1	125.9 ± 7.3	129.9 ± 6.3	131.3 ± 5.5	129.8 ± 4.7	133.8 ± 4.4	138 ± 4.1
BU15	141 ± 15.9	368.7 ± 28.7	369.5 ± 22.1	373.2 ± 18.7	367.1 ± 14.5	374 ± 12.7	370.4 ± 11.6	380.4 ± 10.0	384 ± 9.8
BCIII	**41.1 ± 8.8**	**12.9 ± 1.5**	**13.9 ± 1.1**	**11.5 ± 0.8**	**9.9 ± 0.5**	**10.6 ± 0.4**	**10.8 ± 0.4**	**11.0 ± 0.4**	**12.5 ± 0.4**
BU16	10.1	5.7 ± 1.6	4.9 ± 0.9	3.9 ± 0.5	3.3 ± 0.3	3.8 ± 0.3	4 ± 0.3	3.9 ± 0.2	4.1 ± 0.2
BU17	18.2 ± 2.4	11.3 ± 0.7	11.1 ± 0.6	8.1 ± 0.4	7.7 ± 0.3	8.3 ± 0.3	8.4 ± 0.3	8.7 ± 0.2	8.8 ± 0.2
BU18	81.3	32.2 ± 6.9	35.3 ± 4.8	35.9 ± 4.1	35.6 ± 3.7	36.4 ± 2.9	37.3 ± 2.5	37.8 ± 2.3	40.4 ± 2.1
BU19		9.9 ± 1.7	11.2 ± 0.9	11.1 ± 0.8	10.4 ± 0.8	10.8 ± 0.7	11 ± 0.7	10.9 ± 0.6	10.9 ± 0.5
BCIV	**152.8 ± 34.7**	**129.0 ± 11.6**	**170.3 ± 7.8**	**187.5 ± 10.3**	**177.9 ± 7.4**	**195.3 ± 6.6**	**200.5 ± 5.7**	**205.2 ± 4.8**	**212.3 ± 4.4**
BU20	51.3 ± 3.3	137.7 ± 86.5	189.3 ± 30.6	206 ± 29.9	200.2 ± 18.7	217.6 ± 16.4	227.2 ± 13.7	225.5 ± 10.8	232.5 ± 9.6
BU21	13.2	113.1 ± 14.6	126.4 ± 11.9	124.5 ± 9.7	118.6 ± 7.4	121.8 ± 6.4	124.2 ± 5.0	122.7 ± 4.4	121.5 ± 4.1
BU22	191.4 ± 49.9	186.7 ± 27.1	220.2 ± 16.8	207.9 ± 14.4	204.7 ± 11.9	196.3 ± 12.3	199 ± 10.7	209.4 ± 8.9	208.4 ± 8.5
BU23	55 ± 18.1	66.4 ± 7.5	70.4 ± 5.2	67.6 ± 4.9	61.9 ± 4.0	62.8 ± 3.4	62.5 ± 3.0	62.9 ± 2.7	63.7 ± 2.4
BCV		**110.6 ± 26.9**	**108.4 ± 18.2**	**108.4 ± 18.2**	**105.9 ± 16.2**	**107.9 ± 14.7**	**125 ± 12.7**	**128.7 ± 10.4**	**127.4 ± 10.1**
BU24		110.6 ± 26.9	108.4 ± 18.2	108.4 ± 18.2	105.9 ± 16.2	107.9 ± 14.7	125 ± 12.7	128.7 ± 10.4	127.4 ± 10.1
BCVI	**116.1 ± 13.1**	**147.1 ± 2.5**	**151.8 ± 1.4**	**152.0 ± 1.3**	**149.0 ± 1.2**	**151.5 ± 1.0**	**153.4 ± 0.9**	**157.0 ± 0.8**	**159.9 ± 0.8**
BU25	88.4 ± 8.6	123.2 ± 8.6	136.5 ± 6.2	135.9 ± 6.4	131.7 ± 5.8	134 ± 4.8	139.8 ± 4.5	142.3 ± 4.3	141.7 ± 4.1
BU26	134.9 ± 19.1	153.3 ± 2.8	156.4 ± 1.7	156.5 ± 1.5	155.1 ± 1.4	157.1 ± 1.1	157.9 ± 1.0	161.7 ± 0.9	165.3 ± 0.9
BU27		175.9 ± 52.4	147.7 ± 23.4	147.6 ± 20.7	147.6 ± 20.7	146 ± 14.7	148.9 ± 12.2	165.1 ± 13.2	173.7 ± 13.3
BU28		116.4 ± 68.3	174.5 ± 31.3	167.4 ± 25.7	167.4 ± 25.7	135.5 ± 16.7	138.9 ± 14.2	153.8 ± 12.7	156.4 ± 11.4
BCVII	**77.5 ± 20.4**	**120.6 ± 4.7**	**121.8 ± 3.4**	**118.4 ± 2.7**	**115.5 ± 2.3**	**117.6 ± 1.6**	**116.7 ± 1.4**	**119.4 ± 1.3**	**121.2 ± 1.2**
BU29	77.5 ± 21.8	120.7 ± 4.6	121.8 ± 3.1	118.4 ± 2.5	115.6 ± 2.2	117.6 ± 1.7	116.8 ± 1.5	119.5 ± 1.5	121.4 ± 1.4
BU30		82.6 ± 36.7	129.3 ± 27.1	122.7 ± 23.9	103.9 ± 17.7	106.1 ± 16.6	103.8 ± 15.5	103.8 ± 15.5	99.7 ± 12.2

Table 5. Time evolution of calculated average primary energy use for DHW for the different building categories and building uses in the NR Dbase.

| BC | | | | Consecutive Time Bands | | | | | |
BC	Upto2011	Upto2012	Upto2013	Upto2014	Upto2015	Upto2016	Upto2017	Upto2018	Upto2019
BU				EUIp,DHW $\pm$ Standard Error of the Mean (kWh/m^2)					
BCI	**62.4 ± 8.5**	**63.9 ± 2.1**	**46.8 ± 1.1**	**45.7 ± 1.0**	**44.3 ± 0.9**	**43.8 ± 0.7**	**41.9 ± 0.6**	**41.0 ± 0.5**	**40.5 ± 0.5**
BU1	105.2 ± 25.9	90.5 ± 6.9	73.3 ± 4.2	75.1 ± 3.6	73.3 ± 3.2	71.3 ± 2.5	68.8 ± 2.1	69 ± 1.9	70.2 ± 1.7
BU2	50.3 ± 3.7	50.1 ± 1.9	34.9 ± 1.1	32.4 ± 0.9	30.6 ± 0.8	28.9 ± 0.7	26.7 ± 0.6	25.4 ± 0.5	24.2 ± 0.5
BU3	197.3 ± 8.9	85.6 ± 6.3	71.6 ± 3.7	69.2 ± 3.2	65.8 ± 2.9	61.5 ± 2.4	62.7 ± 2.0	64.2 ± 1.9	64 ± 1.7
BU4	60.3 ± 26.9	34.6 ± 2.1	26.3 ± 1.1	25.3 ± 0.9	25.6 ± 0.9	23.2 ± 0.7	22.7 ± 0.6	22.2 ± 0.5	22 ± 0.5
BU5			21.0	21.4 ± 5.5	21.4 ± 5.5	21.4 ± 5.5	21.4 ± 5.5	22.9 ± 2.8	22.9 ± 2.8
BU6	40.9	60.8 ± 18.1	45.9 ± 10.9	38.9 ± 9.2	38.7 ± 8.7	40.3 ± 9.0	69.9 ± 7.4	67.2 ± 6.6	63.4 ± 5.8
BCII	**245.6 ± 35.7**	**119.2 ± 3.4**	**113.0 ± 2.1**	**111.0 ± 1.9**	**105.5 ± 1.7**	**99.4 ± 1.4**	**96.2 ± 1.2**	**102.8 ± 1.2**	**101.2 ± 1.1**
BU7	217 ± 20.3	185.9 ± 3.4	175.4 ± 2.2	173.5 ± 2	171 ± 1.9	168.3 ± 1.6	166.7 ± 1.4	166.3 ± 1.3	165.8 ± 1.2
BU8	57.8 ± 3.4	52.3 ± 0.8	50.7 ± 0.6	50.4 ± 0.5	50 ± 0.5	49.8 ± 0.4	49.6 ± 0.4	49.7 ± 0.4	49.8 ± 0.3
BU9	56.7 ± 7.3	55.9 ± 1.4	53.9 ± 1.1	53.6 ± 1.0	53.7 ± 1.0	53.6 ± 0.8	53.4 ± 0.8	53.4 ± 0.8	53.5 ± 0.7
* BU10									
* BU11									
* BU12									
* BU13									
* BU14									
BU15	698.8 ± 5.8	253.8 ± 34.4	236.8 ± 23.8	237.8 ± 19.0	207.1 ± 14.4	205.8 ± 12.0	198.7 ± 10.6	199.3 ± 8.4	196 ± 7.9
BCIII									
* BU16									
* BU17									
* BU18									
* BU19									
BCIV	**53.9 ± 10.2**	**41.2 ± 3.6**	**30.1 ± 1.6**	**29.0 ± 1.1**	**26.9 ± 0.9**	**27.5 ± 0.9**	**24.9 ± 0.7**	**26.4 ± 0.7**	**27.7 ± 0.6**
BU20	7.9	27.5 ± 19.6	25.1 ± 6.5	29 ± 5	27.7 ± 3.3	27.6 ± 2.7	24.9 ± 2.1	27.2 ± 1.8	28.7 ± 1.8
BU21	41.8	19.1 ± 3.3	17.2 ± 1.9	15.7 ± 1.4	16.0 ± 1.1	16.3 ± 0.9	13.8 ± 0.8	13.5 ± 0.7	13.6 ± 0.6
BU22	63.9 ± 19.9	56.8 ± 11.3	44.1 ± 4.9	40.7 ± 4.5	38.9 ± 3.9	38.2 ± 3.8	39.6 ± 3.5	37.8 ± 2.9	37.6 ± 2.8
BU23	58.5 ± 7.0	34.2 ± 3.4	27.5 ± 2.1	25.9 ± 1.7	23.8 ± 1.4	22.7 ± 1.2	21.6 ± 1	21.5 ± 1	20.6 ± 0.8
BCV									
* BU24									
BCVI		*1.0 ± 0.4*	*1.8 ± 0.4*	*1.6 ± 0.3*	*1.3 ± 0.3*	*1.5 ± 0.3*	*1.7 ± 0.2*	*1.7 ± 0.2*	*1.9 ± 0.2*
* BU25									
* BU26									
BU27		120.7 ± 30.4	223.4 ± 26.0	231.2 ± 29.3	231.2 ± 29.3	245.1 ± 24.7	238.7 ± 21.7	261.2 ± 21.0	268.2 ± 19.1
BU28		70.3 ± 17.0	67.4 ± 6.1	67.4 ± 5.0	67.4 ± 5.0	66.4 ± 3.8	67.9 ± 3.1	68.2 ± 2.3	66.4 ± 2.4
BCVII									
* BU29									
* BU30									

* BUs that have no DHW in the calculations according to the national regulation.

Table 6. Time evolution of calculated average primary energy use for lighting for the different building categories and building uses in the NR Dbase.

BC	Upto2011	Upto2012	Upto2013	Upto2014	Upto2015	Upto2016	Upto2017	Upto2018	Upto2019
					Consecutive Time Bands				
BU					$EUIp,L \pm$ Standard Error of the Mean (kWh/m^2)				
BCI	**127.5 ± 15.3**	**164.0 ± 3.5**	**177.1 ± 2.3**	**176.2 ± 2.0**	**175.3 ± 1.8**	**179.7 ± 1.6**	**179.9 ± 1.3**	**174.2 ± 1.2**	**170.3 ± 1.1**
BU1	141 ± 29.0	216.7 ± 11.4	231.2 ± 7.5	221.3 ± 6.6	221 ± 5.9	221.3 ± 4.8	222.5 ± 4.4	210 ± 4.0	203.3 ± 3.6
BU2	129 ± 9.6	118.2 ± 4.7	140.5 ± 2.5	144.2 ± 2.2	144.4 ± 2.1	147.2 ± 1.9	148.2 ± 1.5	146.0 ± 1.4	143.9 ± 1.3
BU3	324.2 ± 214.9	214.9 ± 8.2	233.4 ± 6.0	234.2 ± 5.1	231.1 ± 4.8	231.0 ± 4	229.6 ± 3.4	226.7 ± 3.1	224.4 ± 3.0
BU4	110.3 ± 23.9	158.3 ± 5.5	167.3 ± 3.2	167.6 ± 2.9	165.2 ± 2.8	165.7 ± 2.2	165.2 ± 1.9	162.7 ± 1.7	159.7 ± 1.6
BU5			132.0	130.3 ± 0.6	130.3 ± 0.6	130.3 ± 0.6	130.3 ± 0.6	146.2 ± 9.2	146.2 ± 9.2
BU6	85.6	179.2 ± 24.4	229.8 ± 35.4	214.2 ± 29.2	203.8 ± 26.5	195.3 ± 18	230.8 ± 10.7	234.8 ± 9.5	238.6 ± 8.3
BCII	**141.3 ± 11.8**	**128.3 ± 2.7**	**132.7 ± 2.0**	**134.4 ± 1.8**	**136.0 ± 1.7**	**133.2 ± 1.5**	**132.5 ± 1.3**	**133.9 ± 1.2**	**130.9 ± 1.2**
BU7	180.8 ± 24.3	146.7 ± 4.6	155.3 ± 3.3	155.8 ± 3.1	155.1 ± 2.9	156.5 ± 2.6	154.1 ± 2.3	146.9 ± 2.1	140.2 ± 2.0
BU8	130.1 ± 23.9	163.3 ± 4.8	174.2 ± 3.4	175.5 ± 3.1	174.9 ± 2.9	175.5 ± 2.4	176 ± 2.2	175.4 ± 2.0	174.9 ± 1.9
BU9	35.8 ± 11.7	49.8 ± 4.3	49.8 ± 2.8	48.4 ± 2.6	48.4 ± 2.5	47.4 ± 2.2	46.4 ± 2	45.9 ± 1.9	45.4 ± 1.9
BU10		49.8 ± 12.9	74.9 ± 21.9	91.6 ± 19.0	101.5 ± 20.6	79.9 ± 14.8	72.4 ± 12.2	75.2 ± 10.9	74.3 ± 9.6
BU11	84.6	56.4 ± 8.8	53.2 ± 5.1	58.7 ± 5.8	64.2 ± 5.3	66.0 ± 4.3	66.7 ± 4	70.2 ± 4.1	68.1 ± 3.8
BU12		121.0	85.8 ± 13.8	85.8 ± 13.8	80.8 ± 9.3	81.4 ± 7.9	81.7 ± 8.4	107.6 ± 17.3	105.1 ± 15.2
BU13	123.8 ± 46.5	121.3 ± 22.4	126.6 ± 14.4	127.0 ± 11.4	119.0 ± 9	116 ± 6.4	115.5 ± 4.7	113.8 ± 4.5	114.4 ± 4.5
BU14		81.4 ± 6.4	82.8 ± 5.0	87.4 ± 4.5	86.5 ± 4.0	84.6 ± 3.3	86.9 ± 3.1	83.8 ± 2.7	83.1 ± 2.4
BU15	155.7 ± 19.7	168.1 ± 8.5	165.4 ± 6.6	174 ± 6.5	183.4 ± 6.1	182.4 ± 5.6	183.9 ± 5.7	185.3 ± 4.8	182.6 ± 4.6
BCIII	**77.3 ± 9.1**	**62.5 ± 2.0**	**68.5 ± 2.4**	**66.8 ± 1.7**	**61.6 ± 1.1**	**62.1 ± 1.0**	**62.7 ± 1.0**	**62.4 ± 0.8**	**64.2 ± 0.8**
BU16	57.5	50.6 ± 3.0	44.7 ± 3.2	45.2 ± 2.0	49.8 ± 1.9	50.0 ± 1.4	49.6 ± 1.4	49.0 ± 1.3	49.6 ± 1.3
BU17	55.2 ± 3.5	54.4 ± 1.3	57.0 ± 1.3	55.6 ± 1.0	54.5 ± 0.7	54.8 ± 0.6	55.1 ± 0.6	55.5 ± 0.6	55.6 ± 0.6
BU18	116.0	148.3 ± 8.2	146.7 ± 12.9	143.4 ± 11.2	141.1 ± 10.2	138.9 ± 8.2	139.1 ± 7.0	132.1 ± 6.2	126.0 ± 5.0
BU19		82.1 ± 7.8	76.1 ± 3	75.9 ± 2.7	75.4 ± 2.5	75.6 ± 2.2	76.2 ± 1.9	77.0 ± 1.9	75.9 ± 1.8
BCIV	**137.1 ± 14.9**	**132.0 ± 10.0**	**167.8 ± 10.3**	**162.3 ± 5.5**	**162.9 ± 4.5**	**186.1 ± 3.7**	**195.6 ± 3.4**	**196.7 ± 3.0**	**197.0 ± 2.7**
BU20	72.2 ± 19.2	109.0 ± 38.4	164.9 ± 30.6	163.9 ± 15.4	173.3 ± 10.8	196.1 ± 9.7	209 ± 8.9	205.7 ± 7.2	205.4 ± 6.5
BU21	122.9	129.2 ± 10.6	138.7 ± 8.1	141.3 ± 6.6	153.3 ± 5.7	156.6 ± 5.4	159.2 ± 4.2	160.1 ± 3.9	157.5 ± 3.6
BU22	162.4 ± 19.6	201.6 ± 17.9	254.8 ± 18.0	246.6 ± 14.9	242.3 ± 12.7	231.7 ± 10.9	243.6 ± 10.6	252.3 ± 9.2	256.9 ± 9.5
BU23	38.4 ± 10.5	56.9 ± 4.1	61.6 ± 2.9	62.9 ± 2.5	63.5 ± 2.0	64.2 ± 1.7	63.5 ± 1.6	63.9 ± 1.4	63.8 ± 1.3
BCV		**215.8 ± 51.0**	**206.2 ± 35.1**	**206.2 ± 35.1**	**223 ± 35.3**	**222 ± 31.6**	**279.5 ± 33.8**	**280.6 ± 27**	**278.5 ± 24**
* BU24		215.8 ± 51.0	206.2 ± 35.1	206.2 ± 35.1	223 ± 35.3	222 ± 31.6	279.5 ± 33.8	280.6 ± 27	278.5 ± 24
BCVI	*92.4 ± 15.4*	*139.9 ± 2.0*	*149.6 ± 1.1*	*150.3 ± 0.9*	*150.1 ± 0.9*	*150.8 ± 0.7*	*150.7 ± 0.6*	*149.5 ± 0.6*	*148.6 ± 0.5*
BU25	19.1 ± 16.3	126.7 ± 11.5	148.7 ± 6.5	148.5 ± 5.4	148.7 ± 4.9	147.6 ± 4.1	146.1 ± 3.3	148.3 ± 3.2	149.4 ± 3
BU26	142.3 ± 14.2	143.9 ± 1.8	149.8 ± 1.1	150.7 ± 1.0	150.5 ± 0.9	151.6 ± 0.7	151.9 ± 0.6	150.0 ± 0.6	148.3 ± 0.5
BU27		106.2 ± 16.1	148.3 ± 15.9	155.6 ± 15.6	155.6 ± 15.6	165.1 ± 12.4	173.9 ± 11.4	166.8 ± 10.2	164.1 ± 9.6
BU28		123.3 ± 17.1	147.4 ± 11.4	149.3 ± 9.5	149.3 ± 9.5	154.6 ± 7.8	150.7 ± 6.0	149.4 ± 6.4	146.3 ± 5.7
BCVII	**105.5 ± 11.7**	**128.7 ± 2.6**	**134.4 ± 1.7**	**135.4 ± 1.4**	**134.8 ± 1.3**	**137.5 ± 1.0**	**137.2 ± 0.9**	**135.1 ± 0.9**	**132.8 ± 0.8**
BU29	105.5 ± 12.3	128.8 ± 2.7	134.8 ± 1.7	135.8 ± 1.5	135.3 ± 1.4	137.9 ± 1.1	137.6 ± 1	135.4 ± 0.9	133.2 ± 0.9
BU30		99.9 ± 4.9	80.9 ± 7.2	90.5 ± 11.4	81.4 ± 9.2	78.9 ± 8.1	81.4 ± 7.9	81.4 ± 7.9	81.4 ± 6.5

* BUs that have no DHW in the calculations according to the national regulation.

Table 7. Time evolution of calculated average CO_2 emissions intensity for the different building categories and building uses in the NR Dbase.

				Consecutive Time Bands					
BC	Upto2011	Upto2012	Upto2013	Upto2014	Upto2015	Upto2016	Upto2017	Upto2018	Upto2019
BU	$CO_2 \pm$ Standard Error of the Mean (kg/m^2)								
BCI	**131.1 ± 1.1**	**130.9 ± 0.1**	**127.9**	**127.3**	**126.5**	**129.6**	**129.5**	**127.9**	**127.9**
BU1	181.9 ± 17.9	186.4 ± 7.0	183.9 ± 4.7	179.3 ± 4.2	178.8 ± 3.7	177.5 ± 3.0	178.1 ± 2.7	173.1 ± 2.5	172.9 ± 2.2
BU2	77.5 ± 4.8	88.5 ± 2.4	93.3 ± 1.4	94.2 ± 1.2	93.6 ± 1.1	94.4 ± 1.0	96.2 ± 0.8	96.3 ± 0.8	97.0 ± 0.7
BU3	236.7 ± 50.7	183.1 ± 5.3	190.9 ± 3.6	192.3 ± 3.4	191.2 ± 3.2	189.8 ± 2.8	189.0 ± 2.4	190.7 ± 2.3	191.1 ± 2.1
BU4	95.3 ± 8.1	101.0 ± 2.5	104.4 ± 1.6	104.1 ± 1.4	103.2 ± 1.3	102.0 ± 1.1	102.4 ± 1.0	103.5 ± 0.9	103.3 ± 0.8
BU5			177.5	123.7 ± 25.5	123.7 ± 25.5	123.7 ± 25.5	123.7 ± 25.5	128.5 ± 12.0	128.5 ± 12.0
BU6	175.9	175.8 ± 27.7	177.3 ± 16.8	178.9 ± 15.0	169.6 ± 15.2	182.6 ± 17.1	181.8 ± 8.0	188.2 ± 7.4	194.9 ± 7.1
BCII	**233.6 ± 2.4**	**218.3 ± 0.1**	**216.7 ± 0.1**	**216.2**	**213.7**	**209.3**	**206.8**	**218.9**	**220.2**
BU7	269.2 ± 29.4	260.5 ± 3.9	256.8 ± 2.6	255.3 ± 2.4	253.6 ± 2.3	253.5 ± 2.0	251.5 ± 1.7	254.9 ± 1.7	257.1 ± 1.6
BU8	266.9 ± 35.5	260.5 ± 5.0	261.8 ± 3.6	261.8 ± 3.3	260.8 ± 3.1	261.2 ± 2.5	261.8 ± 2.2	270.3 ± 2.3	277.6 ± 2.3
BU9	76.0 ± 8.6	100.9 ± 3.4	101.1 ± 2.4	100.2 ± 2.2	100.2 ± 2.1	100.8 ± 1.8	101.0 ± 1.7	102.3 ± 1.7	103.5 ± 1.7
* BU10		121.3 ± 10.0	108.0 ± 7.8	124.8 ± 10.8	125.7 ± 10.3	119.1 ± 7.5	122.2 ± 6.8	123.8 ± 6.2	126.8 ± 5.8
* BU11	69.6	112.0 ± 8.6	102.7 ± 8.0	105.3 ± 7.6	105.6 ± 6.3	100.3 ± 5.4	101.4 ± 4.6	103.4 ± 4.1	102.6 ± 3.8
* BU12		120.7	97.8 ± 23.6	97.8 ± 23.6	97.4 ± 16.2	92.4 ± 14.6	91.8 ± 10.2	105.0 ± 12.4	104.0 ± 11.5
* BU13	86.4 ± 19.8	89.2 ± 10.2	91.6 ± 6.4	92.4 ± 5.5	93.4 ± 5.4	89.1 ± 4.3	86.5 ± 3.2	86.5 ± 3.1	85.5 ± 3.0
* BU14		95.2 ± 7.3	96.5 ± 4.5	95.7 ± 3.7	98 ± 3.3	98.9 ± 2.8	100.2 ± 2.5	103.1 ± 2.5	106.9 ± 2.5
BU15	297.8 ± 13.9	309.5 ± 13.6	302.4 ± 11.4	312.4 ± 10.5	303 ± 8.5	307.3 ± 7.4	305.5 ± 7.0	319.6 ± 6.8	318.6 ± 6.6
BCIII	**63.5 ± 1.9**	**52.1 ± 0.1**	**53.7 ± 0.1**	**54.8 ± 0.1**	**50.3**	**49.7**	**50.1**	**52.2**	**53.3**
* BU16	54.3	57.6 ± 8.2	48.3 ± 5.4	47.5 ± 3.7	45.7 ± 2.2	47.7 ± 2.3	47.0 ± 2.3	51.3 ± 2.3	50.5 ± 2.0
* BU17	49.9 ± 6.6	49.3 ± 1.5	49.6 ± 1.4	51.1 ± 1.1	47.7 ± 0.8	46.6 ± 0.7	46.8 ± 0.7	49.2 ± 0.7	49.2 ± 0.7
* BU18	87.4	81.8 ± 5.8	83.6 ± 7.3	81.3 ± 6.2	79.9 ± 5.7	82.3 ± 4.4	83.0 ± 3.8	82.7 ± 3.4	82.7 ± 2.8
* BU19		57.3 ± 4.9	50.4 ± 2.1	50.9 ± 2.0	51.3 ± 1.8	52.4 ± 1.7	51.9 ± 1.4	52.8 ± 1.4	53.6 ± 1.4
BCIV	**220.3 ± 10.1**	**164.2 ± 1.5**	**172.4 ± 0.5**	**179.9 ± 0.4**	**172.9 ± 0.3**	**188.2 ± 0.2**	**190.4 ± 0.2**	**198.1 ± 0.1**	**205.6 ± 0.1**
BU20	76.2 ± 6.3	147.7 ± 71.6	172.5 ± 22.0	187.9 ± 19.2	187.5 ± 12.5	201.6 ± 10.9	206.2 ± 8.9	210.6 ± 7.6	218.9 ± 6.6
BU21	61.0	143.2 ± 12.9	143.8 ± 8.0	142.9 ± 6.9	143.2 ± 5.2	142.6 ± 4.6	138.5 ± 3.5	140.8 ± 3.1	141.3 ± 3.1
BU22	275.1 ± 42.3	249.1 ± 21.5	244.7 ± 13.5	234.1 ± 11.7	226.7 ± 10.0	215.4 ± 9.7	224.0 ± 9.1	234.1 ± 7.9	235.4 ± 7.7
BU23	66.5 ± 8.5	76.9 ± 4.2	79.8 ± 3.1	79.8 ± 2.7	77.7 ± 2.4	76.7 ± 2.1	75.3 ± 1.9	76.3 ± 1.8	75.9 ± 1.7
BCV		**176.0 ± 32.7**	**177.9 ± 23.3**	**177.9 ± 23.3**	**178.2 ± 20.5**	**173.8 ± 18.9**	**189.8 ± 14.2**	**188.2 ± 11.6**	**182.6 ± 11.2**
* BU24		176.0 ± 32.7	177.9 ± 23.3	177.9 ± 23.3	178.2 ± 20.5	173.8 ± 18.9	189.8 ± 14.2	188.2 ± 11.6	182.6 ± 11.2
BCVI	**90.1 ± 1.1**	**133.6**	**137.1**	**136.8**	**134.5**	**137.6**	**139.0**	**143.4**	**147.6**
* BU25	51.7 ± 3.2	107.9 ± 5.9	119.2 ± 4.0	116.4 ± 3.6	114.3 ± 3.3	115.1 ± 2.8	116.2 ± 2.4	118.4 ± 2.4	118.6 ± 2.3
* BU26	116.3 ± 8.9	140.3 ± 1.6	141.8 ± 0.9	141.9 ± 0.8	141.1 ± 0.8	144.1 ± 0.6	145.8 ± 0.5	150.8 ± 0.6	155.5 ± 0.6
BU27		161.8 ± 36.6	196.7 ± 13.9	201.6 ± 13.2	201.6 ± 13.2	207.8 ± 11.6	207.7 ± 9.4	225.6 ± 11.4	237.0 ± 11.9
BU28		130.4 ± 38.7	187.7 ± 21.4	185.5 ± 17.6	185.5 ± 17.6	183.6 ± 11.4	182.2 ± 10.2	194.5 ± 9.4	195.7 ± 8.9
BCVII	**86.4 ± 11.4**	**108.6 ± 2.4**	**172.4 ± 0.5**	**109.7 ± 1.4**	**108.3 ± 1.2**	**111.8 ± 1**	**112.2 ± 0.9**	**113.6 ± 0.9**	**115.4 ± 0.8**
* BU29	86.4 ± 11.4	108.6 ± 2.4	110.8 ± 1.6	109.8 ± 1.4	108.4 ± 1.2	111.9 ± 1	112.3 ± 0.9	113.7 ± 0.9	115.5 ± 0.8
* BU30		106.4 ± 17.2	104.8 ± 9.9	104.7 ± 8.4	96.6 ± 6.7	94.2 ± 6.5	92.5 ± 6.3	92.5 ± 6.3	94.9 ± 7.1

* BUs that have no DHW in the calculations according to the national regulation.

Table 8. Results of the applied statistical tests and failure coding (F-codes) for the various energy use intensities and emissions for BU2–hotel (summer), across the eight pairs of time bands.

BU2		Pairs of Consecutive Time Bands *							
		T1	T2	T3	T4	T5	T6	T7	T8
EUI$_p$	t-test/p	−1.22/0.22	−0.35/0.73	−0.10/0.92	0.74/0.46	−0.23/0.82	−0.76/0.45	0.20/0.84	−0.33/0.74
	M–W/p	−1.91/0.06	−0.94/0.35	−0.11/0.91	−0.98/0.33	−0.51/0.61	−1.00/0.32	−0.30/0.76	−0.25/0.80
	Levene's/p	1.25/0.26	7.18/0.01	0.05/0.83	0.00/1.00	0.48/0.49	0.72/0.39	0.31/0.58	0.16/0.69
	Conover/p	−1.98/0.05	−2.40/0.02	−0.07/0.94	−0.18/0.86	−0.68/0.50	−0.74/0.46	−0.27/0.78	−0.55/0.58
	K–S/p	1.44/0.03	1.01/0.26	0.23/1.00	0.64/0.81	0.46/0.98	0.82/0.52	0.31/1.00	0.26/1.00
	F-codes	011	010	000	000	000	000	000	000
EUI$_{p,H}$	t-test/p	−3.14/0.00	2.14/0.03	0.46/0.64	−0.12/0.90	0.08/0.94	−0.82/0.41	−0.08/0.94	−1.28/0.20
	M–W/p	−2.41/0.02	−1.55/0.12	−0.62/0.54	−0.55/0.58	−0.12/0.9	−1.25/0.21	−0.54/0.59	−1.44/0.15
	Levene's/p	8.71/0.00	9.83/0.00	0.00/0.98	0.24/0.62	0.00/0.99	0.00/0.99	0.69/0.41	0.62/0.43
	Conover/p	−3.49/0.00	−4.60/0.00	−0.01/1.00	−0.52/0.60	−0.45/0.65	−0.19/0.85	−0.94/0.35	−1.11/0.27
	K–S/p	1.40/0.04	1.10/0.18	0.79/0.55	0.59/0.87	0.27/1.00	0.95/0.33	0.62/0.84	0.88/0.42
	F-codes	111	110	000	000	000	000	000	000
EUI$_{p,C}$	t-test/p	−2.76/0.01	0.54/0.59	0.18/0.86	0.66/0.51	0.01/1.00	−1.56/0.12	−1.33/0.18	−1.96/0.05
	M–W/p	−3.61/0.00	−0.08/0.94	−0.16/0.88	−0.59/0.55	−0.21/0.84	−1.52/0.13	−1.43/0.15	−2.01/0.04
	Levene's/p	3.16/0.08	2.62/0.11	0.15/0.69	0.41/0.52	0.24/0.62	0.43/0.51	0.48/0.49	1.11/0.29
	Conover/p	−2.61/0.01	−1.71/0.09	−0.21/0.84	−0.86/0.39	−0.48/0.63	−0.64/0.53	−0.74/0.46	−1.14/0.26
	K–S/p	2.49/0.00	0.98/0.29	0.27/1.00	0.59/0.88	0.32/1.00	0.69/0.73	0.82/0.52	0.94/0.34
	F-codes	111	000	000	000	000	000	000	000
EUI$_{p,DHW}$	t-test/p	0.03/0.98	7.18/0.00	1.67/0.09	1.47/0.14	1.52/0.13	2.24/0.03	1.67/0.09	1.56/0.12
	M–W/p	−0.13/0.89	−7.04/0.00	−1.48/0.14	−1.39/0.16	−1.57/0.12	−2.07/0.04	−2.02/0.04	−1.78/0.07
	Levene's/p	2.95/0.09	0.01/0.90	3.19/0.07	2.18/0.14	1.67/0.20	4.46/0.03	0.30/0.58	0.81/0.37
	Conover/p	−1.53/0.13	−0.41/0.68	−2.36/0.02	−1.97/0.05	−2.14/0.03	−2.58/0.01	−1.01/0.31	−1.67/0.09
	K–S/p	0.81/0.52	3.89/0.00	0.99/0.28	0.85/0.47	0.79/0.55	1.10/0.18	0.96/0.31	0.76/0.61
	F-codes	000	100	010	010	010	110	100	000
EUI$_{p,L}$	t-test/p	0.88/0.38	−4.46/0.00	−1.11/0.27	−0.07/0.94	−1.01/0.31	−0.41/0.68	1.06/0.29	1.08/0.28
	M–W/p	−1.46/0.15	−5.91/0.00	−0.94/0.35	−0.15/0.88	−1.00/0.32	−0.60/0.55	−1.52/0.13	−1.36/0.17
	Levene's/p	3.42/0.07	3.17/0.08	0.17/0.68	0.11/0.74	0.23/0.63	1.25/0.26	0.51/0.47	0.08/0.77
	Conover/p	−2.98/0.00	−1.53/0.13	−1.10/0.27	−0.35/0.73	−0.49/0.62	−0.86/0.39	−0.82/0.41	−0.10/0.92
	K–S/p	1.90/0.00	3.17/0.00	0.74/0.64	0.48/0.97	0.51/0.96	0.92/0.37	0.99/0.28	0.83/0.49
	F-codes	011	101	000	000	000	000	000	000
CO$_2$	t-test/p	−2.04/0.05	−1.78/0.08	−0.55/0.59	0.41/0.68	−0.55/0.58	−1.43/0.15	−0.05/0.96	−0.67/0.50
	M–W/p	−1.97/0.05	−2.27/0.02	−0.56/0.58	−0.69/0.49	−0.81/0.42	−1.75/0.08	−0.04/0.97	−0.55/0.58
	Levene's/p	4.97/0.03	3.77/0.05	0.00/1.00	0.00/0.99	0.24/0.63	0.18/0.67	0.43/0.51	0.36/0.55
	Conover/p	−2.86/0.00	−1.93/0.05	−0.23/0.82	−0.19/0.85	−0.44/0.66	−0.22/0.82	−0.32/0.75	−0.77/0.44
	K–S/p	1.31/0.06	1.50/0.02	0.42/1.00	0.57/0.90	0.60/0.86	1.05/0.22	0.27/1.00	0.41/1.00
	F-codes	110	101	000	000	000	000	000	000

* T1:upto2011-upto2012, T2:upto2012-upto2013, T3:upto2013-upto2014, T4:upto2014-upto2015, T5:upto2015-upto2016, T6:upto2016-upto2017, T7:upto2017-upto2018, T8:upto2018-upto2019

Table 9. Failures coding (F-codes) for primary energy uses and CO_2 intensities, for the different building uses in the NR Dbase.

	EUIp						EUIp,H						EUIp,C						EUIp,DHW						EUIp,L						CO_2					
Pairs of Consecutive Time Bands																																					
	T3	T4	T5	T6	T7	T8	T3	T4	T5	T6	T7	T8	T3	T4	T5	T6	T7	T8	T3	T4	T5	T6	T7	T8	T3	T4	T5	T6	T7	T8	T3	T4	T5	T6	T7	T8
BU1	000	000	000	000	000	000	000	000	000	000	000	000	000	000	000	000	000	000	000	000	000	000	000	000	000	000	000	000	101	000	000	000	000	000	000	000
BU2	000	000	000	000	000	000	000	000	000	000	000	000	000	000	000	000	000	000	100	010	010	010	110	100	000	000	000	000	000	000	000	000	000	000	000	000
BU3	000	000	000	000	000	000	000	000	000	000	000	000	000	000	000	000	000	000	000	000	000	000	000	000	000	000	000	000	000	000	000	000	000	000	000	000
BU4	000	000	000	000	000	000	000	000	000	000	000	000	000	000	000	000	000	000	000	000	110	000	000	000	000	000	000	000	000	000	000	000	000	000	000	000
BU5	000	000	000	000	000	000	000	000	000	000	000	000	000	000	000	000	000	000	010	000	000	000	000	000	000	000	010	000	000	000	000	000	000	000	000	000
BU6	000	000	000	000	000	000	000	000	000	100	000	000	000	000	000	000	000	010	000	000	000	000	110	000	000	000	000	000	000	000	000	000	000	000	000	000
BU7	000	000	000	000	000	000	000	000	000	000	010	000	000	000	000	000	000	000	000	010	000	000	000	000	000	000	000	000	101	101	000	000	000	000	000	000
BU8	000	000	000	000	000	000	000	000	000	000	000	000	000	000	000	000	000	000	000	010	000	010	010	010	000	000	000	000	000	000	000	000	000	000	110	110
BU9	000	000	000	000	000	000	000	000	000	000	000	000	000	000	000	000	000	000	000	000	000	000	000	000	000	000	000	000	000	000	000	000	000	000	000	000
BU10	000	000	000	000	000	000	000	000	000	000	000	000	000	000	000	000	000	000							000	000	010	000	000	000	000	000	000	000	000	000
BU11	000	000	000	000	000	000	000	000	000	000	000	000	000	000	000	000	000	000							000	000	000	000	010	000	000	000	000	000	000	000
BU12	000	000	000	000	000	000	000	000	000	000	000	000	000	000	000	000	000	000							000	000	000	000	000	000	000	000	000	000	000	000
BU13	000	000	000	000	000	000	000	000	000	000	000	000	000	000	000	000	000	000							000	000	000	000	000	000	000	000	000	000	000	000
BU14	000	000	000	000	000	000	000	000	000	000	000	000	000	000	000	000	000	000							000	000	000	000	000	000	000	000	000	000	000	000
BU15	000	000	000	000	000	000	000	000	000	000	000	000	000	000	000	000	000	000	000	000	000	000	000	000	000	000	000	000	000	000	000	000	000	000	000	000
BU16	000	000	000	000	000	000	000	000	000	000	000	000	000	000	000	000	000	000							000	000	000	000	000	000	000	000	000	000	000	000
BU17	000	100	000	000	100	000	100	000	000	000	110	000	101	010	000	000	000	000							000	000	000	000	000	000	000	100	000	000	100	000
BU18	000	000	000	000	000	000	000	000	000	000	000	000	000	000	000	000	000	000							000	000	000	000	000	000	000	000	000	000	000	000
BU19	000	000	000	000	000	000	000	000	000	000	000	000	000	000	000	000	000	000							000	000	000	000	000	000	000	000	000	000	000	000
BU20	000	000	000	000	000	000	000	000	000	000	000	000	000	000	000	000	000	000	000	000	000	000	000	000	000	000	000	000	000	000	000	000	000	000	000	000
BU21	000	010	010	000	000	000	000	000	000	000	000	000	000	010	000	000	000	000	010	110	000	000	000	000	010	010	000	000	000	000	010	010	000	000	000	000
BU22	000	000	000	000	000	000	000	000	000	000	000	000	000	000	000	000	000	000	000	000	000	000	000	000	000	000	000	000	000	000	000	000	000	000	000	000
BU23	000	000	000	000	000	000	000	000	000	000	000	000	000	000	000	000	000	000	000	000	000	000	000	000	000	000	000	000	000	000	000	000	000	000	000	000
BU24	000	000	000	000	000	000	000	000	000	000	000	000	000	000	000	000	000	000							000	000	000	000	000	000	000	000	000	000	000	000
BU25	000	000	000	000	000	000	000	000	000	000	000	000	000	000	000	000	000	000							000	000	000	000	000	000	000	000	000	000	000	000
BU26	000	000	110	100	111	111	000	000	110	110	111	111	000	000	000	000	110	110							010	000	010	010	111	111	000	000	110	000	111	111
BU27	000	000	000	000	000	000	000	000	000	000	000	000	000	000	000	000	000	000	000	000	000	000	000	000	000	000	000	000	000	000	000	000	000	000	000	000
BU28	000	000	000	000	000	000	000	000	000	000	000	000	000	000	000	000	000	000	000	000	000	000	000	000	000	000	000	000	000	000	000	000	000	000	000	000
BU29	000	000	100	000	000	000	000	000	110	000	000	010	000	000	000	000	000	000							000	000	010	000	101	101	000	000	100	000	000	010
BU30	000	000	000	000	000	000	000	000	000	000	000	000	000	000	000	000	000	000							000	000	000	000	000	000	000	000	000	000	000	000

T3:upto2013-upto2014, T4:upto2014-upto2015, T5:upto2015-upto2016, T6:upto2016-upto2017, T7:upto2017-upto2018, T8:upto2018-upto2019; Empty cells refer to BUs that have no DHW in the calculations according to the national regulation.

Looking at the four end uses, the average values of energy use intensities don't differentiate significantly across the consecutive time bands in more than half of the building uses. Overall, 13 out of 30 building uses exhibit failures of the average or/and variance tests in one or more of the end uses, which might indicate that there is not yet homogeneity of the respective building uses.

Depending on the results of the statistical tests, an indicator (e.g., EUIp, EUIp,H) can be characterized as stabilized (S), almost stabilized (AS), and nonstabilized (NS), according to the discussion in Section 3.5. Based on the characterization of the respective end uses, the overall status of a building use is considered stabilized when all its end uses are stabilized, nonstabilized when at least one end use is nonstabilized, and almost stabilized in all other cases. The stabilization status of the specific indicators, as well as the overall status of all BUs are summarized in Table 10. The total primary energy use and CO_2 intensity were not taken into account for the overall characterization of a BU, since these may be the result of counteracting variations among the different end uses. Nevertheless the characterization of these two specific parameters was included in Table 10 only for comparison.

Table 10. Stabilization status for all building uses.

BU	Specific Parameters						Overall
	EUIp	CO$_2$	EUIp,H	EUIp,C	EUIp,DHW	EUIp,L	
BU1	S	S	S	S	S	NS	NS
BU2	S	S	S	NS	NS	S	NS
BU3	S	S	S	S	S	S	S
BU4	S	S	S	S	NS	S	NS
BU5	S	S	S	AS	S	AS	AS
BU6	S	S	NS	AS	NS	S	NS
BU7	S	S	AS	S	AS	NS	NS
BU8	S	NS	S	S	NS	NS	NS
BU9	S	S	S	S	S	S	S
BU10	S	S	S	S		AS	AS
BU11	S	S	S	S		AS	AS
BU12	S	S	S	S		S	S
BU13	S	S	S	S		S	S
BU14	S	S	S	S		S	S
BU15	S	S	S	S	S	S	S
BU16	S	S	S	S		S	S
BU17	NS	NS	NS	NS		S	NS
BU18	S	S	S	S		S	S
BU19	S	S	S	S		S	S
BU20	S	S	S	S	S	S	S
BU21	AS	AS	S	AS	NS	AS	NS
BU22	S	S	S	S	S	S	S
BU23	S	S	S	S	S	S	S
BU24	S	S	S	S		S	S
BU25	S	S	S	S		S	S
BU26	NS	NS	NS	NS		NS	NS
BU27	S	S	S	S	S	S	S
BU28	S	S	S	S	S	S	S
BU29	NS	NS	NS	S		NS	NS
BU30	S	S	S	S		S	S

Empty cells refer to BUs that have no DHW in the calculations according to the national regulation.

Overall, a total of 10 BUs (33%) are nonstabilized (Table 10), indicating that the analysis results of available data should be used with caution since they might not yet be sufficiently representative of the respective building population. Among them, BU26 is the most problematic, since all of its end uses are nonstabilized, followed by BU29 and BU17, for which two out of their three end uses have not yet reached steady-state behavior. From the analysis of the available data from the NR Dbase, there were no evident reasons for nonstabilization of the specific building uses. Probably, in a second phase,

Energies 2020, 13, 2100

having access and analyzing the raw data of the corresponding EPCs would allow a more detailed investigation of the driving forces for this behavior.

A similar approach was followed for the six building categories, excluding BCV since it only refers to a single building use (BU24) and was not considered again in the analysis. The temporal evolution of the weighted average primary energy use and CO_2 intensities and their respective standard errors for the building categories, are also in Tables 2–7. The results are listed for the nine time bands from 2011 to 2019. For the building categories, only the two parametric statistical tests were applied, and the resulting failures of the null hypothesis are summarized in Table 11. This time, a two-digit convention is used; the first digit refers to equality of averages test and the second to equality of the variances test. The coding interpretation is the same as the one used for BUs. Again, "0" represents retention of the null hypothesis, while "1" signifies its rejection. Failures (rejections) are illustrated with shaded cells, in order to facilitate the interpretation of the results. Again, the first two pairs of time bands (T1 and T2) are not included in Table 11. Finally, the derived stabilization status is summarized in Table 12, following similar characterization rules as in the building use level. Overall, assessing the currently available data at the building category level, for the different end uses as well as for the overall behavior, the results indicate that they are nonstabilized yet. This level of detail is very demanding, so there is a need to accumulate additional data in the future.

Table 11. Failures coding (F-codes) for primary energy uses and CO_2 intensities, for the different building categories in the NR Dbase.

BC	EUIp T3	T4	T5	T6	T7	T8	EUIp,H T3	T4	T5	T6	T7	T8	EUIp,C T3	T4	T5	T6	T7	T8	EUIp,DHW T3	T4	T5	T6	T7	T8	EUIp,L T3	T4	T5	T6	T7	T8	CO₂ T3	T4	T5	T6	T7	T8
BCI	00	00	00	00	00	00	00	00	00	01	00	00	00	00	00	00	10	10	00	01	00	11	00	00	00	01	01	01	10	10	00	00	00	00	00	00
BCII	00	00	10	00	11	00	00	00	00	00	10	00	00	01	00	00	10	00	00	10	11	00	10	00	00	00	00	00	01	00	00	00	00	00	11	00
BCIII	00	10	00	00	10	00	11	01	00	00	11	00	01	00	00	01	11	11							01	11	01	00	00	00	00	11	00	00	10	00
BCIV	01	01	00	00	00	00	00	01	00	00	10	10	01	01	00	00	01	01	01	00	01	10	00	00	00	01	11	01	00	00	01	01	10	00	01	00
BCVI	00	10	11	00	11	11	00	10	10	01	11	11	01	00	00	00	10	10	00	00	00	01	01	01	00	00	00	00	00	00	00	10	10	00	11	11
BCVII	00	00	10	00	00	00	00	01	10	00	00	10	01	00	01	00	00	00							00	00	00	00	00	00	00	00	11	00	00	00

T3:upto2013-upto2014, T4:upto2014-upto2015, T5:upto2015-upto2016, T6:upto2016-upto2017, T7:upto2017-upto2018, T8:upto2018-upto2019; Empty cells refer to BUs that have no DHW in the calculations according to the national regulation; BCV was not included in the analysis, since it includes only the buildings of BU24.

Table 12. Stabilization status for all building categories.

BC	EUIp	CO₂	EUIp,H	EUIp,C	EUIp,DHW	EUIp,L	Overall
BCI	S	S	AS	NS	NS	NS	NS
BCII	NS	NS	NS	NS	NS	AS	NS
BCIII	NS	NS	NS	NS		NS	NS
BCIV	AS	NS	NS	NS	NS	NS	NS
BCVI	NS	NS	NS	NS	NS	S	NS
BCVII	NS	NS	NS	AS		S	NS

Empty cells refer to BUs that have no DHW in the calculations according to the national regulation. BCV was not included in the analysis, since it includes only the buildings of BU24.

5. Discussion and Conclusions

This paper examined the energy performance of existing Hellenic NR buildings, by analyzing the data from about 30,000 EPCs for whole buildings that have been issued up to the end of 2019. The data were clustered into seven building categories, according to the definitions in the national technical guidelines, and into a total of thirty building uses based on similarities between building types. The calculated primary energy use and CO_2 emission intensities were analyzed, taking into account heating, cooling, DHW and lighting (heating and cooling include mechanical ventilation).

The first objective of this work was to derive practical baselines for the energy use and CO_2 emission intensities for different NR buildings. Although the number of buildings in the NR Dbase is rather small, the total floor area of the audited buildings reaches a significant percentage of the national building stock, justifying the use of the findings as baselines, at least for most of the building uses.

The calculated primary energy use intensity for the Hellenic NR sector averages 448.0 kWh/m^2, ranging from 167.2 kWh/m^2 to 1003.9 kWh/m^2 for the different buildings uses. Lighting and space cooling are the most consuming services, contributing about 33% each to the total primary energy use, following by space heating with 29% and finally by DHW with only 5%. The average calculated CO_2 emissions for NR buildings is 147.5 kg/m^2.

The relatively low energy performance of NR buildings is also reflected by their poor energy rating. They average an energy class -D by 30%, while only 1.3% are ranked in energy class -B+ or better towards nZEB. Apparently, there is a great potential for significant energy conservation by renovating the existing building stock, comparing the average total primary energy use intensity to the corresponding values for the reference buildings (indicating a "good" building) as illustrated in Figure 9.

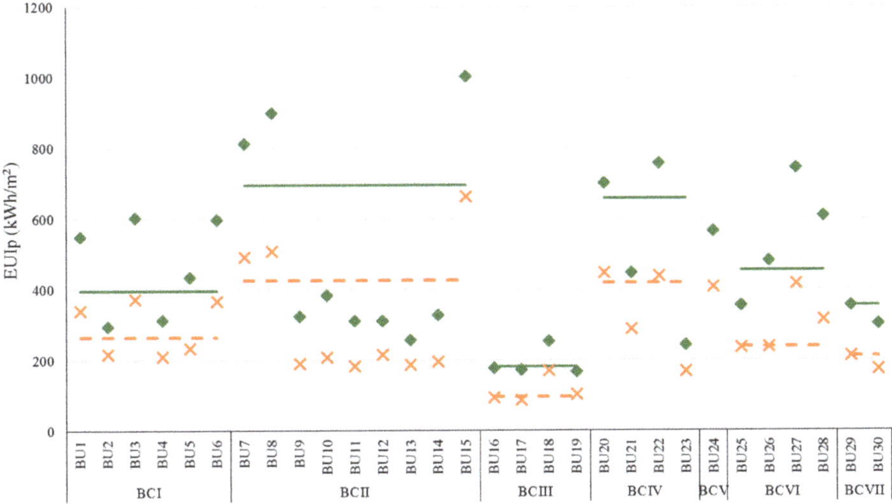

Figure 9. Calculated average total primary energy use intensity for the various building uses (diamond) and for the reference building of each BU (cross). The weighted average for the various building categories is presented with the green solid line and for the reference buildings with the orange dash line.

The primary energy use and the carbon emissions were estimated using the national conversion factors per energy carrier, which were defined in 2010. This may impact the results for electricity since they do not reflect the recent evolution of the energy mix used for power generation. This is expected to change even more in the near future, given the national efforts to decarbonize power generation and the national policy to abandon lignite power plants by 2028, according to the proposed NECP [37]. As a result, the calculated average intensities are much higher than they should be. For example, in NECP it is reported that in 2016 for electricity the conversion factor to primary energy was 2.17 and to CO_2 emissions 0.606. The use of these updated factors would result in a reduction of about 25% in the primary energy use for lighting and cooling and about 39% in CO_2 emissions. The total primary energy use would also be reduced by a smaller yet significant percentage, since lighting and cooling are important end uses for NR buildings.

The second objective of this paper was to analyze the time evolution of the average primary energy use and CO_2 emission intensities (arranged in nine consecutive time bands) and conclude about their stabilization status The cumulative datasets for first two time bands (up to 2011 and up to 2012) included very few cases and for this reason were not taken into account in the analysis.

Accordingly, for the time bands after 2013, the majority of the building uses (i.e., 19 out of 30 building uses) could be considered as stabilized, based on the test results performed on their end uses. For some of them (i.e., 11 building uses identified in Table 10) more EPCs should be gathered in order to constitute a more representative sample. Investigating only the stabilization of the EUIp (e.g., in case of lack of more detailed data), a total of 26 BUs could be considered as stabilized, indicating that when less detailed data are used, the results can be misleading. It is interesting to note that although the stabilized BUs are more than the nonstabilized ones, they represent only 28% of the total heated area, indicating that the NR Dbase as a whole is nonstabilized.

The variations in the last two pairs of consecutive time bands (T7:upto2017-upto2018 and T8:upto2018-upto2019) could be attributed to the updates incorporated in the national calculation engine in late 2017, which have affected—to some extent—the calculations of EPCs issued in 2018 and 2019. Generally, regular updating of the methodology and the software results to disruptions in the continuity of the time series, nullifying the possibility for assessing the degree of representativity of the dataset. Another limitation of the followed procedure is that for datasets with relative small number of cases, the statistical tests may not be so accurate, thus, in these cases, even if they are characterized as stabilized, more data will be required.

As expected, most of the building categories are nonstabilized, since they included heavy, nonstabilized BUs, covering between 44% and 99% of their heated area. On the other hand, BCIV is characterized nonstable, even though its only unstable building use is BU21, which corresponds to only 8% of the category's heated area. One possible cause for this is that different variations from the four building uses comprising this category may not be statistically significant at the BU level, but, when aggregated at the BC level, they became significant. This is another example of results obtained from less detailed data having the potential to be misleading, albeit at the opposite direction than the case of EUIp discussed above. In any case, even though for many BUs the population may be sufficient, when it comes to BCs, more data are required.

Furthermore, it would be desirable to evaluate and analyze the impact of the end use on EUIp in conjunction with the construction period and the climate at the location of the buildings. The national regulation recognizes three construction periods and four climate zones across Greece. However, the population of the currently available dataset is not yet sufficient to support such a discretization of the data.

As a general benefit, this work presented a well-structured integrated approach for clustering the building stock, screening the data, performing statistical tests and overall data assessment. The proposed methodology has numerous advantages, since it is a modular approach with sequential steps that can also be implemented independently, can be extended and adopted for other types of buildings (i.e., residential buildings), and can be replicated in other countries with their national certificate databases. Even if building uses and building categories are not yet all stabilized and representative, the analysis of the available data comprises a first step for bridging the gap of knowledge about the Hellenic NR building uses. The derived practical baselines on the energy use and CO_2 emission intensities, elaborated for the first time at this level of detail, provide new insight for practically all NR building uses defined in Greece for gaining a deeper understanding of national building stock modeling.

Author Contributions: Conceptualization, C.A.B. and K.G.D.; methodology, K.G.D. and C.A.B.; formal analysis, K.G.D., S.K., S.L., E.G.D. and C.A.B.; investigation, K.G.D.; data curation, K.G.D. and S.K.; writing—original draft, K.G.D.; writing—review and editing, C.A.B., S.K., S.L., E.G.D. and A.A.A.; visualization, K.G.D. All authors have read and agreed to the published version of the manuscript.

Funding: This research received no external funding.

Acknowledgments: This research is part of K.G.D. dissertation work as doctorate candidate at the Department Physics, University of Patras, Greece. The national EPC repository (buildingcert) has been developed and maintained by the Hellenic Ministry of Environment & Energy (YPEN) in collaboration with the Centre for Renewable Energy Sources. The authors wish to acknowledge YPEN for allowing access to the EPC database. The analysis presented herein does not necessarily reflect the opinion of the Ministry.

Conflicts of Interest: The authors declare no conflict of interest.

Nomenclature

BC	Building Category
BU	Building Use
DHW	Domestic Hot Water
EED	Directive on Energy Efficiency
EPBD	Energy Performance of Building Directive
EPC	Energy performance certificate
EU	European
EUIp	Total Primary Energy Use per unit floor area
EUIp,H	Primary Energy Use per unit floor area for space heating
EUIp,C	Primary Energy Use per unit floor area for space cooling
EUIp,DHW	Primary Energy Use per unit floor area for domestic hot water
EUIp,L	Primary Energy Use per unit floor area for lighting
HDD	Heating Degree Days
IQR	Interquartile Range
MFH	Multifamily Houses
NECPs	National Energy and Climate Plan
NR	Nonresidential
NR BS	Nonresidential Building Stock
NR Dbase	Screened data base for NR whole-buildings from the certificates
SFH	Single-Family Houses

References

1. European Commission. Energy Performance of Buildings. Available online: https://ec.europa.eu/energy/en/topics/energy-efficiency/energy-performance-of-buildings/energy-performance-buildings-directive (accessed on 29 January 2020).
2. Organisation for Economic Co-Operation and Development. Built-up Area and Built-up Area Change in Countries and Regions. Available online: https://stats.oecd.org/Index.aspx?DataSetCode=BUILT_UP (accessed on 29 January 2020).
3. EU Energy in Figures. Statistical Pocketbook. 2019. Available online: https://ec.europa.eu/energy/en/data-analysis/energy-statistical-pocketbook (accessed on 29 January 2020).
4. Bertoldi, P.; Atanasiu, B. An In-Depth Analysis of the Electricity End-Use Consumption and Energy Efficiency Trends in the Tertiary Sector of the European Union. *Int. J. Green Energy* **2011**, *8*, 306–331. [CrossRef]
5. Capros, P.; de Vita, A.; Tasios, N.; Siskos, P.; Kannavou, M.; Petropoulos, A.; Evangelopoulou, S.; Zampara, M.; Papadopoulos, D.; Nakos, C.; et al. *EU Reference Scenario 2016 Energy, Transport and GHG Emissions to 2050*; Publications Office of the European Union: Luxembourg, 2016. Available online: https://ec.europa.eu/energy/sites/ener/files/documents/20160713%20draft_publication_REF2016_v13.pdf (accessed on 29 January 2020).
6. ELSTAT. Buildings Census 2011. Athens: Hellenic Statistical Authority. 2015. Available online: www.statistics.gr/census-buildings-2011 (accessed on 29 January 2020).

7. Gaglia, A.G.; Balaras, C.A.; Mirasgedis, S.; Georgopoulou, E.; Sarafidis, Y.; Lalas, D.P. Empirical Assessment of the Hellenic Non-Residential Building Stock, Energy Consumption, Emissions and Potential Energy Savings. *Energy Convers. Manag.* **2007**, *48*, 1160–1175. [CrossRef]

8. Hellenic Ministry of Environment & Energy. *Long Term Strategy Report Mobilizing Investments in the Renovation of Residential and Commercial Buildings, Public and Private, of the National Building Stock*; Pursuant to Article 4 of Directive 2012/27/EU; Hellenic Ministry of Environment, Energy and Climatic Change: Athens, Greece, 2014.

9. European Commission. The Amending Directive. (2018/2002). Available online: https://ec.europa.eu/energy/en/topics/energy-efficiency/targets-directive-and-rules/energy-efficiency-directive#content-heading-0 (accessed on 29 January 2020).

10. DIRECTIVE (EU) 2018/844. Official Journal of the European Union. 2018. Available online: https://eur-lex.europa.eu/legal-content/EN/TXT/PDF/?uri=CELEX:32018L0844&from=EN (accessed on 29 January 2020).

11. European Commission. Clean Energy for All Europeans Package. Available online: https://ec.europa.eu/energy/en/topics/energy-strategy-and-energy-union/clean-energy-all-europeans (accessed on 29 January 2020).

12. Maldonado, E. (Ed.) *Implementing the Energy Performance of Buildings Directive (EPBD)*; ADENE: Lisbon, Portugal, 2015.

13. Attanasio, A.; Piscitelli, M.S.; Chiusano, S.; Capozzoli, A.; Cerquitelli, T. Towards an Automated, Fast and Interpretable Estimation Model of Heating Energy Demand: A Data-Driven Approach Exploiting Building Energy Certificates. *Energies* **2019**, *12*, 1273. [CrossRef]

14. Dall'O, G.; Sarto, L.; Sanna, N.; Tonetti, V.; Ventura, M. On the use of an energy certification database to create indicators for energy planning purposes: Application in northern Italy. *Energy Policy* **2015**, *85*, 207–217. [CrossRef]

15. Gangolells, M.; Casals, M.; Forcada, N.; Macarulla, M.; Cuerva, E. Energy mapping of existing building stock in Spain. *J. Clean. Prod.* **2016**, *112*, 3895–3904. [CrossRef]

16. Streicher, K.N.; Padey, P.; Parra, D.; Bürer, M.C.; Patel, M.K. Assessment of the current thermal performance level of the Swiss residential building stock: Statistical analysis of energy performance certificates. *Energy Build.* **2018**, *178*, 360–378. [CrossRef]

17. Dascalaki, E.G.; Kontoyiannidis, S.; Balaras, C.A.; Droutsa, K.G. Energy Certification of Hellenic Buildings: First findings. *Energy Build.* **2013**, *65*, 429–437. [CrossRef]

18. D'Agostino, D.; Cuniberti, B.; Bertoldi, P. Energy consumption and efficiency technology measures in European non-residential buildings. *Energy Build.* **2017**, *153*, 72–86. [CrossRef]

19. Gangolells, M.; Casals, M.; Ferré-Bigorra, J.; Forcada, N.; Macarulla, M.; Gaspar, K.; Tejedor, B. Energy Benchmarking of Existing Office Stock in Spain: Trends and Drivers. *Sustainability* **2019**, *11*, 6356. [CrossRef]

20. Hjortling, C.; Björk, F.; Berga, M.; af Klintberg, T. Energy mapping of existing building stock in Sweden—Analysis of data from Energy Performance Certificates. *Energy Build.* **2017**, *153*, 341–355. [CrossRef]

21. Armitage, P.; Godoy-Shimizu, D.; Steemers, K.; Chenvidyakarn, T. Using Display Energy Certificates to Quantify Public Sector Office Energy Consumption. *Build. Res. Inf.* **2014**, *43*, 691–709. [CrossRef]

22. Godoy-Shimizu, D.; Armitage, P.; Steemers, K.; Chenvidyakarn, T. Using Display Energy Certificates to Quantify Schools' Energy Consumption. *Build. Res. Inf.* **2011**, *39*, 535–552. [CrossRef]

23. Dascalaki, E.G.; Balaras, C.A.; Gaglia, A.G.; Droutsa, K.G.; Kontoyiannidis, S. Energy performance of buildings—EPBD in Greece. *Energy Policy* **2012**, *45*, 469–477. [CrossRef]

24. Ministerial Decision No 85251/242/2018—FEK 5447/B/5-12-2018 (in Greek). Available online: https://www.e-nomothesia.gr/kat-periballon/upourgike-apophase-upendepea-85251-242-2018.html (accessed on 29 January 2020).

25. Droutsa, K.G.; Balaras, C.A.; Dascalaki, E.G.; Kontoyiannidis, S.; Argiriou, A.A. Energy Use Intensities for Asset Rating of Hellenic Non-Residential Buildings. *Glob. J. Energy Technol. Res. Updates* **2018**, *5*, 19–36. [CrossRef]

26. Arcipowska, A.; Anagnostopoulos, F.; Mariottini, F.; Kunkel, S. *Energy Performance Certificates across the EU*; Buildings Performance Institute Europe (BPIE): Brussels, Belgium, 2014; Available online: http://bpie.eu/wp-content/uploads/2015/10/Energy-Performance-Certificates-EPC-across-the-EU.-A-mapping-of-national-approaches-2014.pdf (accessed on 29 January 2020).

27. Pasichnyi, O.; Wallin, J.; Levihn, F.; Shahrokni, H.; Kordas, O. Energy performance certificates—New opportunities for data-enabled urban energy policy instruments? *Energy Policy* **2019**, *127*, 486–499. [CrossRef]
28. Report on Quality of Statistical Data from EPCs (in Greek). Available online: http://ypeka.gr/Default.aspx?tabid=907&language=el-GR (accessed on 29 January 2020).
29. Mathew, P.A.; Dunn, L.N.; Sohn, M.D.; Mercado, A.; Custudio, C.; Walter, T. Big-data for building energy performance: Lessons from assembling a very large national database of building energy use. *Appl. Energy* **2015**, *140*, 85–93. [CrossRef]
30. Tukey, J.W. Exploratory Data Analysis. Addison-Wesley Publishing Company Reading, Mass.—Menlo Park, Cal., London, Amsterdam, Don Mills, Ontario, Sydney 1977, XVI, 688 S. *Biom. J.* **1981**, *23*, 413–414. [CrossRef]
31. Skew and Kurtosis. 2 Important Statistics Terms You Need to Know in Data Science. Available online: https://codeburst.io/2-important-statistics-terms-you-need-to-know-in-data-science-skewness-and-kurtosis-388fef94eeaa (accessed on 29 January 2020).
32. Sprent, P. *Applied Nonparametric Statistical Methods*, 2nd ed.; Chapman and Hall: London, UK, 1993.
33. Homogeneity of Variance Tests. Available online: www.unistat.com/guide/homogeneity-of-variance-tests/ (accessed on 29 January 2020).
34. Levene, H. *Contributions to Probability and Statistics: Essays in Honor of Harold Hotelling*; Olkin, I., Ed.; Stanford University Press: Redwood City, CA, USA, 1960; pp. 278–292.
35. Devore, J. *Probability and Statistics for Engineering and the Sciences*, 3rd ed.; Brooks/Cole: Pacific Grove, CA, USA, 1991.
36. Droutsa, K.G.; Kontoyiannidis, S.; Dascalaki, E.G.; Balaras, C.A. Mapping the Energy Performance of Hellenic Residential Buildings from EPC (energy performance certificate) Data. *Energy* **2016**, *98*, 284–295. [CrossRef]
37. National Plan for Energy and Climate (in Greek). Available online: www.opengov.gr/minenv/?p=10155 (accessed on 29 January 2020).

Article

Investigating Primary Factors Affecting Electricity Consumption in Non-Residential Buildings Using a Data-Driven Approach

Sooyoun Cho [1], Jeehang Lee [2], Jumi Baek [1], Gi-Seok Kim [3] and Seung-Bok Leigh [1,*]

[1] Department of Architectural Engineering, Yonsei University, 50 Yonsei-ro, Seodaemun-gu, Seoul 03722, Korea; suyouncho@yonsei.ac.kr (S.C.); jumi100@yonsei.ac.kr (J.B.)

[2] Department of Bio and Brain Engineering, KAIST, 291 Daehak-ro, Yuseong-gu, Daejeon 34141, Korea; jeehanglee@gmail.com

[3] Center for Sustainable Buildings, Yonsei University, 50 Yonsei-ro, Seodaemun-gu, Seoul 03722, Korea; giseok_kim@yonsei.ac.kr

* Correspondence: sbleigh@yonsei.ac.kr

Received: 6 September 2019; Accepted: 21 October 2019; Published: 24 October 2019

Abstract: Although the latest energy-efficient buildings use a large number of sensors and measuring instruments to predict consumption more accurately, it is generally not possible to identify which data are the most valuable or key for analysis among the tens of thousands of data points. This study selected the electric energy as a subset of total building energy consumption because it accounts for more than 65% of the total building energy consumption, and identified the variables that contribute to electric energy use. However, this study aimed to confirm data from a building using clustering in machine learning, instead of a calculation method from engineering simulation, to examine the variables that were identified and determine whether these variables had a strong correlation with energy consumption. Three different methods confirmed that the major variables related to electric energy consumption were significant. This research has significance because it was able to identify the factors in electric energy, accounting for more than half of the total building energy consumption, that had a major effect on energy consumption and revealed that these key variables alone, not the default values of many different items in simulation analysis, can ensure the reliable prediction of energy consumption.

Keywords: feature selection; prediction of energy consumption; electricity consumption; machine learning; non-residential buildings

---

1. Introduction

As carbon emissions and energy problems have become key issues in major cities worldwide [1], there has been an increasing awareness regarding the need for energy and carbon emission reductions in the urban development industry (including construction). After the Paris Climate Conference in 2015, in contrast to the arrangements under the Kyoto Protocol, cities around the world must now propose their own reduction goals, the target ranges must be expanded, and the established reduction goals must be actively presented. Therefore, the importance of energy reduction in the building industry, which accounts for at least 25% of all energy usage, must be recognized, and detailed stage-based energy reduction strategies have become an urgent requirement [2]. Energy consumption in the building sector has significantly increased in the last few decades. Energy is an essential part of our lives, and almost all things in some way are associated with electricity [3,4]. According to a report issued by the US Energy Information Administration (EIA), 28% growth in the global energy demand may occur until 2040 [5]. Because of improper usage, a tremendous amount of energy is wasted annually; thus, energy wastage can be avoided by the efficient utilization of energy. In addition, the third largest use of electricity among total energy sources [6] may have a direct impact on the aforementioned CO_2 savings. Most studies that have analyzed the energy performance of buildings have focused on an analysis or estimation of the energy consumption by each area of a building, on the quantitative prediction of energy savings, and on verification of the efficiency of heating and cooling systems. Rather than relying on actual, measured data, these studies have provided estimates by modeling the physical characteristics of buildings (e.g., building surfaces and volumes), the number and behavior of occupants, and the functioning of the heating/cooling systems [7].

An issue arising from such dynamic simulation is that the succession of estimates obtained from calculation formulas can cause an increasing discrepancy between the estimated and actual values of energy consumption, saving, and efficiency. This limitation has been pointed out in earlier studies [8]. Another drawback is that the reliability of the results of dynamic simulation depends on the skill of the expert who performs the simulation. Because an increasing number of buildings are operated automatically, the number of studies in which sets of data from buildings are analyzed by means of analytical tools initially devised for other fields is also increasing.

Machine learning algorithms have recently been used in academic fields including medicine, as well as industry [9]. Estimates obtained through machine learning have contributed to resolving the discrepancy between estimated and measured values [10]. In addition, such algorithms provide objective results that are independent of the operator's skills (i.e., the same estimate is produced by any operator when the same process is performed). This research aims to identify the major variables that affect the energy consumption in highly energy-demanding buildings, using the clustering method. With respect to the prediction of major variables through the machine learning algorithm, an estimate that takes all variables of the dynamic simulation and all parameters into account can be expected to eliminate the discrepancy between estimated and actual values which arises from the use of calculated values. It can also be expected to enhance the reliability of such estimates (regardless of the operator performing the analysis).

This study focused on electric energy consumption in office buildings [11], which has been increasing [12], and identified the factors contributing to electric energy consumption. Energy consumption prediction based on traditional dynamic simulation methods produces values predicted by default values, which cannot exactly reflect the physical conditions prior to a building's design or construction, e.g., building area, window-to-wall ratio, envelope thermal efficiency, and number of occupants. This study identified which factors in the measured data of electric energy consumed by users in an actual building had the largest effect on electric energy consumption. The variables that could predict electric energy consumption most accurately were identified among other data measured from an office building using data-driven clustering, and it was confirmed that the combinations of these major variables predicted an electric energy consumption similar to that which occurred when all kinds of information were used together. Based on the assumption that only the electric energy of the building is used, it is sufficient to measure only the major sensors and measuring instruments related to the electric energy consumption derived by clustering. This is expected to greatly reduce the cost and time in conventional energy or consumption analysis. The key significance of this study is that a comparison of the results predicted using all information and the results predicted using only major variables to test the research methods showed no significant difference.

Two additional procedures have also been adopted for validation of the results to improve the reliability of the study. The significance of our study lies in the ability of the proposed method to identify the fundamental cause of excessive energy consumption in buildings with diverse functions, not only in office buildings, which are the primary subject of the study.

The rest of the paper is organized as follows. Related work is given in Section 2, and a detailed explanation of the limitations of previous studies that have analyzed the building energy performance using engineering methods for energy building analyses is first presented. Then, after a review of the recent literature on building energy analysis conducted by machine learning, the novel approach used in this study is presented. In Section 3, for 11 non-residential office buildings located across South Korea, the clustering analysis method of machine learning is employed to identify the major variables that affect the energy usage. Section 4 shows the results of the main variables of building electricity energy usage with features selected by machine learning. Section 5 validates the prediction results using only the main variables and the prediction results using all variables in two different ways. The paper's conclusion and future work are discussed in Section 6.

2. Related Works

Methodologies for building energy analyses have been developed over the last 50 years. Because the results obtained by these methods vary, based on their suitability, accuracy, sensitivity, and purpose [13], it is important to identify the correct method for research purposes, target applications, and the environment. Over the last 20 years, steady-state data-based simulation and dynamic simulation-based engineering analysis methods have typically been used for building energy analyses in Korea [14]. These methods estimate the energy usage from the environmental conditions of a building and physical data such as the building envelope, building heat-cooling-ventilation system, and a thermodynamic equation. They also predict the building's energy consumption and performance of the equipment system [15]. Dynamic simulation programs such as BLAST, ESP, EnergyPlus, ESP-r, and TRNSYS are widely used, as they are accurate and can be utilized without restrictions of purpose and usage.

However, there are disadvantages to using these programs, as usage becomes more complex as the number of variables increases. In contrast, data-driven (machine learning-based) prediction method approaches in conjunction with machine learning techniques use data covering the building's entire history to predict the amount of energy that will be used in the future under detailed, but limited, conditions. However, the applicability of such methods is often hindered by malfunctions and defects in the equipment system [16]. Popular algorithms that are used for such methods include linear regression, artificial neural networks, and support vector regression [17]. It has been reported that the prediction accuracy can be improved using ensemble-algorithms, which have an improved accuracy and reliability compared with a single-algorithm prediction model under the same building energy usage and system performance conditions.

2.1. Engineering Analysis Method and Its Threshold

Dynamic simulations, which are widely used in industry as well as in academia, use a calculation method that treats heat movement by considering indoor and outdoor conditions and assumes that heat movement occurring indoors will vary with time. BLAST, EnergyPlus, ESP-r, and TRNSYS are some of the main dynamic simulation applications [18]. Although the calculations are relatively accurate, and these methods are widely applicable, they do have limitations because of the difference between the calculations performed in the simulation and the actual measurements pertaining to the building. This occurs because of the increasing amount of data and variables and because there is a drastic difference in the results, depending on the user's understanding of the variables (i.e., user bias) [19]. The energy performance gap (EPG) is a representative example of the difference between dynamic simulation results and the actual energy consumption in buildings. Table 1 indicates the differences between the energy consumption estimated by simulation and the actual consumption measured after the building was constructed.

Regarding the characteristics of the majority of engineering analysis methods used in studies and reports published outside Korea, a potential common disadvantage is that the results may vary according to the user, as well as by time and money requirements. Another limitation seen in international settings is related to the supply of low-energy buildings and sustainable buildings. Research results for the initial design concerning energy usage and the performance of key technologies used in these types of buildings are different from the results obtained after the buildings are actually built. This phenomenon is referred to as the EPG [20]. In some of these studies, the difference between the values predicted by the engineering analysis method and the values measured in the actual building differ significantly [21], suggesting that the fundamental cause of these EPGs differs, according to the design–construction–operation method, as well as occupants [22]. Moreover, the difference in the results also results from errors in the energy modeling program used in the engineering analysis method. The second cause is the construction quality of the actual building under study. The EPG often occurs as a result of not having a detailed understanding regarding the implementation of certain technologies (e.g., envelope construction, window sealing, and ventilation system) regarding construction, which is due to a lack of awareness regarding eco-friendly buildings and a dearth of skilled workers. Furthermore, EPGs also occur when a detailed performance inspection is not properly conducted during the construction process, or when the construction is completed, but the performance was not fully guaranteed. Lastly, EPGs are caused by a delay in verification techniques and performance measurements for each specific technology once the building is constructed.

Table 1. Published papers on the building energy performance.

Author	Building Use	Target/Energy Source	Evaluations
Yu et al. [16]	Domestic	Gas, Electricity	Data Mining
Wilde et al. [20]	Domestic	Gas, Electricity	Monitoring
Menezes et al. [21]	Non-Domestic	Ventilation	Post-Occupancy Evaluation (POE)
Olivia et al. [22]	Hospital, School	Indoor Comfort	POE, Monitoring
Choi et al. [23]	Non-Domestic	Ventilation	POE
Hossein et al. [24]	Non-Domestic	Electricity	POE
Salehi et al. [25]	Non-Domestic	Ventilation	Dynamic Simulation
Niu et al. [26]	Domestic	Ventilation	POE
Herrando et al. [27]	Non-Domestic	Ventilation	Dynamic Simulation
Min et al. [28]	Non-Domestic	Air Handling Unit	Facility Management Review

In summary, because the energy consumption analysis or prediction studies traditionally used in the building field have been performed by experts who are highly experienced in the field and are time- and labor-intensive, it would be difficult to rely on the analysis results without expertise. Like the aforementioned EPG results, this research found a large difference in the reliability and resulting values of the analyzed results. For this reason, analysis methods have recently been developed that identify, analyze, and diagnose phenomena using data from only the actual building, instead of results calculated by default values embedded in the program.

2.2. Data-Driven Analysis Method

In recent years, many studies employing building energy prediction and analysis methods have incorporated machine learning, which helps to address the limitations of engineering analysis. This involves the extensive use of building control and energy management systems (BEMS) and BEMS data, whose use is linked to increases in the amount of energy data derived from buildings and the number of analysis variables that must be considered. Machine learning is a technique employed for discovering models, patterns, and rules within data [9]. It can be used to extract previously unknown but useful knowledge, by recognizing complex patterns in data and building statistical models based on those findings [29]. As a result, it is utilized not only in engineering, but also in various other fields, such as medicine [30]. There are also studies that have applied machine learning to the field of building energy. Building energy forecasting methodologies based on machine learning use historical data to predict the future energy use under specific constraints. These methodologies also involve analyzing building energy systems to detect malfunctions or defects [31].

A single prediction model is run using one learning algorithm and training a monolithic model throughout the model development process [32]. Various machine learning algorithms, such as Multiple Linear Regression (MLR) [33], Artificial Neural Networks (ANNs) [34], decision trees [35], and Support Vector Regression (SVR) [36], have been introduced to building energy prediction and have provided promising prediction results during the past two decades. These algorithms can process continuous real-time data derived from buildings to predict the energy use and the performance of certain equipment systems, or to detect malfunctions. Figure 1 shows a schematic diagram of a typical single prediction model. The historical data recorded from the building are divided into training (60%), verification (20%), and testing (20%) sets, after removing outliers during preprocessing. The procedures are repeated until the final result is obtained.

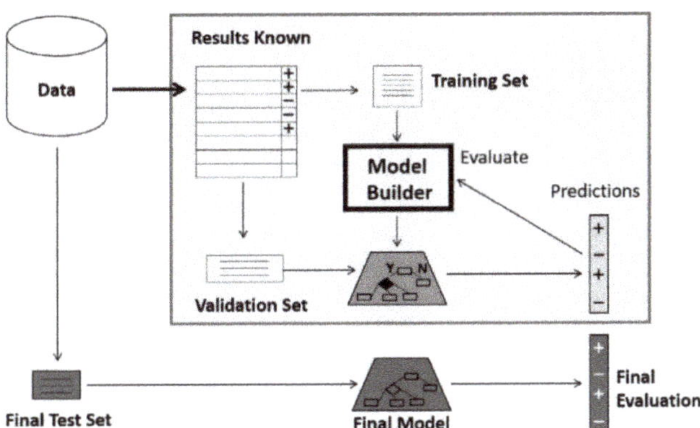

Figure 1. The process of machine learning analysis. Source: https://blog.algorithmia.com/page/50 [37].

2.2.1. Analysis of Building Energy Consumption Using Machine Learning

Although there have not been as many studies utilizing machine learning in the building energy sector as there have been in other fields of research, there has been much research on building energy consumption predictions and analytical studies that utilize machine learning. Studies by Paudel et al., [38] Yildiz et al., [39] and Rahman et al. [40] used machine learning to predict the energy consumption in buildings. Table 2 shows the target buildings and evaluation indexes for predicting building energy use. Although similarities exist with respect to utilizing actual energy usage data, predicting the usage volume, and presenting the mean absolute percentage error (MAPE) and RMSE as evaluation indicators, there is a lack of research on the analysis of the causes that induce a specific energy usage, which is one of the objectives of this study. Moreover, there has not been specific and persuasive evidence on the standard by which the variables of datasets were selected and utilized by the machine learning algorithms. The variables selected for predicting the energy use of buildings should not be based on researchers' experience; instead, they should be chosen to ensure the reliability of the results and consistency of the research process using statistical methods. Therefore, the use of machine learning for variable selection remains very important.

Table 2. Prediction of building energy consumption through machine learning.

Authors	Research Topic	Target Building	Algorithm	Evaluation Indices
Paudel et al. [38]	Prediction of building energy consumption All data vs. relevant data compared	Low-energy buildings	SVR	RMSE, R^2
Yildiz et al. [39]	Building electricity consumption prediction	Commercial buildings, educational facilities	ANN, SVR, Regression tree	MAPE, RMSE, R^2
Rahman et al. [40]	Prediction of building's fuel use Hourly data used over one year	Office buildings, supermarkets, restaurants	MLR, ANN, SVR, GP	RMSE
Moon et al. [41]	Prediction of building's electricity use	University buildings	ANN, SVR	MAPE, RMSE
Seong et al. [42]	Energy optimization model for buildings	Office buildings	ANN	MBE, CVRMSE
Support Vector Regression (SVR) Artificial Neural Network (ANN) Multi Liner Regression (MLR)			Root Mean Square Error (RMSE) Mean Absolute Percentage Error (MAPE) Coefficient of Variation of the Root Mean Square Error (CVRMSE)	

2.2.2. Analysis of Building Energy Consumption by the Clustering Method

Clustering is a common technique in unsupervised learning, which is the machine learning method of identifying a specific pattern in data without the ground level. Clustering refers to a group of methodologies that classify data with similar attributes into a number of groups [29]. Few studies have used the clustering analysis method for the analysis of building energy [43–47]. Most of them have focused on the analysis of consumption patterns of specific buildings. However, they only mention the possible causes of transmission and distribution, instead of providing details through experimental results, indicating that they only used the method with the aim of studying the methodology itself.

Table 3 outlines the results of recent studies that employed the clustering method and classifies them in terms of the algorithms used and their evaluation indices. Recent studies on clustering that are related to this study tend to use centroid-based K-means algorithms, and most studies were not studies looking for key variables that contribute to energy consumption, but were only used to analyze building energy consumption patterns with clustering algorithms.

Table 3. Building energy analysis using the clustering method.

Authors	Research Topic	Target Building	Algorithm	Evaluation Indices
Naganathan et al. [43]	Identifying the loss of energy during transmission and distribution	105 buildings	K-means	-
Ko et al. [44]	Improving the estimation accuracy of building energy consumption	Office buildings	K-means	R^2
Yang et al. [45]	Utilized K-shape clustering for analyzing building energy use patterns	Educational facilities	K-Shape SVR	RMS
Moon et al. [46]	Analysis of cooling and heating energy consumption patterns in office buildings	Office buildings	K-medoids	-
Hwang et al. [47]	Building energy demand predictions using hierarchical clustering	No information on target buildings	Hierarchical Clustering	APE, R^2

To summarize the literature review, there are studies that have used machine learning methods to predict the energy consumption in buildings, and some studies have used clustering methods to analyze energy usage patterns. However, there are no studies that have used machine learning and clustering at the same time to determine the cause of energy consumption.

3. Approach

The purpose of this study was to identify the power consumption of buildings and determine the variables that contribute to their use. This study confirms that key variables obtained using clustering in machine learning strongly correlate with the energy consumption of buildings. For this purpose, it is important to first identify building energy consumption patterns. This was done based on a total of 16 variables related to building energy usage, in addition to the electricity consumption of the target building. Then, the variables that best define the energy consumption patterns were identified. The major elements affecting consumption are not identified by simulation, or by the analyst's experience, but through a machine learning process. The structure of the research work is described below and Figure 2 shows simplify the process of this study from the number one to six.

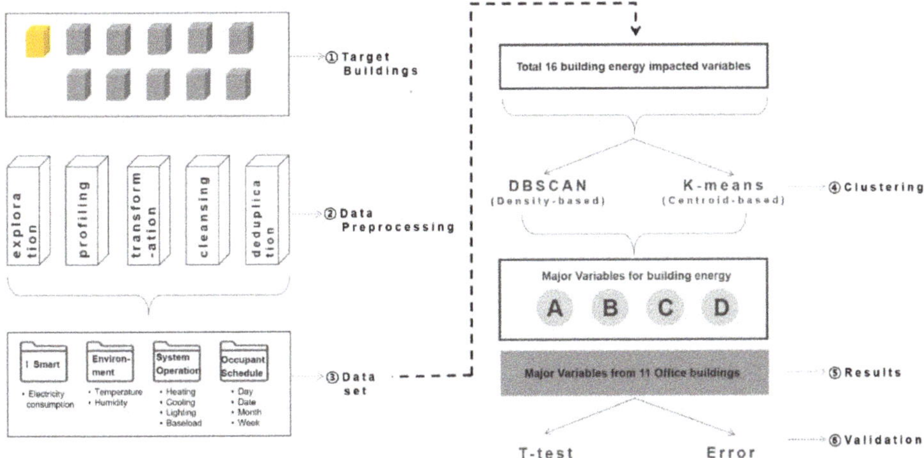

Figure 2. Schematic representation of the research process.

(1) This study targets 11 office buildings in eight different regions of South Korea. Their heating and cooling systems use electricity. The 11 buildings all have different areas, building designs, and heating and cooling systems, and are used for different purposes. In the data analysis, the physical qualities are referred to as nominal variables. In this study, only the continuous data that change in real time as a result of the actions of the occupants were used for the analysis. Four categories of variables were identified for the analysis, i.e., the amount of electric energy used by the building, external environmental data, building system data, and occupancy schedules. In total, 16 continuous variables were selected;

(2) Before analyzing the large amount of data used in machine learning, preprocessing is a necessary step that prepares data by processing missing values, removing noise, or removing outliers. In data preprocessing, cleaning is a process that fills in or removes missing values, corrects noisy data, identifies outliers, and therefore ensures the consistency of data outcomes [48];

(3) The data set in this research contains time variables (such as the year, month, day, day of the week, and hour), external environment variables (including temperature, humidity, sunlight, airflow, wind speed, and soil temperature), and energy variables that track the energy consumed during the building operating hours (in 2016, use of electricity consumption hourly data of 11 offices) [49];

(4) To determine the most important variables, unsupervised learning, which does not require a ground truth, was used. This involved using clustering to extract the most important variables that control the building energy usage and cluster similar variables together;

(5) In addition, using the *t*-test, we aimed to verify the validity of the variables, derived from machine learning, that affect the building energy consumption;

(6) For each target building, the first 16 variables related to energy consumption, and five variables derived from machine learning, were compared with the same verification logic.

As stated in Section 3, this study aims to use energy data derived from a building and identify important variables that control most of the energy consumption. Two representative clustering methodologies used for this purpose are the Density-Based Spatial Clustering of Applications with Noise (DBSCAN) and K-means.

3.1. Clustering: DBSCAN and K-Means

Density-based spatial clustering of applications with noise, one of the most widely used density-based clustering methods, explores core objects that have a high density in a neighborhood [50]. Centroid-based clustering methods of K-means predetermine the K value and the number of clusters, and then assign each datum to one of the K clusters to minimize dispersion. Centroid-based algorithms have the advantages of clustering many data points quickly and easily, and spherical data have a relatively better reliability [28,50].

Table 4 shows a sequential list of operations performed through DBSCAN and K-means using the open-source program R 3.6.1. [51] To summarize the clustering process for extracting key variables, first, similar characteristics (density center, distance-based) in the various data were assessed to create the first cluster. Although N clusters were formed, it was impossible to determine which clusters were meaningful. Then, results were extracted only for clusters containing a large amount of data. Based on these results, the data were classified into high-energy-consumption and low-energy-consumption data clusters. Lastly, for the two clusters with large amounts of high-energy-consumption data, box plots were used to confirm which of the 16 variables related to building energy were the most important.

Table 4. Clustering process.

<table>
<tr><th>Order</th><th>R-Code and Clusters</th><th>Contents</th></tr>
<tr><td>1</td><td>> #1-1. DBSCAN clustering
> # dbscan
> library(dbscan)
> db <- dbscan(pcaiscores[, 1:2], eps = 0.1, minPts = 5)
> plot(pcaiscores, col = (db$cluster+1), pch = 19, main = "DBSCAN")
> pairs(pcaiscores[, 1:5], col = (db$cluster + 1))
>
> table(db$cluster)

0 1 2 3 4 5 6 7 8 9 10 11 12 13 14 15 16 17 18 19 20 21 22 23
189 23 3032 37 675 9 2475 52 115 7 8 5 28 6 9 5 24 9 6 8 7 1111 5 7
30 31 32 33 34 35 36 37
4 21 11 8 5 5 6 5

> t <- table(db$cluster)
> names(t) <- paste(names(t), "(", rep(palette(), 10)[1:length(t)], ")", sep = "")
> t
0(black) 1(red) 2(green3) 3(blue) 4(cyan) 5(magenta) 6(yellow) 7(gray) 8(black) 9(red)
189 23 3032 37 675 9 2475 52 115 7
12(cyan) 13(magenta) 14(yellow) 15(gray) 16(black) 17(red) 18(green3) 19(blue) 20(cyan) 21(magenta)
28 6 9 5 24 9 6 8 7 1111
24(black) 25(red) 26(green3) 27(blue) 28(cyan) 29(magenta) 30(yellow) 31(gray) 32(black) 33(red)
58 48 157 396 7 36 4 21 11 8
36(cyan) 37(magenta)
6 5</td><td>First cluster is formed based on similar characteristics in the data (density center). It is unable to distinguish which cluster data are significant among n clusters through seven repeated colors.</td></tr>
<tr><td>2</td><td>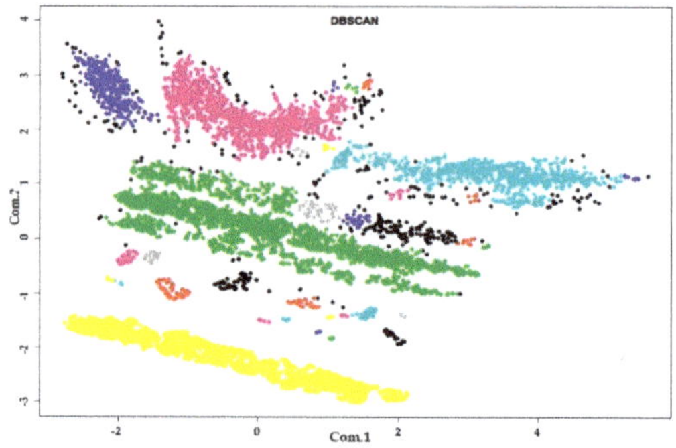
</td><td>Results of re-executed DBSCAN code for n clusters sorted by the amount of data in the clusters in descending order.</td></tr>
</table>

Table 4. *Cont.*

Order	R-Code and Clusters	Contents
3	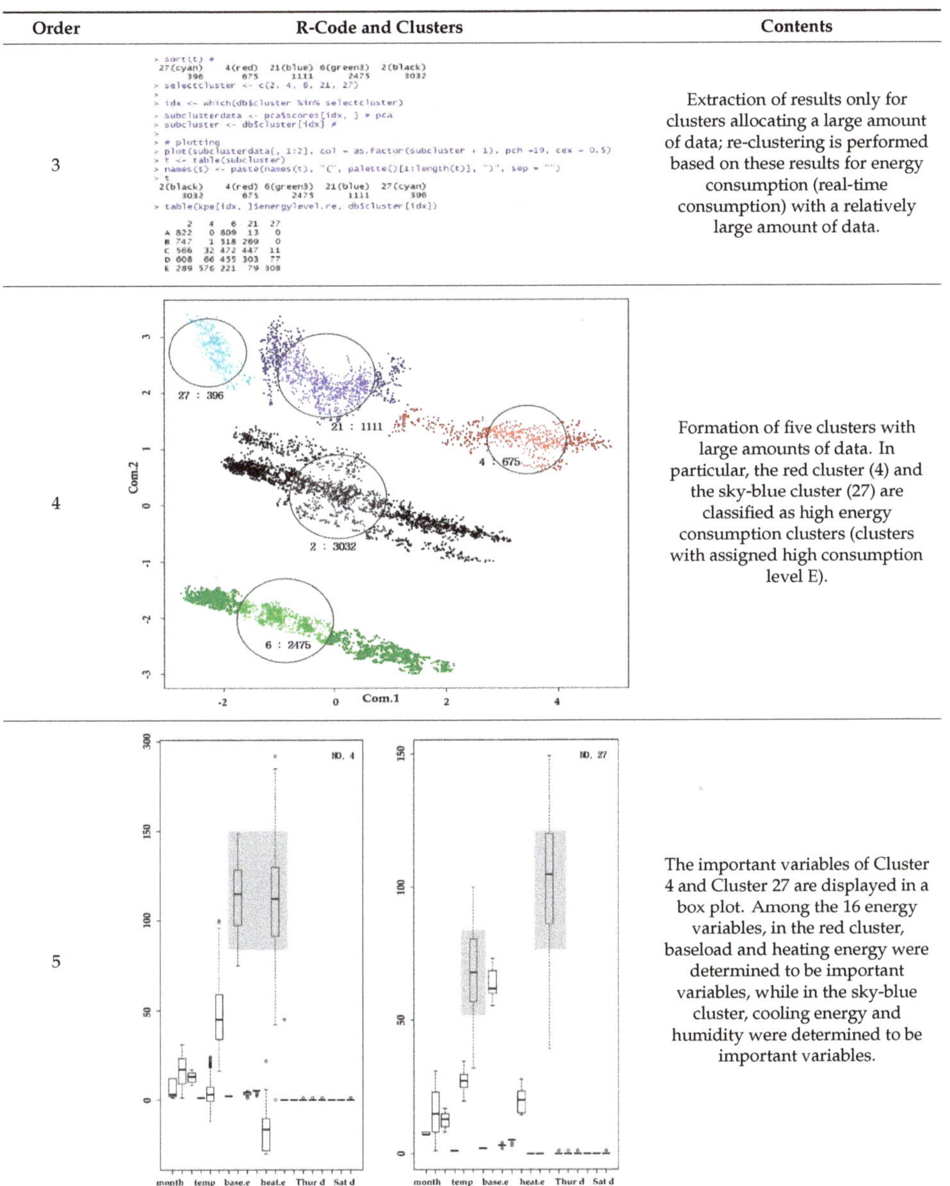	Extraction of results only for clusters allocating a large amount of data; re-clustering is performed based on these results for energy consumption (real-time consumption) with a relatively large amount of data.
4		Formation of five clusters with large amounts of data. In particular, the red cluster (4) and the sky-blue cluster (27) are classified as high energy consumption clusters (clusters with assigned high consumption level E).
5		The important variables of Cluster 4 and Cluster 27 are displayed in a box plot. Among the 16 energy variables, in the red cluster, baseload and heating energy were determined to be important variables, while in the sky-blue cluster, cooling energy and humidity were determined to be important variables.

Figure 3 shows the results after clustering using the density-based DBSCAN code to derive the clusters for the 11 buildings, where two parameters were required for DBSCAN: epsilon (ε) and the minimum number of points required to form a cluster (minPts). Epsilon is a distance parameter that defines the radius used to search for nearby neighbors. In Figure 4, the distance-based K-means results are shown. The consumption data were categorized based on energy usage and divided into five classes (energy consumption level A to E), with A representing the lowest energy consumption and E representing the highest. The energy consumption cluster results (e.g., size and shape, density of clusters, and number) of the buildings for each region are different. This occurs because the characteristics of the 11 regions' building envelope, heating and cooling systems, and building use are different.

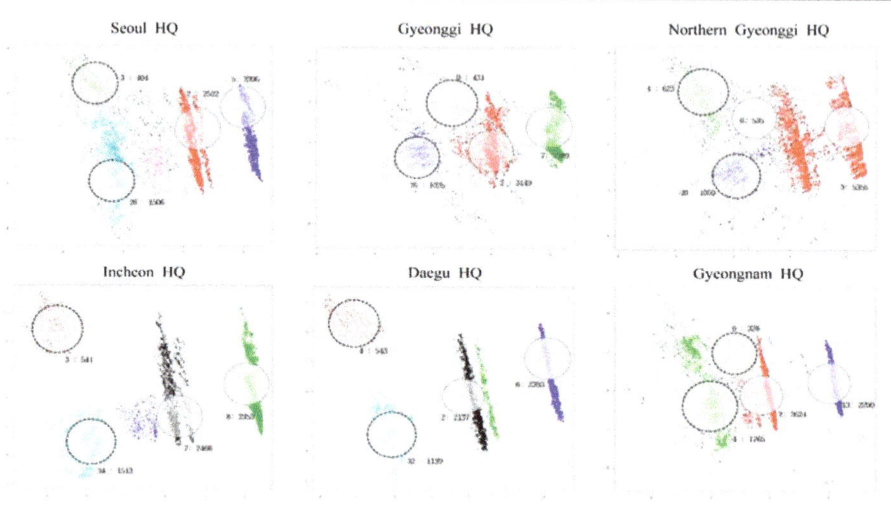

Figure 3. Result of DBSCAN clustering.

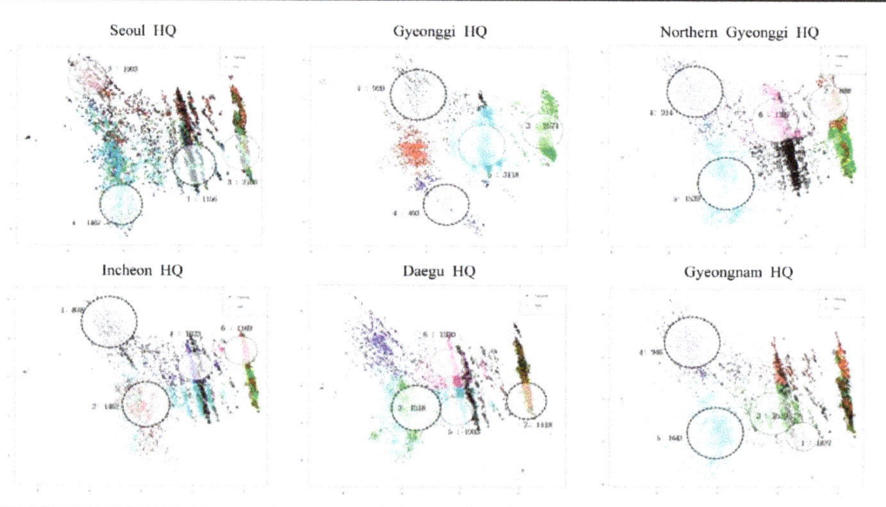

Figure 4. Result of K-means clustering.

3.2. Clustering Result

This study identified important variables affecting the energy consumption of 11 regional buildings. It classified them into high and low energy consumption categories to identify the clusters. Data obtained from a building are a mix of continuous and categorical data; hence, it is difficult to determine the unique features of the building using only the building energy data. Therefore, two different clustering algorithms were employed. A comparison was made to determine which of the two clustering algorithms correctly analyzes building energy data characteristics and energy consumption patterns. The four qualitative evaluation items are as follows [7,52]: the shape of a cluster and density of color, slope of an inverse model, sensitivity of an outlier (determined by the mean of the absolute values of the standardized variables), and grades that are accurately distributed against the amount of energy consumption when it is graded based on relative criteria. Table 4 was developed to determine which of the two clustering algorithms can accurately analyze the data features and energy consumption patterns of the building by weighting the above-mentioned four qualitative evaluation indices. The selected clustering algorithm was then used to categorize buildings only with variables contributing to the energy consumption, which were identified from the buildings; this result is expected to lead to a consistent analysis and provide meaningful results.

Table 5 shows the results of four qualitative indicators for assessing the two aforementioned clustering algorithms. The cluster density using DBSCAN was 2.53, which was 1.03 times higher than that using K-means based on converting data into the energy consumption based on the outdoor temperature. This result means that the use of DBSCAN rather than K-means can lead to meaningful results for deriving the key variables that directly affect consumption among the 16 variables that affect the building energy consumption.

Table 5. Comparison of DBSCAN and K-means.

| | DBSCAN | | | | K-Means | | | |
Criteria	Level of Clustering	Inverse Model *	Outlier	Energy Consumption Grading	Level of Clustering	Inverse Model	Outlier	Energy Consumption Grading
Weighted value	1.00	0.75	0.50	0.25	1.00	0.75	0.50	0.25
Seoul	3.00	2.25	1.50	0.75	1.00	2.25	1.50	0.75
Gyeonggi	3.00	2.25	1.50	0.50	2.00	2.25	1.00	0.50
Northern Gyeonggi	3.00	2.25	1.00	2.00	1.00	2.25	1.50	0.75
Incheon	3.00	1.50	1.50	0.75	2.00	2.25	1.50	0.75
Daegu	3.00	2.25	1.00	0.75	1.00	1.50	1.50	0.75
Gyeongnam	3.00	2.25	1.50	0.50	1.00	1.50	1.50	0.50
Pusan	3.00	0.75	1.50	0.50	1.00	1.50	1.50	0.50
Jeonbuk	3.00	1.50	1.00	0.50	1.00	1.50	1.50	0.75
Gwangju	3.00	2.25	1.00	0.25	1.00	1.50	1.50	0.75
Chung cheong	3.00	1.50	1.00	0.50	1.00	1.50	1.50	0.25
Kangwon	3.00	1.50	1.00	0.25	1.00	1.50	1.00	0.50
mean	3.00	1.84	1.23	0.66	1.18	1.77	1.41	0.61

* A graph showing both winter heating energy consumption (left slope) and summer cooling energy consumption (right slope) compared with outdoor air temperature, with the interim period without heating or cooling in the middle.

4. Research Results

This section discusses the result of clustering (DBSCAN) of the main variables affecting building energy consumption in 11 buildings in eight regions.

Figure 5 shows the results of classifying the energy consumption (A to E) of each building and deriving the variables that affect the high-energy-use clusters. First, the data were sorted into energy consumption levels A to E by hour, from a low to high energy consumption. Clusters with more levels D and E than A and B can be considered high-energy-consumption clusters. The important variables (abbreviated I-V) are arranged by rank among the high-energy-consumption clusters. We conducted a level-of-importance analysis for those clusters with large amounts of high-energy-consumption data to obtain the important variables, which were then ranked by order of importance (Figure 5). For example, for the Seoul headquarters, clusters 3 and 26 contain large amounts of high-energy-consumption data. Through an analysis of the cluster attributes, we found the important variables to be the heating energy, the intermediate energy, and the lighting energy, in order of importance. January and March were identified as the months with large effects on the energy consumption.

ation

Seoul HQ

level	5	2	3	26
A	747	791	0	0
B	777	761	0	0
C	802	640	0	2
D	55	184	14	796
E	15	146	390	708
D+E	70	330	404	1,504
AtoE	2,396	2,522	404	1,506
D,E %	2.92%	13.08%	100%	99.87%
I-V	humidity		heating	
I-V	time		intermediate season	
I-V	temperature		lighting	
I-V	Operation, Sat		month, day	
month	Jan		Jan, Mar	

Gyeonggi HQ

level	3	7	0	25
A	673	874	12	8
B	827	781	16	9
C	844	600	49	42
D	507	115	118	733
E	298	19	236	233
D+E	805	134	354	966
AtoE	3,149	2,389	431	1,025
D,E %	25.56%	5.61%	82.13%	94.24%
I-V	time		intermediate season	
I-V	operation		temperature	
I-V	Sun		base load	
I-V	energy level		humidity	
month	Jan		Jan, Apr	

Northern Gyeonggi HQ

level	0	3	4	20
A	35	1,563	0	1
B	13	1,626	1	6
C	64	1,503	1	52
D	211	451	297	479
E	212	212	324	521
D+E	423	663	621	1,000
AtoE	535	5,355	623	1,059
D,E %	79.07%	12.38%	99.68%	94.43%
I-V	humidity		heating	
I-V	base load		intermediate season	
I-V	operation		temperature	
I-V	time		lighting	
month	jan		Jan, Apr	

Incheon HQ

level	2	8	3	34
A	754	781	0	4
B	644	833	0	10
C	777	552	0	6
D	246	176	12	808
E	45	11	529	685
D+E	291	187	541	1,493
AtoE	2,466	2,353	541	1,513
D,E %	11.80%	7.95%	100%	98.68%
I-V	time		heating	
I-V	humidity		intermediate season	
I-V	temperature		lighting	
I-V	operation, Sunday		base load	
month	Jan		Jan, Apr	

Daegu HQ

level	2	6	4	32
A	670	687	0	0
B	599	733	0	2
C	506	766	0	17
D	207	173	198	465
E	155	34	345	655
D+E	362	207	543	1,120
AtoE	2,137	2,393	543	1,139
D,E %	16.94%	8.65%	100%	98.33%
I-V	humidity		intermediate season	
I-V	time		heating	
I-V	temperature		lighting	
I-V	operation, Sat		base load	
month	Jan		Jan, Apr	

Gyeongnam HQ

level	2	13	0	4
A	441	957	23	6
B	675	709	7	4
C	903	496	10	7
D	552	118	128	669
E	53	10	160	1,079
D+E	605	128	288	1,748
AtoE	2,624	2,290	328	1,765
D,E %	23.06%	5.59%	87.80%	99.04%
I-V	humidity		intermediate season	
I-V	temperature		lighting	
I-V	operation		month	
I-V	saturday		time	
month	Jan		Jan	

Pusan HQ

level	2	6	28	39
A	483	944	17	0
B	682	696	137	0
C	944	335	144	0
D	359	144	722	70
E	542	176	106	296
D+E	901	320	828	366
AtoE	3010	2295	1126	366
D,E %	29.93%	13.94%	73.53%	100%
I-V	humidity		intermediate season	
I-V	time		lighting	
I-V	temperature		lighting-v	
I-V	Saturday		operation	
month	Jan		march, July	

Jeonbuk HQ

level	2	4	0	41
A	647	897	10	0
B	810	765	12	11
C	902	521	35	6
D	513	159	158	470
E	26	3	196	919
D+E	539	162	354	1389
AtoE	2898	2345	411	1406
D,E %	18.60%	6.91%	86.13%	98.79%
I-V	humidity		intermediate season	
I-V	time		lighting	
I-V	temperature		month	
I-V	operation, Sat		energy level	
month	Jan		Jan, Apr	

Gwangju HQ

level	2	7	3	32
A	532	950	0	6
B	692	816	1	14
C	887	348	91	136
D	269	80	319	729
E	568	48	331	122
D+E	837	128	650	851
AtoE	2948	2242	742	1007
D,E %	28.39%	5.71%	87.60%	84.51%
I-V	time		heating	
I-V	humidity		intermediate season	
I-V	temperature		lighting	
I-V	base load		month	
month	Jan		Jan, Apr	

Chungcheong HQ

level	2	6	22	3
A	519	996	10	0
B	778	773	9	0
C	908	291	270	3
D	299	129	879	128
E	406	165	208	513
D+E	705	294	1087	641
AtoE	2910	2354	1376	644
D,E %	24.23%	12.49%	79.00%	99.53%
I-V	humidity		intermediate season	
I-V	time		heating	
I-V	temperature		base load	
I-V	operation, Sat		lighting	
month	Jan		Apr	

Kangwon HQ

level	3	7	0	5
A	650	894	17	0
B	916	663	27	0
C	1022	505	68	11
D	1175	186	106	137
E	967	110	185	319
D+E	2142	296	291	456
AtoE	4730	2358	403	467
D,E %	45.29%	12.55%	72.21%	97.64%
I-V	temperature		heating	
I-V	humidity		lighting	
I-V	operation		date	
I-V	month		base load	
month	Jan		Jan	

A: the lowest energy consumption
E: the highest energy consumption
D + E: relatively high energy consumption
A to E: total energy consumption
D, E%: high energy consumption/total energy consumption
I–V: Important variables by consumption section

Figure 5. Result of the clustering of 11 buildings.

Figures 6 and 7 show the location of the buildings and the classification of the important variables derived from the high-energy clusters of the 11 regions, respectively. The temperature and humidity, which can be called environmental variables, were excluded because of duplicate derivation from the 11 regions. In the Seoul, Gyeonggi, Incheon, Daegu, and Gyeongnam buildings, lighting and heating energy were the most influential variables, followed by baseload energy and intermediate energy. It can be concluded that occupancy is closely related to building energy consumption. Moreover, the difference in energy consumption between winter heating and summer cooling was also found to be an important variable for office buildings. Lighting and baseload energy were also found to be important variables (in that order) in the six regions of North Gyeonggi, Busan, Jeonbuk, Gwangju, Chungcheong, and Kangwon. Occupancy and related lighting energy variables, which can be considered characteristics of office buildings, were found to be variables affecting high energy consumption for all 11 regions, rather than regional variables related to the building's location (i.e., latitude and topography). In addition, baseload energy, which is involved in the operation of the building's systems and the outlet load, is also considered to be related to occupancy and was found to have the greatest effect on energy consumption in office buildings. These results indicate that the most important variables for energy usage in non-residential buildings are related to lighting and baseload energy consumption, based on the building's occupancy, rather than to regional and topographical characteristics based on the building's location.

To sum up, the beginning of this study assumed that "only key data can be used to effectively analyze building electricity consumption". This study focused on operational buildings, i.e., not buildings that are yet to be built or recently completed buildings. In the case of existing office buildings, this study aimed to determine which of the continuous energy data derived from the buildings had more influence on the electricity consumption. In general, the U-value of building materials was treated as a categorical variable because it is a set value that does not vary over time. Based on these results, we concluded that the electricity of a building can be effectively managed if only four or five major energy variables derived from the building are controlled or used as key points of operation [53,54].

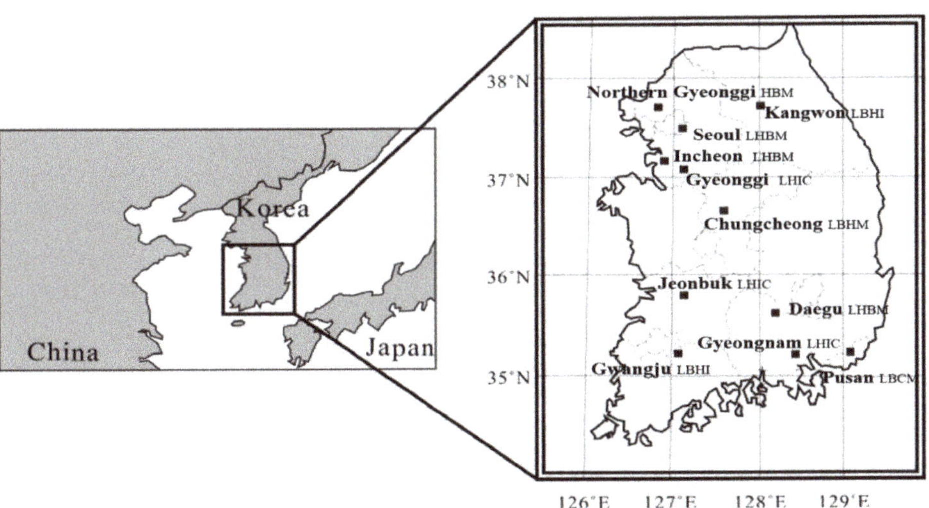

Figure 6. Office location of 11 regions. (L: Lighting energy; H: Heating energy; B: Baseload energy; C: Cooling energy; I: Intermediate energy; M: Month-time).

Figure 7. Results of the clustering of 11 regions and important variables.

5. Validation

5.1. Validation of Variables for High and Low Energy Consumption with a t-Test

Although a clustering algorithm was used to categorize the building energy data into high- and low-energy-consumption clusters, the *t*-test was used as a method to determine whether the difference between the averages of these two clusters was significant. The important variables of each cluster derived by the boxplot can also be compared using the averages of the absolute values. This comparison, however, does not reflect the range and variance of the actual values. For this reason, the *t*-test was performed to analyze whether the difference between the averages of the two different clusters, which considers the variations between them, was significant.

$$t = \frac{\overline{X_1} - \overline{X_2}}{SE_{\overline{X_1} - \overline{X_2}}} \quad SE_{\overline{X_1} - \overline{X_2}} = \sqrt{\frac{S_1^2}{n_1} + \frac{S_2^2}{n_2}} \tag{1}$$

The *t*-test explored whether the difference between the averages of the two different clusters (high and low energy consumption) from the building under analysis was significant. Table 6 shows the *t*-test determined six explanatory variables that led to high or low energy consumption of the building under analysis: month, temp (temperature), humi (humidity), lighting.e (energy), base.e (baseload energy), and heat.e (energy). The mean and variance of the day of the week and date variables in the two clusters were almost the same, so they were considered unnecessary for an analysis of the energy consumption.

Table 6. Result of the *t*-test of the 11 regions.

Region	Seoul	Gyeonggi	N-Gy	Incheon	Daegu	G-Nam	Pusan	Jeonbuk	Gw-Ju	Ch-Ch	Ka-W
month	0.000	0.001	0.000	0.000	0.000	0.620	0.000	0.000	0.000	0.000	0.000
date	0.803	0.616	0.728	0.219	0.024	0.531	0.450	0.484	0.288	0.177	0.739
hour	0.992	0.563	0.980	0.021	0.015	0.551	0.428	0.008	0.573	0.258	0.161
temp	0.000	0.098	0.000	0.000	0.000	0.000	0.000	0.000	0.000	0.000	0.000
humi	0.000	0.000	0.001	0.000	0.000	0.000	0.000	0.000	0.000	0.000	0.000
base.e	0.000	0.006	0.000	0.000	0.000	0.544	0.000	0.170	0.000	0.000	0.000
lit.e	0.000	0.000	0.280	0.000	0.000	0.000	0.000	0.000	0.000	0.000	0.000
heat.e	0.000	0.000	0.000	0.000	0.000	0.000	0.500	0.000	0.000	0.000	0.000
inter.e	0.318	0.000	0.000	0.500	0.500	0.000	0.318	0.000	0.000	0.318	0.000
cool.e	0.500	0.000	0.016	0.500	0.500	0.000	0.000	0.000	0.500	0.500	0.000
Mon	0.056	0.359	0.492	0.673	0.487	0.012	0.801	0.069	0.219	0.011	0.162
Tue	0.563	0.310	0.967	0.825	0.212	0.015	0.801	0.768	0.395	0.145	0.892
Wed	0.473	0.071	0.532	0.129	0.005	0.045	0.326	0.195	0.781	0.209	0.508
Thu	0.551	0.278	0.005	0.454	0.005	0.023	0.903	0.036	0.477	0.818	0.226
Sat	0.500	0.500	0.318	0.500	0.500	0.500	0.500	0.500	0.500	0.500	0.500
Sun	0.500	0.318	0.157	0.500	0.500	0.500	0.500	0.500	0.500	0.500	0.500

5.2. Regression Results

A regression analysis was performed using only the 16 major variables from each building, derived from the analysis process, and five or six variables selected by machine learning, to verify that the observed values, i.e., the Y variable, were well-described. Two validations were carried out for this purpose.

The coefficient of determination (R^2) of the 16 original variables was 0.8356 in the Figure 8. whereas that of the important variables was 0.8261 from the Figure 9. This suggests that the important variables alone are able to explain most of the observed variance, as the performance of the regression equation built using only the important variables is practically identical to that obtained using all the original variables.

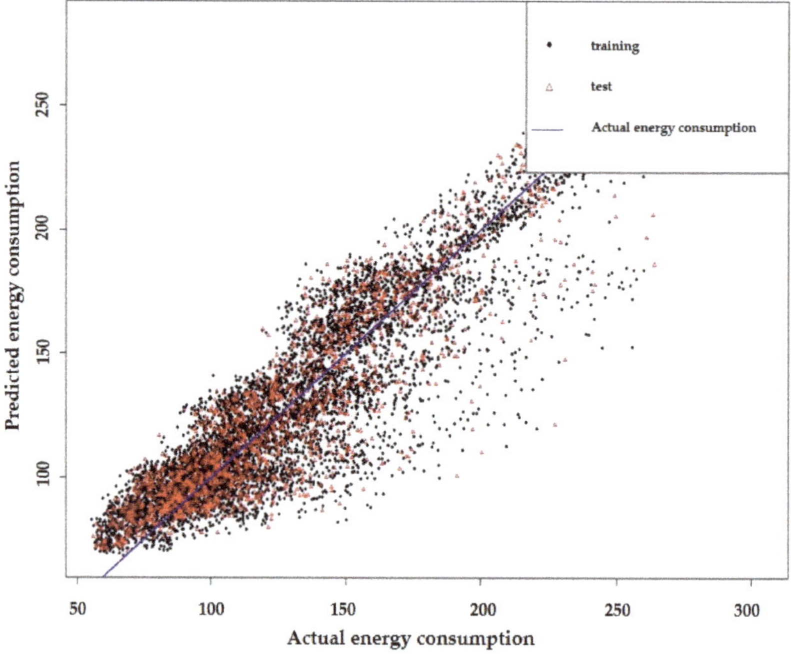

Figure 8. Results of a regression analysis of the 16 original variables.

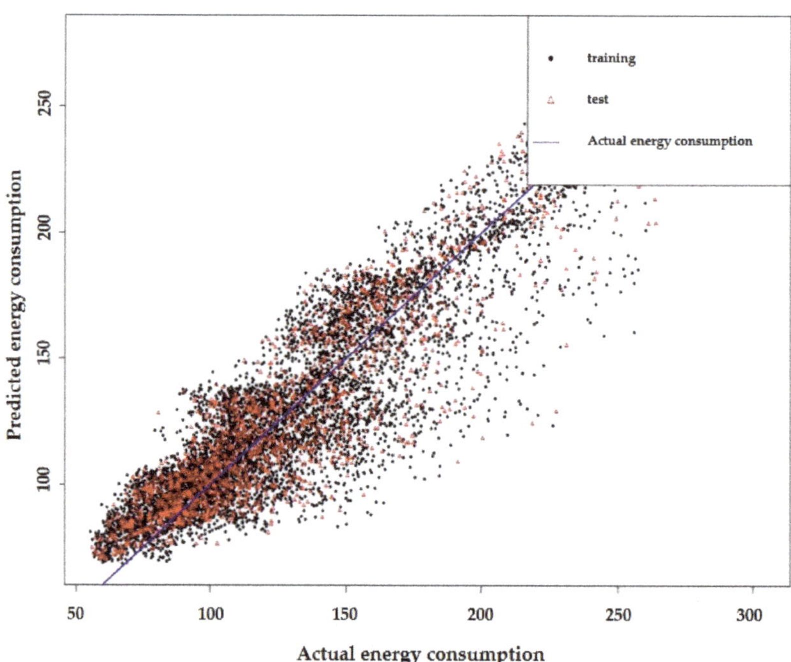

Figure 9. Results of a regression analysis of the major variables.

5.3. Mean Squared Error (MSE) and Mean Absolute Percentage Error (MAPE)

The MSE takes the square root of the average of the squares of the errors from an observation and compares it to deviation between the observed values. As a measure that generalizes standard deviations, MSE is used to validate the amount of difference between the actual value and estimated value using a regression equation. Table 7 shows the results were obtained from the two regression analyses performed for the 16 original variables and important variables based on the computation of their MSE and MAPE. The MSE and MAPE results of these two regression analyses show that even though the MSE and MAPE of the important variables were only slightly higher than those of the 16 original variables, and Table 8 shows the regression analysis error and coefficient of determination of the key variables were as significant as those of the 16 original variables, which were larger in number.

Table 7. Comparison of Mean Squared Error (MSE) and Mean Absolute Percentage Error (MAPE) associated with the regression analyses.

Evaluation Index	Original 16 Variables	Major Variables	Differences
MSE Training set	360.946	381.8843	21.938
MSE Test set	337.726	365.9697	28.243
MAPE Training set	11.65511	12.06266	0.40
MAPE Test set	11.65511	11.77404	0.12

Table 8. Verification of error rates with three algorithms.

Algorithms	Original 16 Variables	Major Variables	Differences
Regression R^2	0.835	0.826	0.009
Regression CvRMSE	13.84	14.60	0.8
SVM CvRMSE	7.69	9.03	1.34
Random Forest CvRMSE	6.74	7.20	0.46

Figures 10 and 11 show comparisons of energy consumption predictions, where one predicted energy consumption uses the initial 16 variables and the other prediction only uses the major variables derived by machine learning with the testbed as a subject. Apart from the R^2 value, which is the coefficient of determination, and the MSE, which is an indicator of the error rate of prediction, the results using the major variables showed error rates ranging from 7.2% to 14.6% compared with the predictions using 16 variables. The error rate using the major variables was lower than the ASHRAE Guideline 14 [55] criteria of 30%. Moreover, the random forest method showed the lowest error rates among the three different methodologies (simple linear regression, support vector machine, and random forest).

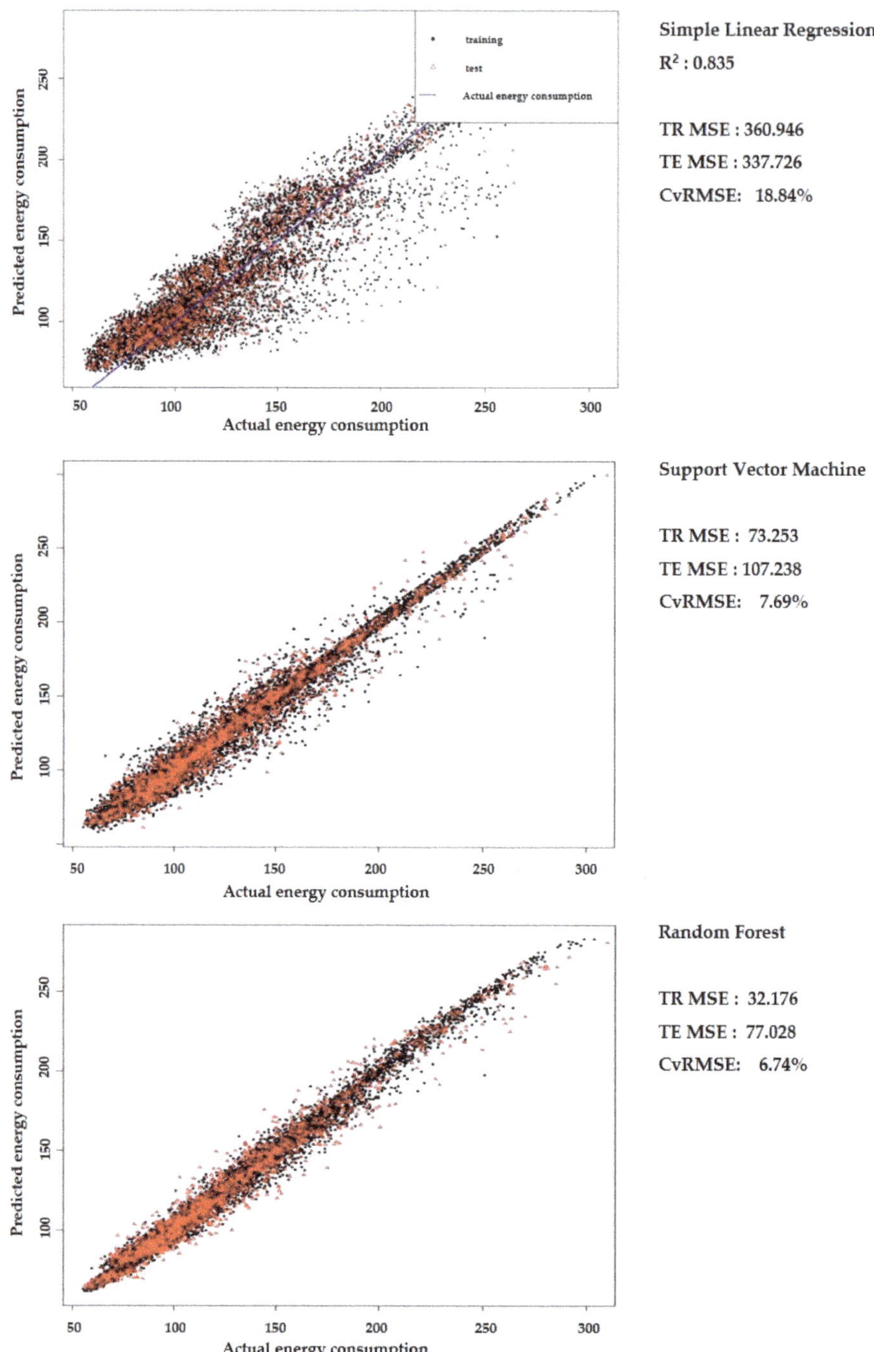

Figure 10. Results of the simple linear regression, support vector machine, and random forest analyses using the original variables.

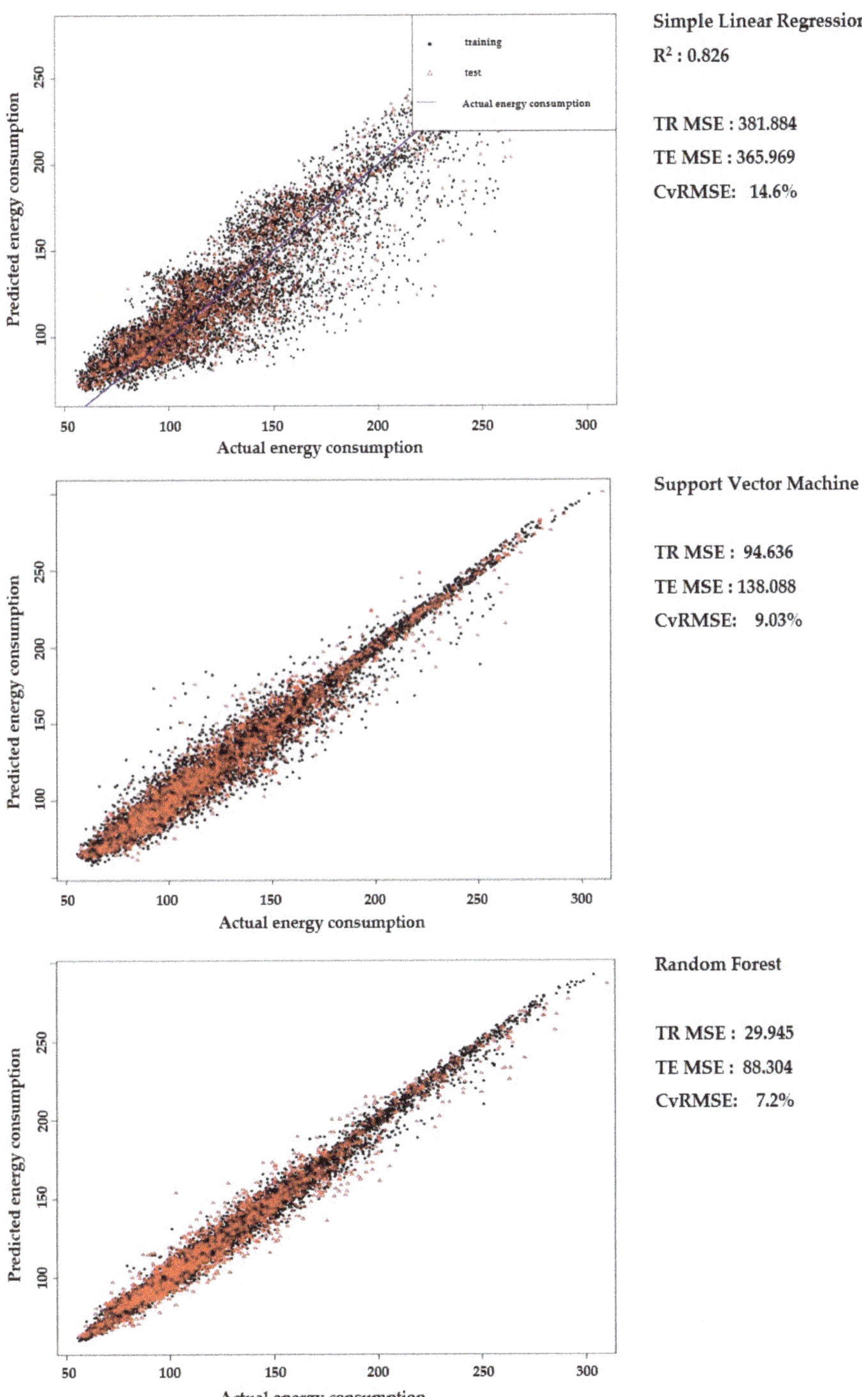

Figure 11. Results of the simple linear regression, support vector machine, and random forest analyses using only the major variables.

6. Conclusions

6.1. Research Conclusions

The purposes of this study were to examine the factors contributing to electric energy use, which accounts for more than 65% of the total building energy consumption, using clustering in machine learning based on data measured from the actual building, and then to find which variables were identified and which had a strong correlation with energy consumption.

Machine learning of the previous studies only predicted the energy usage of buildings in the building sector or confirmed the patterns of usage through clustering methods. However, this study found the key variables that affect the consumption of buildings with a high energy consumption through two different characteristics of the clustering methodology.

The results suggest that the energy consumption predicted by the major variables alone is reliably accurate and that the method can be expected to reduce energy consumption more when these major variables are controlled, or their efficiency is improved. The results can be summarized as follows:

1. The important variables for building energy consumption were derived based on machine learning clustering and were clustered into high- and low-energy-consumption clusters;
2. Based on the clustering, the energy consumption of 11 regional buildings was analyzed according to changes in the outdoor air temperature, which can reveal the building energy features;
3. *T*-tests were performed on the results of the buildings categorized into similar clusters to determine the explanatory variables that led to a high or low energy consumption;
4. Lastly, the important variables identified from this methodology were validated.

    -   Comparison of R^2 values
    -   Validation of the two regression equations for the 16 original variables and important variables by obtaining the MSE and MAPE.

    With respect to Conclusion 3, the important variables that had a decisive effect on the energy consumption of a single building under analysis and the 11 regional buildings (12 buildings in total) were found to be two environmental variables (temperature and humidity), lighting energy, heating energy, and a time variable (month);
5. This study determined the key variables affecting the electrical electricity consumption of buildings, especially non-residential buildings. Except for the external environment (geographic location, temperature, and humidity), the studied building's electricity consumption was found to be as important as its physical characteristics, such as an increased cooling energy, lighting energy, and baseload, due to the working conditions of the occupants. Internal heat gains varied according to occupancy time and density.

Buildings may appear similar, but vary in electricity consumption and patterns, depending on their use, size, and occupants. Therefore, the results of this study cannot necessarily be applied to all buildings. Since this study had a specific application (office building, commercial building), it can be used as an example of variables affecting the electricity consumption of a non-residential building. However, the main influencing factor may be different for a residential building or buildings with other uses (e.g., where the working hours of the occupants are not common).

6.2. Significance and Application

The significance of this study relates to three aspects. First, the analysis of building energy use by existing engineering methods was used to derive absolute values by load factors using a simulation program and to predict reductions by subloads (cooling, heating, and ventilation). Although it may be possible to predict the detailed reduction volume by energy loads or performance, the data were not based on actual data, and the reliability of the results has thus been questioned. However, predicting building energy use using major variables derived from machine learning (as in this study) utilizes the actual total energy volume of the target building, which helps to overcome the problem of reliability of the resulting value. This might occur because of differences in the actual consumption or performance versus the simulated amounts. Second, the prediction of building energy consumption applied with machine learning leads to the same results, despite the lack of experience or intuition by the researcher. Consistent results can be provided with no direct influence from an expert.

It is possible to derive major variables influencing the building energy consumption (high consumption/low consumption), identify which energy source is most responsible for consumption, and view how they influence consumption.

The major variables causing building energy consumption can be used to identify the status of energy consumption of the building and as an indicator of post hoc maintenance.

The machine-learning methodology can be widely applied to various buildings with different uses or located in different climates, and buildings can be classified by major variables influencing building energy consumption.

This method can be used for selecting the variables affecting building energy consumption and can be applied without constraints to non-residential buildings. Furthermore, if there is a collectible data set, such as real-time information regarding a building service, variables that have an impact on energy consumption can be identified without constraints.

From an economic perspective, the fourth significance is as follows. Utilizing this study's methodology in a practical work setting, it would be possible to present monitoring points for building maintenance for aging buildings. Selecting and using energy measurement sensors also make it possible to find the most significant measurement points because the data from actual equipment systems can be used to obtain clustering results (for correlation analysis and the extraction of major variables).

Author Contributions: Conceptualization, S.C. and G.-S.K.; methodology, S.C.; validation, S.C.; writing—original draft preparation, S.C.; writing—review and editing, S.C., and J.L.; visualization, S.C., and J.B.; supervision, S.-B.L.

Funding: This research was sponsored by the Korean Institute of Energy Technology Evaluation and Planning (KETEP) of the Republic of Korea (No. 2018201060010A).

Conflicts of Interest: The authors declare no conflict of interest.

References

1. Climate Change, Sustainable Development Goals. Available online: https://www.un.org/sustainabledevelopment/climate-change-2/ (accessed on 5 April 2019).
2. Choi, M.S.; Choi, D.Y. *Building Energy Consumption Sampling Survey*; Korea Energy Economics Institute: Ulsan, Korea, 2014.
3. Fayaz, M.; Kim, D. Energy Consumption Optimization and User Comfort Management in Residential Buildings Using a Bat Algorithm and Fuzzy Logic. *Energies* **2018**, *11*, 161. [CrossRef]
4. Selin, R. *The Outlook for Energy: A View to 2040*; ExxonMobil: Irving, TX, USA, 2013.
5. Sieminski, A. *International Energy Outlook*; Energy Information Administration: Washington, DC, USA, 2014.
6. KEA. *Statistics on Energy Use and Greenhouse Gas Emissions in the Industrial Sector*; KEA: Ulsan, Korea, 2017.
7. Woo, H.J.; Choi, K.W.; Kim, H.S.; Auh, J.S.; Cho, S.Y.; Baek, J.; Kim, G.S.; Leigh, S.B. A Study on Classifying Building Energy Consumption Pattern Using Actual Building Energy Data. *J. Arch. Inst. Korea Plan. Des.* **2016**, *32*, 143–151. [CrossRef]
8. Cho, S.Y.; Leigh, S.B. Comparing methodology of building energy analysis—Comparative analysis from steady-state simulation to data-driven analysis. *KIEAE J.* **2017**, *17*, 77–86. [CrossRef]
9. Moon, H.J.; Yoon, Y.R. *A Case Study on the Use of Machine Learning Technique for Building Energy Analysis*; Korea Institute of Architectural Sustainable Environment and Building Systems: Seoul, Korea, 2017.
10. Seo, W.J.; Ahn, K.U.; Park, C.S. *Utilizing Machine Learning Technology in Building Energy Diagnosis and Facility Control*; Korea Institute of Architectural Sustainable Environment and Building Systems: Seoul, Korea, 2017.

11. Wang, H.; Chiang, P.C.; Cai, Y.; Li, C.; Wang, X.; Chen, T.L.; Wei, S.; Huang, Q. Application of Wall and Insulation Materials on Green Building: A Review. *Sustainability* **2018**, *10*, 3331. [CrossRef]

12. Korea Energy Economics Institute. *Year Book of Energy Statistics 2018*; Korea Energy Economics Institute: Ulsan, Korea, 2019.

13. Crawley, D.B.; Hand, J.W.; Kummert, M.; Griffith, B.T. Contrasting the capabilities of building energy performance simulation programs. *Build. Environ.* **2008**, *43*, 661–673. [CrossRef]

14. Kim, Y.H.; Park, W.J.; Yang, S.H.; Kim, S.J. *Building Energy Consumption Prediction and Evaluation System*; The Society of Air-Conditioning and Refrigerating Engineers of Korea: Seoul, Korea, 2018.

15. Beak, Y.R. Thermal energy analysis program of building. *J. Mech. Sci. Technol.* **2002**, *42*, 20–21.

16. Yu, Y.; Woradechjumroen, D.; Yu, D. A review of fault detection and diagnosis methodologies on air-handling units. *Energy Build.* **2014**, *82*, 550–562. [CrossRef]

17. Fan, C.; Xiao, F.; Yan, C.; Liu, C.; Li, Z.; Wang, J. A novel methodology to explain and evaluate data-driven building energy performance models based on interpretable machine learning. *Appl. Energy* **2019**, *235*, 1551–1560. [CrossRef]

18. Harish, V.; Kumar, A. A review on modeling and simulation of building energy systems. *Renew. Sustain. Energy Rev.* **2016**, *56*, 1272–1292. [CrossRef]

19. Harish, V.S.K.V.; Kumar, A. Techniques used to construct an energy model for attaining energy efficiency in building: A review. In Proceedings of the International Conference on Control, Instrumentation, Energy and Communication (CIEC), Calcutta, India, 31 January–2 February 2014; pp. 366–370.

20. Wilde, P.D. The gap between predicted and measured energy performance of buildings. *Autom. Constr.* **2014**, *41*, 40–49. [CrossRef]

21. Menezes, A.C.; Cripps, A.; Bouchlaghem, D. Predicted vs. actual energy performance of non-domestic buildings using post occupancy evaluation data to reduce the performance gap. *Appl. Energy* **2012**, *97*, 355–364. [CrossRef]

22. Olivia, G.S.; Christopher, T.A. In-use monitoring of buildings: An overview and classification of evaluation methods. *Energy Build.* **2015**, *86*, 176–189. [CrossRef]

23. Choi, J.H.; Loftness, V.; Aziz, A. Post-occupancy evaluation of 20 office buildings as basis for future IEQ standards and guidelines. *Energy Build.* **2012**, *46*, 167–175. [CrossRef]

24. Agha-Hossein, M.; El-Jouzi, S.; Elmualim, A.; Ellis, J.; Williams, M. Post-occupancy studies of an office environment: Energy performance and occupants' satisfaction. *Build. Environ.* **2013**, *69*, 121–130. [CrossRef]

25. Salehi, M.M.; Cavka, B.T.; Frisque, A.; Whitehead, D.; Bushe, W.K. A case study: The energy performance gap of the Center for Interactive Research on Sustainability at the University of British Columbia. *J. Build. Eng.* **2015**, *4*, 127–139. [CrossRef]

26. Niu, S.; Pan, W.; Zhao, Y. A virtual reality integrated design approach to improving occupancy information integrity for closing the building energy performance gap. *Sustain. Cities Soc.* **2016**, *27*, 275–286. [CrossRef]

27. Herrando, M.; Cambra, D.; Navarro, M.; De La Cruz, L.; Millán, G.; Zabalza, I.; Bribian, I.Z. Energy Performance Certification of Faculty Buildings in Spain: The gap between estimated and real energy consumption. *Energy Convers. Manag.* **2016**, *125*, 141–153. [CrossRef]

28. Min, Z.; Morgenstern, P.; Halburd, L.M. Facilities management added value in closing the energy performance gap. *Int. J. Sustain. Built Environ.* **2016**, *5*, 197–209. [CrossRef]

29. Witten, I.H.; Frank, E.; Hall, M.A. *Data Mining: Practical Machine Learning Tools and Techniques*; Elsevier: Amsterdam, The Netherlands, 2013; pp. 45–52.

30. Lee, J.H.; Shin, J.; Realff, M.J. Machine learning: Overview of the recent progresses and implications for the process systems engineering field. *Comput. Chem. Eng.* **2018**, *114*, 111–121. [CrossRef]

31. Molina-Solana, M.; Ros, M.; Ruiz, M.D.; Gómez-Romero, J.; Martin-Bautista, M. Data science for building energy management: A review. *Renew. Sustain. Energy Rev.* **2017**, *70*, 598–609. [CrossRef]

32. Wang, Z.; Wang, Y.; Srinivasan, R.S. A novel ensemble learning approach to support building energy use prediction. *Energy Build.* **2018**, *159*, 109–122. [CrossRef]

33. Catalina, T.; Virgone, J.; Blanco, E. Development and validation of regression models to predict monthly heating demand for residential buildings. *Energy Build.* **2008**, *40*, 1825–1832. [CrossRef]

34. Ekici, B.B.; Aksoy, T.U. Prediction of building energy consumption by using artificial neural networks. *Adv. Eng. Softw.* **2009**, *40*, 356–362. [CrossRef]

35. Yu, Z.; Haghighat, F.; Fung, B.C.M.; Yoshino, H. A decision tree method for building energy demand modeling. *Energy Build.* **2010**, *42*, 1637–1646. [CrossRef]

36. Dong, B.; Cao, C.; Lee, S.E. Applying support vector machines to predict building energy consumption in tropical region. *Energy Build.* **2005**, *37*, 545–553. [CrossRef]

37. Classification: Train, Validatio, Test Split. Available online: https://blog.algorithmia.com/page/50 (accessed on 29 September 2019).

38. Paudel, S.; Elmitri, M.; Couturier, S.; Nguyen, P.H.; BrunoLacarrière, R.; Le Corre, O. A relevant data selection method for energy consumption prediction of low energy building based on support vector machine. *Energy Build.* **2016**, *138*, 240–256. [CrossRef]

39. Yildiz, B.; Bilbao, J.; Sproul, A. A review and analysis of regression and machine learning models on commercial building electricity load forecasting. *Renew. Sustain. Energy Rev.* **2017**, *73*, 1104–1122. [CrossRef]

40. Rahman, A.; Smith, A.D. Predicting fuel consumption for commercial buildings with machine learning algorithms. *Energy Build.* **2017**, *152*, 341–358. [CrossRef]

41. Moon, J.; Jun, S.; Park, J.; Choi, Y.H.; Hwang, E. An Electric Load Forecasting Scheme for University Campus Buildings Using Artificial Neural Network and Support Vector Regression. *KIPS Trans. Comput. Commun. Syst.* **2016**, *5*, 293–302. [CrossRef]

42. Seong, N.C.; Kim, J.H.; Choi, W.; Yoon, S.C.; Nassif, N. Development of Optimization Algorithms for Building Energy Model Using Artificial Neural Networks. *J. Korean Soc. Living Environ. Syst.* **2017**, *24*, 29–36. [CrossRef]

43. Naganathan, H.; Chong, W.O.; Chen, X. Building energy modeling (BEM) using clustering algorithms and semi-supervised machine learning approaches. *Autom. Constr.* **2016**, *72*, 187–194. [CrossRef]

44. Ko, J.H.; Kong, D.S.; Huh, J.H. Baseline building energy modeling of cluster inverse model by using daily energy consumption in office buildings. *Energy Build.* **2017**, *140*, 317–323. [CrossRef]

45. Yang, J.; Ning, C.; Deb, C.; Zhang, F.; Cheong, D.; Lee, S.E.; Sekhar, C.; Tham, K.W. k-Shape clustering algorithm for building energy usage patterns analysis and forecasting model accuracy improvement. *Energy Build.* **2017**, *146*, 27–37. [CrossRef]

46. Moon, H.J.; Jung, S.K.; Ruy, S.H. *Building Cooling and Heating Energy Consumption Pattern Analysis Based on Building Energy Management System Data Using Machine Learning Techniques*; The Society of Air-Conditioning and Refrigerating Engineers of Korea: Seoul, Korea, 2015.

47. Hwang, H.M.; Lee, S.H.; Park, J.B.; Park, Y.G.; Son, S.Y. Load Forecasting using Hierarchical Clustering Method for Building. *Trans. Korean Inst. Electr. Eng.* **2015**, *64*, 41–47. [CrossRef]

48. Shmueli, G.; Petel, N.R.; Bruce, P.C. *Data Mining for Business Intelligence*; Wiley: New York, NY, USA, 2010; pp. 91–97.

49. Cho, S.Y.; Leigh, S.B. A Study of the Possibility of Building Energy Saving through the Building Data: A Case Study of Macro to Micro Building Energy Analysis. *Korean J. Air Cond. Refrig. Eng.* **2017**, *29*, 580–591.

50. Flach, P. *Machine Learning: The Art and Science of Algorithms That Make Sense of Data*, 1st ed.; Intelligent Systems Laboratory, University of Bristol: Bristol, UK, 2012; pp. 295–328.

51. R-3.6.1 for Window. Available online: https://cran.r-project.org/bin/windows/base/ (accessed on 16 August 2019).

52. Cho, K.H.; Oh, J.H.; Kim, S.S.; Lee, B.H.; Yeo, M.S. *An Analysis of Energy Consumption Patterns in University Buildings Using Inverse Modeling*; Architectural Institute of Korea: Seoul, Korea, 2017.

53. Tronchin, L.; Manfren, M.; James, P.A.B. Linking design and operation performance analysis through model calibration: Parametric assessment on a Passive House building. *Energy* **2018**, *165*, 26–40. [CrossRef]

54. Attanasio, A.; Piscitelli, M.S.; Chiusano, S.; Capozzoli, A.; Cerquitelli, T. Towards an Automated, Fast and Interpretable Estimation Model of Heating Energy Demand: A Data-Driven Approach Exploiting Building Energy Certificates. *Energies* **2019**, *12*, 1273. [CrossRef]

55. ASHRAE. *ASHRAE's Guideline 14, Measurement of Energy and Demand Savings*; ASHRAE: Atlanta, GA, USA, 2002.

Article

Examining the Impact of Daylighting and the Corresponding Lighting Controls to the Users of Office Buildings

Lambros T. Doulos [1,2,*], **Aris Tsangrassoulis** [3], **Evangelos-Nikolaos Madias** [2], **Spyros Niavis** [4], **Antonios Kontadakis** [3], **Panagiotis A. Kontaxis** [2,5], **Vassiliki T. Kontargyri** [6], **Katerina Skalkou** [1,7], **Frangiskos Topalis** [2], **Evangelos Manolis** [8], **Maro Sinou** [7] and **Stelios Zerefos** [1]

[1] School of Applied Arts, Hellenic Open University, Parodos Aristotelous 18, 26335 Patras, Greece; katerinaskalkos@yahoo.gr (K.S.); zerefos@eap.gr (S.Z.)

[2] Lighting Laboratory, School of Electrical and Computer Engineering, National Technical University of Athens, Zografou, 15780 Athens, Greece; madias@mail.ntua.gr (E.-N.M.); pkont@uniwa.gr (P.A.K.); fvt@central.ntua.gr (F.T.)

[3] Department of Architecture, University of Thessaly, 38221 Volos, Greece; atsagras@uth.gr (A.T.); kontadakis@uth.gr (A.K.)

[4] Department of Economics, University of Thessaly, 38221 Volos, Greece; spniavis@uth.gr

[5] Lighting Technology Laboratory, Department of Electrical and Electronics Engineering, School of Engineering, University of West Attica, Egaleo, 12241 Athens, Greece

[6] High Voltage Laboratory, School of Electrical and Computer Engineering, National Technical University of Athens, Zografou, 15780 Athens, Greece; vkont@power.ece.ntua.gr

[7] Department of Interior Architecture, School of Applied Arts & Culture, University of West Attica, 12243 Athens, Greece; msinou@uniwa.gr

[8] Capture Visualisation AB, AtlaBase Ltd., 11852 Athens, Greece; vangelis@atlalite.com

* Correspondence: ldoulos@mail.ntua.gr; Tel.: +30-6937-086820

Received: 26 June 2020; Accepted: 24 July 2020; Published: 4 August 2020

Abstract: Daylight utilization significantly contributes to energy savings in office buildings. However, daylight integration requires careful design so as to include variations in daylight availability and maintain a balance between factors such as lighting quality and heat gain or loss. Designers with proper planning can not only improve the visual environment and create higher-quality spaces, but simultaneously minimize energy costs for buildings. The utilization of photosensors can exploit the benefits of daylighting by dimming the lighting system, so that no excessive luminous flux is produced, thus leading to energy savings as well as visual contentment. However, the human factor is crucial for the proper function of a lighting control system. Without its acceptance from the users, energy savings can be minimized or even negligible. The objective of this paper is to present a post-occupancy evaluation regarding occupant satisfaction and acceptance in relation to daylighting in offices equipped with automated daylight controls. In addition, the response of the users was compared with lighting measurements that were performed during the post-occupancy evaluation. Three case studies of office buildings with installed daylight-harvesting systems were examined. The age of the occupants was a crucial factor concerning their satisfaction in relation to the lighting levels. Aged users were more comfortable with lighting levels over 500lx, while young users were satisfied with 300lx. The impact of different control algorithms was outlined, with the integral reset algorithm performing poorly. The acceptance of the users for the closed loop systems maintained the expected energy savings of the daylight harvesting technique. Most of the occupants preferred to use daylight as a light source combined with artificial light but having the control to either override or switch it on and off at will. The results shown that a post-occupancy survey along with lighting measurements are significant for making an office environment a humancentric one.

Keywords: daylight; lighting control; lighting; occupant preferences; occupant satisfaction; photosensor; post-occupancy evaluation; survey

1. Introduction

A significant amount of electricity, namely 17–20%, is consumed by lighting at a global scale [1,2]. In Greece, the respective quotient of the electricity consumed by lighting is 21% [3]. Apart from daylight integration as a means for electric energy reduction [4,5], by examining the daylight zones [6–9] and redirecting solar radiation [10–14], a lot of research has been conducted for (a) the promotion of renewable energy sources in the building sector [15–18] and in electrical energy storage systems [19] and (b) the energy savings in exterior places [20–28]. LED luminaires have a great impact on the lighting market due to their prominent characteristics such as high energy efficiency and long lifespan and are acknowledged as the prevailing lighting solution in the future [29–32]. The use of solid-state lighting can contribute to the reduction of energy consumption due to the increased luminous efficacy when compared to traditional light sources (i.e., fluorescent). Consequently, the global market penetration of LED lamps has increased from 9% in 2011 in the European Union and is expected to exceed 70% in the near future [3]. Decision makers in the lighting sector should take into account circular economy features and use of lighting controls [33–43]. Reviewing the literature, it is evident that the installation of daylight-harvesting systems leads to high energy savings [44–46]. Despite the proven benefits from daylight, daylight controls are not widely installed in buildings due to the additional costs of equipment, labor, studies for lighting control design, potential training for usage and maintenance. Another factor that impedes the proliferation of daylight controls is the misperception among building contractors that these systems are unreliable, although various studies have highlighted their high reliability [47–52]. In order to better understand the operation of photosensor-based daylight-harvesting systems and maximize the benefits of daylight integration, the users' reaction to these systems must be thoroughly studied.

Building facilities have the sole purpose of accommodating the needs of their users. A post-occupancy evaluation study, abbreviated as POE, is defined as a holistic and rigorous procedure of assessing a building's performance with regards to its users after its occupation. It can be described as the documentation and review of user satisfaction, space and resources exploitation of an occupied built facility so as to quantify its efficiency and designate critical occupant and building performance issues. It involves both technological as well as human factors [53]. Therefore, it can be considered as a framework which demonstrates the continuous interaction between buildings and the need of the occupants and suggests appropriate measures to optimize the built environment in order to fulfill these needs [54]. POEs are categorized in three types according to their purpose: indicative, investigative and diagnostic [55,56]. Post-occupancy evaluation is characterized by a plethora of benefits. The most important is that it provides feedback for the actual operation of a building [56]. Other benefits include the effective application of building design skills, improvement of building commissioning process, facilitation of building management, accumulation of knowledge concerning building operation, optimization of building efficiency with regards to a variety of both technical characteristics, e.g., a building's energy or thermal management, its structure and architecture, as well as human characteristics, e.g., occupant's wellbeing and productivity, opportunities to improve dialogue within design teams and their partners, improvement of existing building regulations and establishment of new more effective regulatory frameworks and policies [53,56]. The advantages of POE are applicable not only to a specific building or facility, but extend further to the building sector overall. A POE provides valuable conclusions, which can be utilized for the establishment guidelines and the adoption of measures that improve a buildings' performance in existing, as well as future projects [57,58]. The conduction of POE's enhances knowledge thus steadily improving building indoor environment [59].

Numerous studies have established the correlation between the occupants' satisfaction and their productivity [58,60–62]. Buildings which address efficiently the needs of their users are considered more pleasant [58,63]. Lieman and Bordass [58] have proven that the level of control each user has over his working environment is a critical factor which is able to influence the occupants' satisfaction and productivity. Another finding of their survey was that users were more satisfied when they worked in an office with the preferred operable windows, rather than offices with automatic ventilation in their working environment systems. An extensive post-occupancy evaluation study [64] has accumulated data from various office buildings and has highlighted the importance of indoor environment quality factors such as air quality, temperature, sound, privacy, etc. Thomas [61] has shown that high levels of daylight, glare and noise control in a work environment can positively affect the occupants' satisfaction. Choi et al. [65] have performed on-site measurements of indoor environment quality characteristics, as well as occupant surveys and concluded that thermal quality along with lighting adequacy were highlighted as key factors for user satisfaction and wellbeing. Kim and de Dear [66] have compared the satisfaction of employees working in open-plan as well as private offices and deduced that the users of enclosed private offices expressed a higher level of satisfaction with their work environment. The most important factor affecting an occupant's satisfactions in both types of offices was the amount of individual space available. However, several differences regarding user satisfaction between the two types of offices have been identified. Absence and high level of commotion in open space offices were significant factors that affected employee satisfaction negatively, whereas lighting levels, ease of interaction and comfortable furniture were identified as more significant for the satisfaction of the occupants in private offices. Another POE study by Agha-Hossein et al. [62] has denoted a strong correlation between the productivity and contentment of employees with two groups of variables: (a) interior usage of their workspace which includes office layout and appearance, comfort of furniture, visual and auditory privacy, and (b) physical conditions which include natural and artificial light, air quality and temperature. Filippín et al. [67] have conducted an evaluation study concerning a building in Argentina and quantified its thermal and energy performance which were highly praised by its occupants. Mustafa [59] published a POE study regarding the architecture, accessibility as well as indoor environment quality of a university facility in Iraq. The procedure of POE involved a combination of two sophisticated surveys. The first involved rating by experts and the second was addressed to the users of the building. The results of this study demonstrated that the majority of building environment characteristics, namely a quotient of 88%, is highly associated with the contentment of its occupants. Ponterosso et al. [63] have performed a POE in a "green" building and reached the conclusion that it is important that occupants are able to understand the sustainability features of a building. According to their study, thermal comfort is a profound factor which can affect comfort levels. Moreover, seemingly small details of the indoor environment can induce great impact on the level of comfort experienced by occupants. A large-scale POE study by Park et al. [68], including 1601 workstations in 64 office buildings, utilized measurements of indoor environment quality and occupant surveys. Their objectives were to determine critical indoor environment quality and physical parameters that affect occupant's satisfaction with regards to thermal quality and indicate correlations between the building systems, the measurements of indoor environment factors and user satisfaction in simultaneous time frames. The researchers [68] deduced that the most important indoor environmental factors affecting occupant satisfaction with regards to thermal quality were air temperature and radiant temperature asymmetry which exists between the two sides of the building's wall. Respectively, the most significant technical attributes of the building were the size of each working plane, view and the extent of thermal control. Candido et al. [69] aimed to identify key factors that correlate to employees' contentment, productivity, and wellbeing in open-space office areas while concurrently indicating design similarities in contemporary working environments. The data that were analyzed came from 8827 evaluation studies in office buildings in Australia. The results have proven that maintenance and aesthetics of the building, quality of air, level of sound and commotion, visual comfort were considered among other features as key parameters by the occupants. Göçer et al. [70] documented the results from a dataset comprising of 9794 POE surveys in Australian office buildings. Their conclusions were that open-space office buildings could exploit the advantages of enhanced aesthetics, design of zones that addresses the needs of the work environment, daylight integration and access to an outdoor environment. A recent POE study [71] was conducted in four Swiss green office buildings and employed in situ measurements and occupant surveys in an effort s to demonstrate the factors that maximize users' comfort. The main finding of the study was that although the observed environmental factors adhered to standards, the indoor conditions did not reach the 80% satisfaction threshold by the users. The occupants designated temperature and air quality as crucial parameters.

As energy audits and surveys in buildings have a significant role [72–80], the scope of the proposed research is to emphasize their corresponding role into the lighting systems. The number of POE studies focused only to lighting is limited and this is exactly the basic scope of the proposed method. The collection of POE studies, especially these focusing on the lighting systems, systematically organizes the existing knowledge in the field. In that sense, these studies can provide information on the quality features of the lighting design. In this paper, a post-occupancy evaluation study is performed to document the reactions of the occupants concerning daylight harvesting control systems. The study aims at drawing valuable conclusions concerning the efficiency of lighting systems and daylight controls in an office building. Three office buildings have been examined and a combination of methods has been selected in order to carry out the POE study, namely (a) observation of the lighting system of building, (b) physical measurements regarding the total illuminance on working plane level originating from both daylight and artificial lighting, and (c) an occupant survey was formed that takes into account the occupants' reaction to daylight, windows and lighting control. An occupant survey was delivered to the users of three office buildings with different daylight harvesting control algorithms in each of them, in order to elucidate their preferences. The results of the study can produce a quality profile of the building based on different user opinions that can be utilized as a model for the evaluation of the lighting systems and controls, as well as to determine the correlation between the performance of these controls, the illuminance measurements and the user's overall satisfaction and wellbeing.

2. Materials and Methods

This research is based on the occupants' preferences together with lighting measurements. A questionnaire was designed that can demonstrate user preferences regarding daylight integration and lighting control. Three buildings were selected (A, B and C), each one with different control algorithm (closed loop, open loop and integral-reset control algorithm) in their daylight harvesting control system, in order to investigate their performance. The selection of these buildings was based on two factors: the suitability of the building (i.e., installed lighting control systems) and the availability of users to participate in a survey. The latter is a major problem since by publishing any energy-related results, the value of the buildings might be affected. Thus, there is a reluctance from building owners to participate in surveys or measurement campaigns. A total of 122 people were involved in the questionnaire and an equal number of task areas were measured. Figure 1 shows the methodology used. Lighting measurements combined with the occupants' preferences were statistically analyzed. Presented in the following paragraphs are the examined buildings, the definition of the questionnaire, the lighting measurement and the statistical analysis procedures.

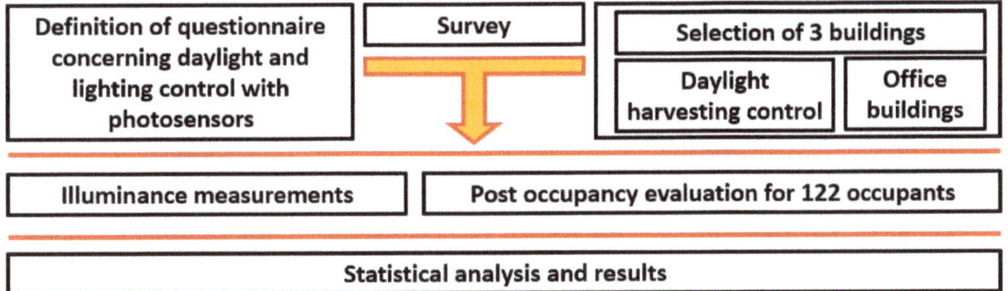

Figure 1. Diagram of the post-occupancy evaluation method.

2.1. Selection of the Buildings

Three office buildings located in Athens, Greece, were selected. In all buildings, a daylight-harvesting system was installed. Building A (Figure 2) is a 2-storey office building of the tertiary sector in an open plan area. On the first floor there are open plan offices, while on the second floor there are closed office spaces. POE was performed in both floors. Building B (Figure 3) is an 8-storey office building. The sky view to the north is partially obstructed by a three-storey building. The users are located mainly in the perimeter of an open plan space. POE was performed on the 3rd and 4th floor. Building C (Figure 4) is a 4-storey office building. The sky view to the north and south is partially obstructed by two similar height buildings. Floors 2 and 3 were selected for this study. A detailed description of the lighting system and daylight control is presented in Table 1 for each building. In all buildings, all examined areas were used as offices. In the rest floors the activities were different, thus were not taken into account.

Figure 2. Interior space (**left**), exterior façade (**middle**) and position of the photosensor (**right**) for Building A.

Figure 3. Interior space (**left**), exterior façade (**middle**) and position of the photosensor (**right**) for Building B.

Figure 4. Interior space (**left**), exterior façade (**middle**) and position of the photosensor (**right**) for Building C.

Table 1. Technical characteristics of lighting system and daylight harvesting equipment.

Building	Type of Luminaire	Placement of Photosensor	Control Algorithm
A	Recessed with parabolic louvres and 4 T8 18W lamps	Interior, aiming the task area, controlling a group of luminaries (Figure 2, right)	Closed loop
B	Downlight with 2 Compact Fluorescent 18W lamps	Exterior, one photosensor controlling the luminaires in perimeter zone of the building (Figure 3, right)	Open loop
C	Recessed with parabolic louvres and 4 T5 14W lamps	Interior, one photosensor for each luminaire installed in the daylight zone of the building (Figure 4, right)	Integral reset

2.2. Survey with Questionnaire

Since the main objective of the study concerned daylight integration and control, the occupants that were working in offices placed in the perimetric zone of each building were chosen for the study. The questionnaires were handed to the users that (a) are located in areas with high daylight penetration, (b) their working station was inside the daylight zone according to EN 15,193 [81] (the maximum depth of the daylight zone was 2.5 times the height between the window lintel above the floor and the height of the task area above the floor) and (c) the corresponding lighting system in the daylight zone was controlled using from a daylight-harvesting system. Due to the general personal data protection regulations the questionnaires were filled anonymously and the detailed position of the occupants' working stations was not recorded, except their age. Figure 5 presents the orientation of the working spaces as a percentage.

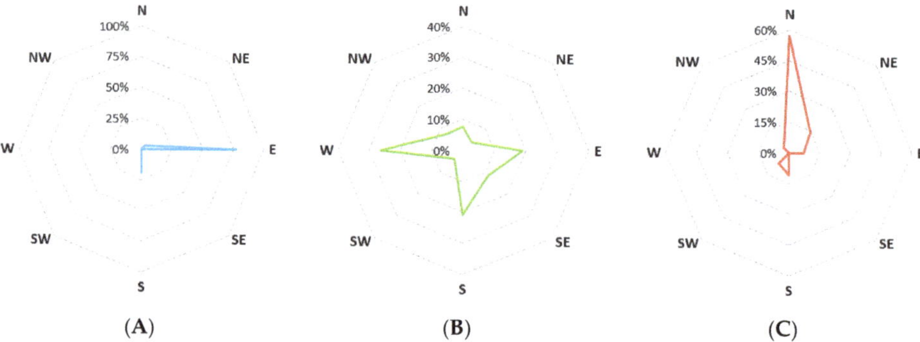

Figure 5. Position of the occupants' working planes concerning the orientation of the building façade where the office spaces were located for each examined building (Building **A**: left, Building **B**: middle, Building **C**: right).

The sample of the occupants that were selected, were between 20–60 years old. (Figure 6, right). The questionnaire was completed by a total number of 122 occupants (53 women, 69 men, Figure 6, left). The questions are presented in Table 2. The format of the questionnaire was based mainly on rating scales. Rating scales are used so as to model the respondents' feedback in a comparative form and are applied in various types of surveys where respondents are expected to assess or classify attributes, such as performance, efficiency, etc. The rating scales are suitable for field research due to their reliability and efficiency for subsequent statistical analysis. Based on the answers, researchers were able to associate a qualitative measure with a certain attribute or feature. The answers were on an ordinal 5-point scale from 'too much' to 'very poor' and a 3- or 4-point scale, while some questions used a tick box (Table 2). The occupants were not trained or informed before filling the questionnaire, in which simple questions were used. This was a prerequisite for the current research. We believe that any training to the occupants prior the questionnaire could affect their responses, especially their awareness to the lighting controls systems, a crucial factor for their successful implementation in to the buildings.

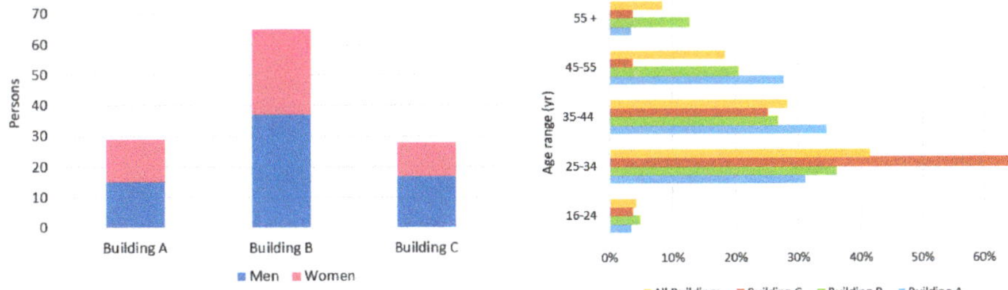

Figure 6. Number of occupants that completed the POE (left), and their age frequency distribution (right).

Table 2. The questionnaire given to the occupants of the buildings.

	Questions	Answer	
1	Which three of the following parameters are the most important for you in order to have a pleasant environment in your work?	O Thermal Comfort O Visual comfort O Ventilation O External openings O Aesthetics of the interior space	O Noise free O Privacy O Large space O View out O Other
2	From 5 (very satisfied) to 1 (very dissatisfied), how satisfied are you with the parameters on the right in your working environment?	O Visual comfort O Noise level O Odor O Ventilation O Thermal Comfort	O Size of external openings O Privacy O Large space O View O Aesthetics of the interior space
3	Do you prefer working in with daylight, artificial light or both?	Daylight, artificial, both	
4	How important is it to you to have a window in your room or immediate work area?	Very, Moderately Not important	
5	Do you ever work using only daylight?	Often, sometimes, only occasionally, never	
6	Is your visual environment pleasing?	Yes, No	
7	When you look up from your working plane does the scene that your see in front of you seem	Too bright 5( ) 4( ) 3( ) 2( ) 1( ) Too dim	
8	Does glare () ever disturb or annoy you?	Frequently 5( ) 4( ) 3( ) 2( ) 1( ) Never	
9	Do you consider yourself as very sensitive to glare?	Yes, No	
10	The amount of total light on your office is:	Too much 5( ) 4( ) 3( ) 2( ) 1( ) Very poor	
11	The amount of total light on your desk is:	Too much 5( ) 4( ) 3( ) 2( ) 1( ) Very poor	
12	The amount of daylight on your desk is:	Too much 5( ) 4( ) 3( ) 2( ) 1( ) Very poor	
13	Is an automated lighting control system installed in your office or in your work environment generally? If yes, it is annoying?	Yes ( ) No ( )	
14	Level of importance with regards to the ability to control the lighting output of the luminaires over your personal office?	Important 5( ) 4( ) 3( ) 2( ) 1( ) Unimportant	
15	What degree of control do you have over the electric lighting above your work plane?	Full control 5( ) 4( ) 3( ) 2( ) 1( ) No control	
16	How satisfied are you with this level of control	Satisfied 5( ) 4( ) 3( ) 2( ) 1( ) Unsatisfied	
17	Is it important to be able to control the lighting of your working plane separately from that of nearby desks?	Important 5( ) 4( ) 3( ) 2( ) 1( ) Unimportant	
18	Do you have any kind of control on the amount of daylight that impinges on your working plane?	Full control 5( ) 4( ) 3( ) 2( ) 1( ) No control	

2.3. Survey with Lighting Measurements

During the survey and the evaluation of the questionnaire from the users (Spring time, from 11:00 to 15:00), illuminance measurements were performed. The illuminance measurements are a crucial parameter in a POE survey, because (a) the occupants' opinion can be verified and (b) the lighting installation is checked as if it is in accordance with EN 12464-1 [82] or the initial lighting design. Special actions should be performed after POE and measurements are concluded. If the space is over illuminated and there are complaints, the redundant luminaires should be removed if possible. On the other hand, a number of luminaires should be added if lighting levels are inadequate. For the measuring procedure, a T-10a Minolta calibrated luxmeter was used (Figure 7). The measuring sensor was placed at a height of 0.8 m, on the working surface of each occupant. The luminaires in the measuring area were all in operation using daylight harvesting control. The measured value was the total illuminance, namely daylight and dimmed artificial lighting levels at the time of the survey.

Figure 7. Illuminance measurement procedure in a lighting survey. A T-10a Minolta calibrated illuminance meter was used for the measurements.

2.4. Statistical Analysis

The survey responses were used to build two types of variables, namely ordinal and dummy. Therefore, the methods for testing the research questions are adjusted on the type of variables incorporated in each one of them. There are two basic methods for testing the associations of variables. When the variables are of ordinal scale, the correlation analysis is conducted. There are two basic correlation coefficients: Pearson and Spearman. Pearson measures the association of variables using their original values while the Spearman coefficient tests the monotonic association based on the rankings of the variables. The first is parametric and is implemented in rather large samples, while the second is a non-parametric one and is implemented on rather small samples [82]. In addition, it can be used when the data are of ordinal scale, which is the case of the present paper [83]. Since, the correlation analyses will consider both the whole sample where observations are quite a lot, as well as on individual buildings, which have smaller samples, the present paper adopts the Spearman correlation analysis. The formula for estimating the Spearman coefficient between two variables y and x is as follows:

$$\rho = 1 - \frac{6 \sum d_i^2}{n(n^2 - 1)},\tag{1}$$

where n is the number of cases and d_i the difference between the ranks of their observation in y and x variables, respectively. The ρ coefficient takes values in the closed interval $[-1,1]$ with values close to 1 denoting a strong positive correlation and values close to -1 a strong negative correlation. Values close to 0 denote a weak correlation of the considered variables [83,84].

On the other hand, when analysis incorporates both ordinal and dummy variables, the non-parametric Mann–Whitney test will be conducted. Mann–Whitney test the null hypothesis that the scores of a variable follow the same distribution over two samples being formulated by the two values of the dummy variable. Similar to Spearman coefficient, the Mann and Whitney statistic U, is computed based on the rankings rather on the actual scores of the variables [83,84]. Then, the statistical significance of the estimation is extracted based on a z normal approximation of the U statistic [85].

3. Results

3.1. Survey

In the first question (Table 2), the occupants ranked the three parameters that are most important for them in making an office environment a pleasant one. In this question, the users indicated what they want mostly and not necessarily what they experience in their office. Visual comfort, thermal comfort and ventilation are ranked as the most desired features with average with average occupants' preferences 24.6%, 23.8% and 21.3%, correspondingly (Figure 8). On the other hand, in the second question, the occupants voted for what they experience in their offices. These comparable values for the individual examined buildings are shown in Figure 9. The results of occupants' satisfaction for Buildings B and C related to the most parameters came out very well. Only visual comfort came out well for Building A, because of the lighting control system with photosensors and occupancy sensors.

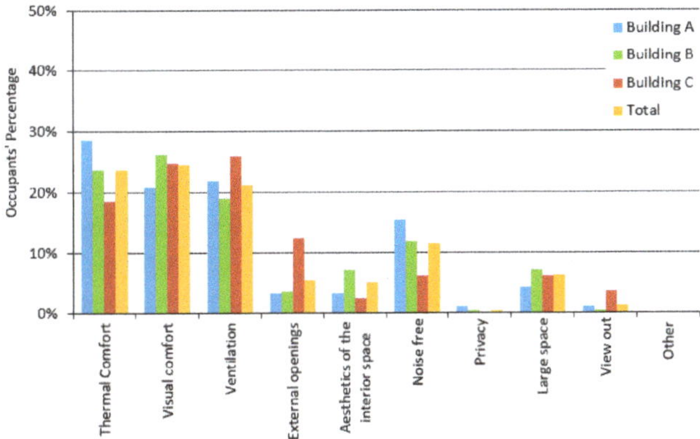

Figure 8. Results of occupants' opinion of what they want most from the main parameters that affect the overall comfort in an office building.

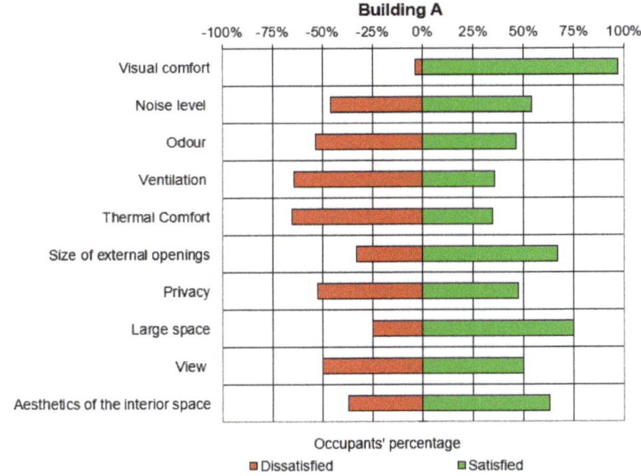

Figure 9. *Cont.*

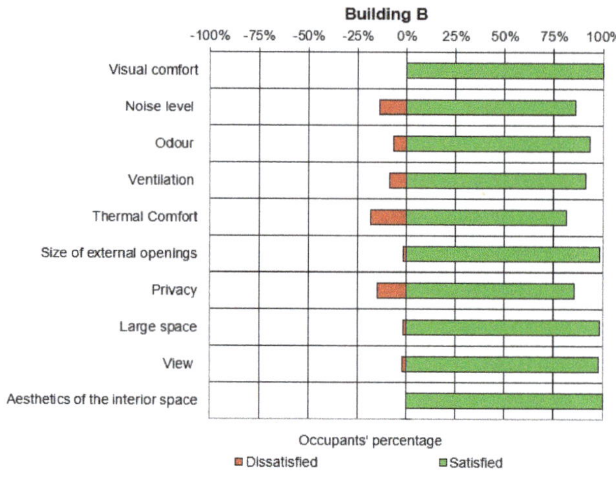

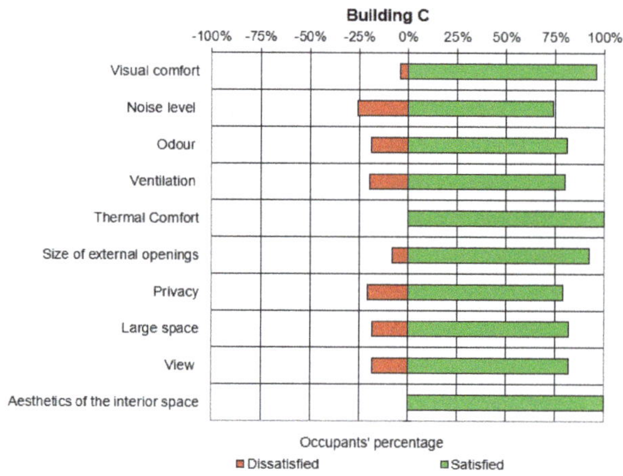

Figure 9. Satisfaction (green color) and dissatisfaction (red color) of different parameters (Visual comfort, Noise level, Odor, Ventilation, Thermal Comfort, Size of external openings, Privacy, Large space, View, Aesthetics of the interior space) for the examined buildings.

As most of the occupants want to work with both daylight and artificial lighting (Figure 10, left) a daylight-harvesting system is preferable than a simple on-off switch. Thus, every user selected that the existence of an external opening is very or moderately important (Figure 10, right). However, only the users of Building C prefer their workplace to be illuminated solely by daylight, while the majority of the employees of Building A oppose it (Figure 11, left). Most of the occupants reported that there is too much light, not only into the entire area of their offices but also in their working planes. Their answers were almost the same either for their space or desk (Figure 11, right, and Figure 12, left). This is due to the overall and general lighting design using a typical grid. None of the buildings had placement of the luminaires according to the task areas. The occupants of Building A, while reporting high illuminance values for both daylight and artificial lighting (Figure 12, left), they also reported low amounts of daylight reaching their working space (Figure 12, right). This means that in many working positions in Building A, the luminaires above them were emitting their maximum light flux. In this building, the control of solar gains is realized by both external and internal shading systems (Figure 2, middle) which are adversely affecting daylight levels. In open-plan spaces, their use can result in gloomy conditions and hence low daylighting levels. A façade renovation can definitely improve the aforementioned conditions [10–14].

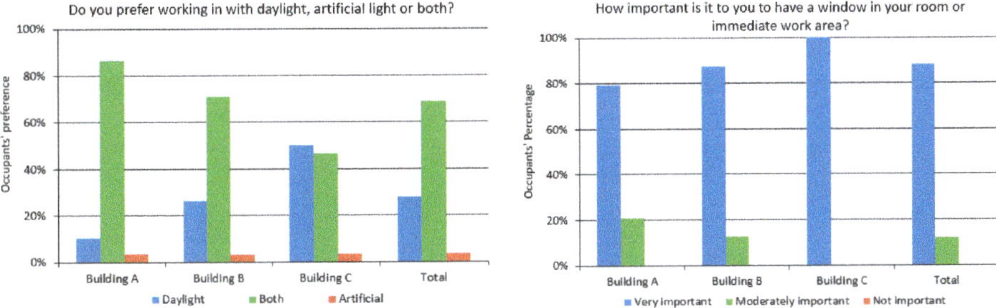

Figure 10. Results for how the users prefer their lighting conditions (**left**) and how important is for them to have a window in their office (**right**).

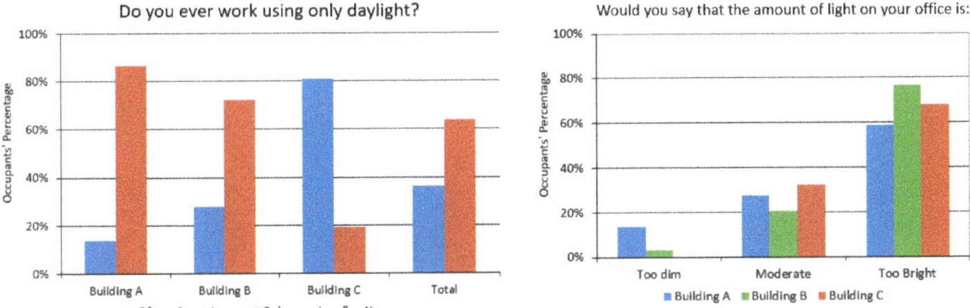

Figure 11. Results of how often the occupants work using only daylight (**left**) and the amount of total light on their space (**right**).

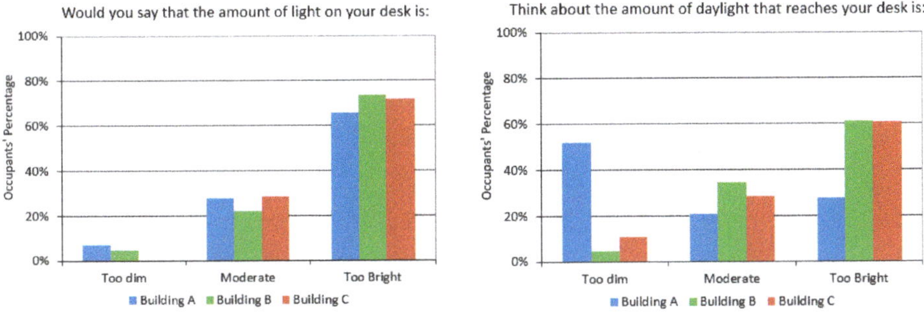

Figure 12. Results about the amount of total light (**left**) and daylight only (**right**) on their desk.

From the total number of the users, 21% of them were not aware of the existence of a daylight harvesting control. The less awareness was recorded in Building C, where 43% of the occupants in that building did not know about the installed photosensors. The users' responses show different reactions for the occupants of the buildings concerning light controls (Figures 13–15). The users of buildings A and B had similar responses concerning the degree of control that the users have over the artificial lighting above their working plane (Figure 13, right). In general, the occupants stated that it is important to be able to control not only the artificial lighting above them (Figure 13, left) but also to control it separately from the advancement areas (Figure 14, right). However, they are highly or moderately satisfied with the degree of control of the lighting system that they have (Figure 14, left).

It should be mentioned that daylight harvesting was in an automated mode in all examined buildings. Moreover, both buildings A and B do not offer their users the ability to override lighting controls or perform manual control (almost 90% of the users in buildings A and B; Figure 13, right). On the contrary, Building C offers that ability. For this reason, over 60% of all the users have expressed a strong desire to control the level of artificial lighting (Figure 13, left). Furthermore, while it could generally be expected that the degree for individual control would be high, the reality of control especially in areas with shared control groups, strengthens these convictions (Figure 14, right). Most of the users, over 50% for all the case studies, have the ability to control the amount of daylight that impinges on their working plane (Figure 15) by using the internal blinds (Figures 2 and 4, left) to prevent direct solar radiation.

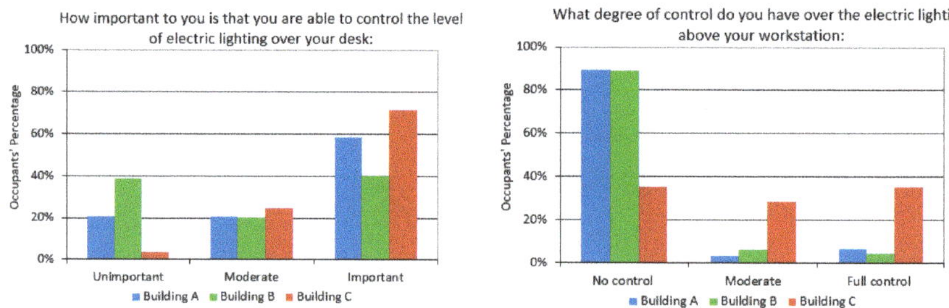

Figure 13. Results of how important is from the occupants to control the level of artificial lighting over their working plane (**left**) and in what degree (**right**).

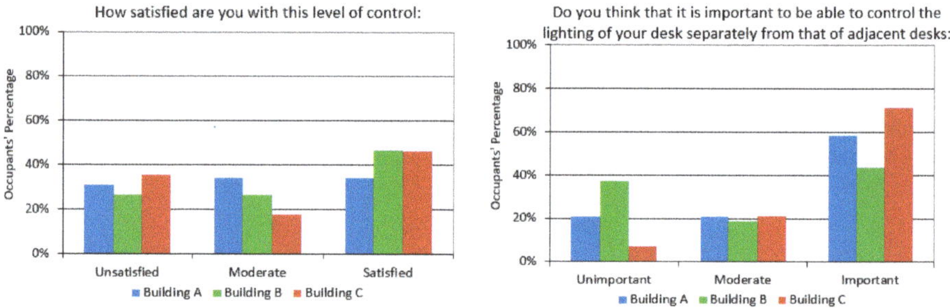

Figure 14. Preference of occupants for how satisfied they are with their degree of control (**left**) and how important is from them to control the artificial lighting of their working plane separately from the others (**right**).

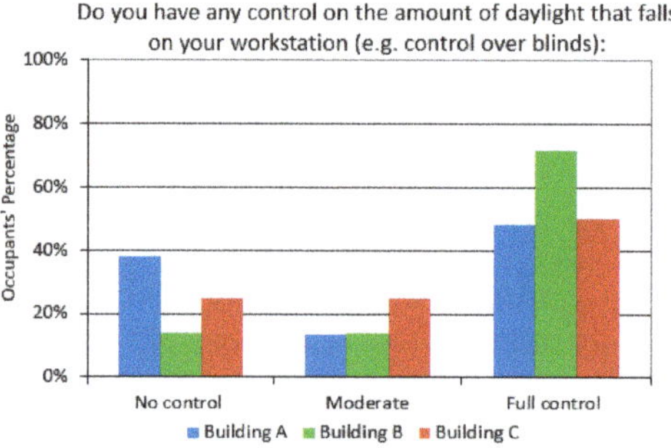

Figure 15. Preference of occupants for the occupants' control on the amount of daylight that impinges on their working plane.

3.2. Lighting Measurements and Occupants' Preferences

According to EN 12464-1 [81], the target illuminance in an office space is 500lx. Figures 16–18 present the lighting measurements concerning the installed power per person, glare, the total amount of light and daylight that experience in their space. With the help of an orange (upper limit) and a purple (lower limit) dashed line, a comfort zone of 500 ± 100lx was highlighted [86]. Most of the occupants (67%) are working in spaces with installed power up to 108W per person (Figure 16), emitting light that results from 150l to 883lx in illuminance values.

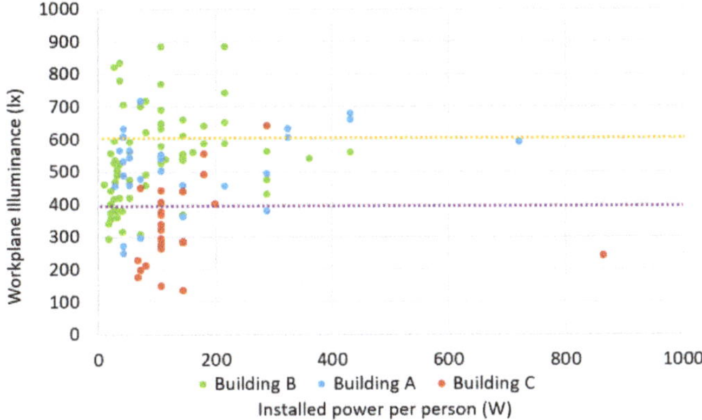

Figure 16. Illuminance measurements vs. installed power per person.

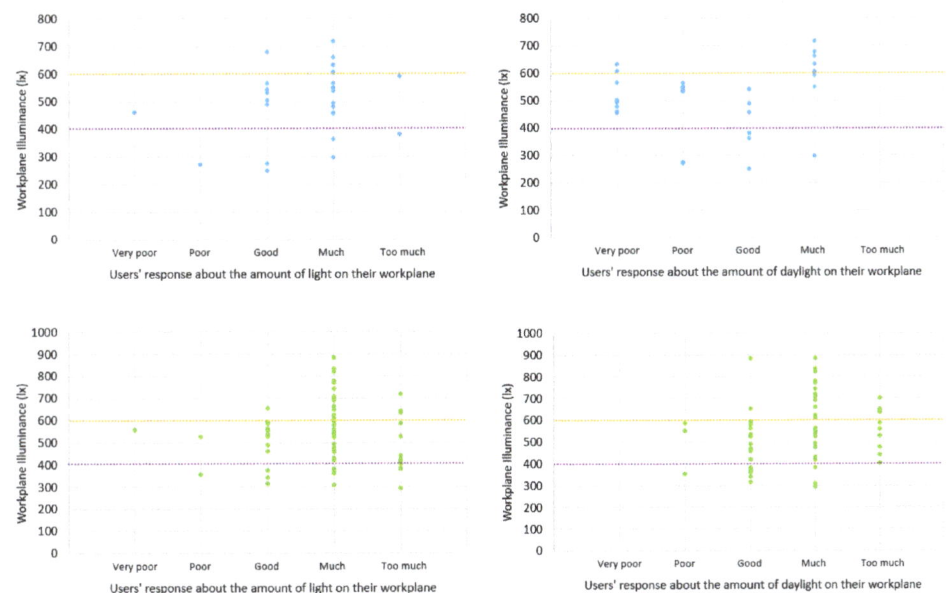

Figure 17. *Cont.*

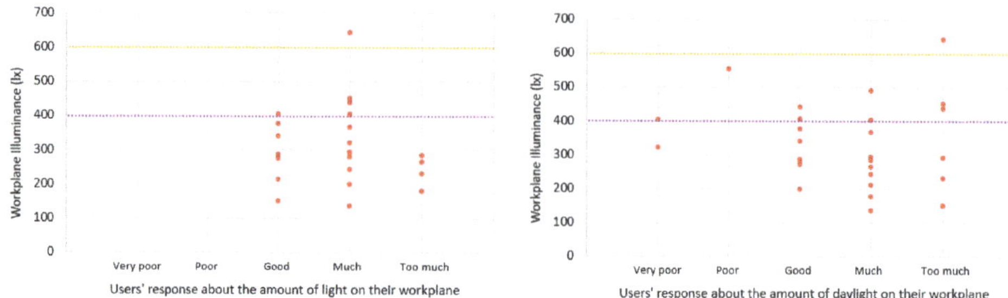

Figure 17. Occupants' response for the amount of total lighting levels (left column) and the corresponding daylight on their working plane (right column) in relation to the measured illuminance (Building A: Upper part, Building B: Middle, Building C: Lower part).

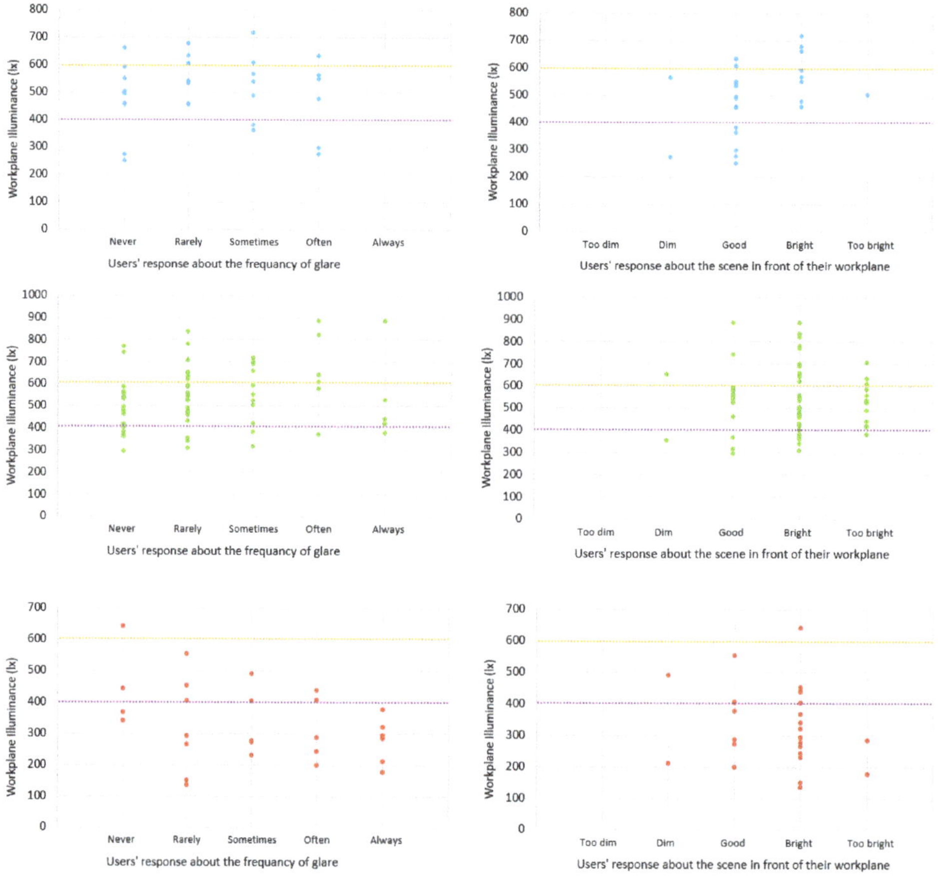

Figure 18. Responses about the frequency of glare (left column) and the brightness of the scene in front of their working plane (right column) in relation to the measured illuminance (Building A: Upper part, Building B: Middle, Building C: Lower part).

In buildings A and B, the similar responses of the users concerning the amount of total light and daylight (Figure 17) on their working plane indicate that artificial light is dimmed as the total light should include high values of daylight. This is an indicator that the daylight-harvesting system is efficient as there were very few task areas (offices) where the measured total lighting levels were below 400lx. For the same buildings (A and B), evaluating the illuminance measurements, the "good" response of the users is coinciding with the comfort zone (400–600lx), while the "much" response is embodying many values higher than 600lx, especially in Building B. On the other hand, in Building C, while the responses of the occupants between the total light levels and daylight are again similar, the corresponding illuminance measurements are below the comfort zone (400–600lx). Although someone should expect that the occupants will describe this situation as "poor" or "very poor", they experienced it as "good" or even "too much".

As glare can be affected by how bright a scene can be, the frequency of glare and the brightness of the external views were examined through the measured lighting levels (Figure 18). There is a debate regarding the selection of an upper illuminance limit above which daylight is not preferable due to glare. While most of the users experienced bright scenes, the frequency of glare was declared from "never" to "sometimes". Even in Building B, where there are high illuminance values, the same users felt, correctly, that the bright environment as the illuminance values were up to 900lx. However, as noted above, while the glare is related with luminance (cd/m^2) distribution rather than illuminance values (lx).

4. Discussion

4.1. Lighting Levels and Age

The occupants' perception regarding the amount of light on their desk did not correlate with the actual measurements of illuminance in their desk. A Spearman correlation analysis both on the whole sample but also on every individual building basis was performed. All analyses returned low ρ scores and non-statistically significant results indicating that the perceived capacity of lighting differs from the actual amount of illuminance. This could be a factor of the occupants' age.

In Building A, 45% of the total lighting measurements, both artificial and daylight, were below 500lx (21% below 400lx) and 24% higher than 600lx. A total of 65% of the users responded (Figure 19) that they like their visual environment (Question 6, Table 2), while comparing the lighting system with the other building parameters (Question 2) their satisfaction for the lighting system was much higher. In the same way, in Building B, 38% of the measurements were below 500lx (17% below 400lx) and 28% higher than 600lx mostly due to daylight. 92% of the users liked their visual environment. Lastly, for Building C, 93% of the working stations were under illuminated, below 500lx (68% below 400lx). The measurements verified the problem of the integral reset algorithm used in Building C (Table 1). Even with a proper commissioning this control algorithm shows erratic behavior [34,36,39,40,44].

Returning to age, it seems that the young age of the Building C users (mainly 25–34) were not affected by the reduced lighting levels (Figure 20) as 93% of them like their visual environment even if it is dimmed below 400lx in most cases. In Building B, where their users of increased age work (+55, Figure 19), a preference for illuminance values between 559 and 794lx was recorded (Figure 20). Figure 20 shows the average measured illuminance values along with the standard deviation per group age in which the occupants' response for liking their visual environment was positive. The measured values of negative responses were not taken into account. These results verify the newer edition of EN12464-1 [87], in which, using the scale of illuminance (from 500 to 750lx) the lighting designer can decide if the illuminance levels should be raised by at least one step when working conditions or age require it. However, a high percentage of older staff should be within the specific working area. This is due to the retinal illuminance that decreases as people getting older and can be attributed to the reduction of the pupil size and higher spectral absorption of the crystalline lens. Higher lighting levels are needed also for persons that underwent presbyopic corrections [88].

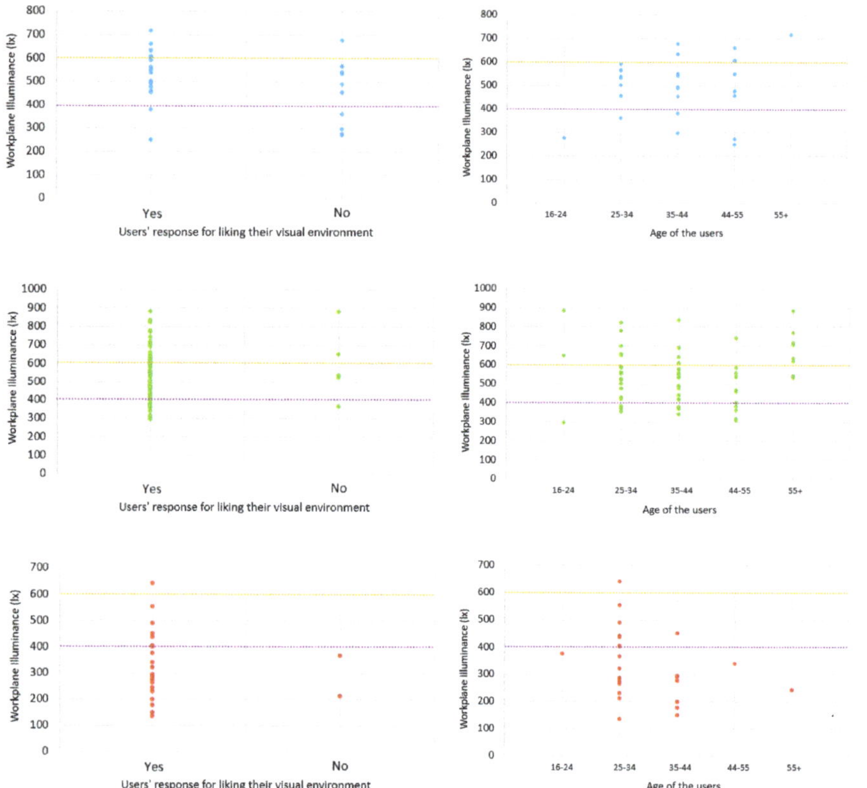

Figure 19. Illuminance measurements vs. the occupants' response concerning liking their visual environment (**left**) and their age (**right**).

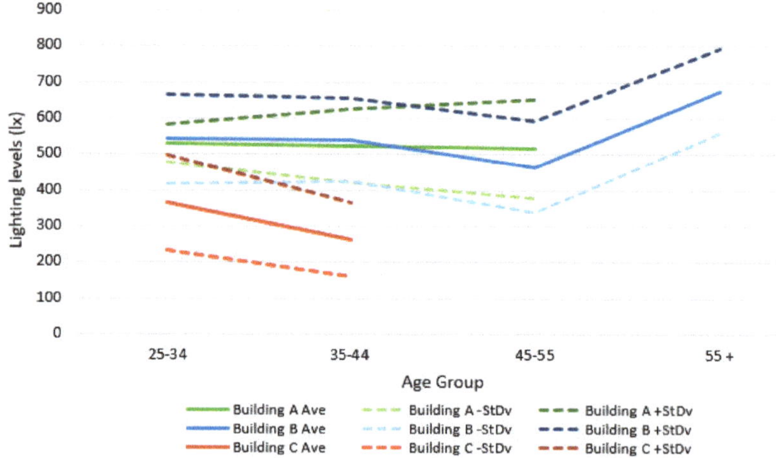

Figure 20. Illuminance measurements on the users' desks in regards their age, in which, the occupants' response for liking their visual environment was positive.

4.2. Total Lighting Levels and Daylight

In Table 3, the estimated Rho coefficients and the respective statistical significance are presented for the three buildings. The results show that the relationship between the two variables is only found statistically significant for Building B. Considering the nature of the two questions (Questions 11 and 12, Table 2) and the fact that Building B is controlled by open loop sensors placed outdoors it is concluded that employees in that building give a positive evaluation to the total quantity of lighting only when they perceive that daylight is sufficient. This relationship is absent in the other buildings, signifying that satisfaction is achieved not only by the provision of daylight but also with artificial light. This raises a question regarding the effectiveness of artificial lighting provision in Building B.

Table 3. The estimated Spearman Rho coefficients between variables of question 11 (total light) and question 12 (daylight) across the three buildings.

Correlation	A	B	C
Rho	0.180	0.537	0.349
Statistical Significance (two tailed)	0.349	0.000	0.131
N	29	64	22

In Building C, the total lighting levels were lower than the corresponding value of 500lx, while there was presence of daylight (Figure 17). As already explained, the integral reset algorithm used in the photosensors of Building C, might be the main reason that artificial lighting was dimmed too much, lower than the necessary lighting levels. Secondly, another reason for the low lighting levels, could be a possible override of the system from the occupants. Due to both reasons, the users of younger age, gave a significant tolerance in visual comfort for achieved lower lighting levels (Figure 20).

4.3. Glare

Most of the users are satisfied with their visual environment with the slight exception of the Building A (Figure 21, left). The use of the external blinds in Building A may dissatisfy some occupants because the deterioration of the view. More than 50% of the occupants consider themselves as insensitive to glare (Figure 22, right). The main sources of glare were the reflections on the monitors either by the artificial lighting systems or the daylight. The employees' satisfaction of their visual environment was tested by taking into account (a) the abundance of light when they look in front of them (Questions 6 and 7, Table 2) and (b) the frequency of glare incidents (Questions 6 and 8). Variable from question 6 (Is your visual environment pleasing?) is a dummy variable and variables from question 7 (When you look up from your working plane does the scene that your see in front of you seem …) and question 8 (Does glare ever disturb or annoy you?) are of ordinal Likert scale, thus a direct correlation analysis is not feasible. Instead, a Mann–Whitney test is preferred. The Mann–Whitney test for questions 6 and 7 returned a z value of −3.176 which denotes statistical significance at the (<0.01) level. This result shows that when employees consider the light in their front view as adequate (inside the field of view of the human eye looking forward), they tend to show greater satisfaction for their visual environment (including their working surface). The relationship across the three buildings was tested and we acquired statistically significant result only for the Building B. In addition, the results of the test for questions 6 and 8 returned a z value of −0.345 with no statistical significance. Therefore, glare does not seem to affect the visual comfort satisfaction level of employees at the measured illuminance levels. To test for any differences across the buildings three tests were also conducted returning no significant results. Hence, the absence of any relationship between visual satisfaction and glare is confirmed for the whole sample. In addition, user sensitivity to glare was tested by estimating the number of glare incidents that encounter for all buildings (Questions 6 and 8). Since the variable of question 9 is a dummy one, the test will rely on the Mann–Whitney method. The tests performed over all the three buildings returned no statistically significant result. Thus, it is concluded that sensitivity to glare does not influence the judgement of employees regarding the frequency of glare incidents in their work.

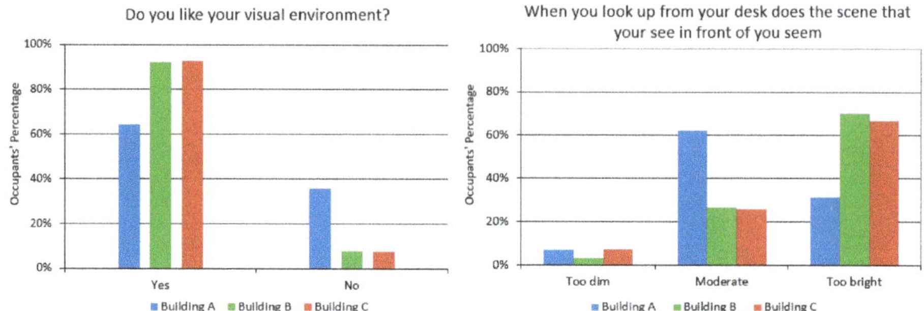

Figure 21. Occupants' preference for their visual environment (**left**) and how dim or bright is the scene in front of them (**right**).

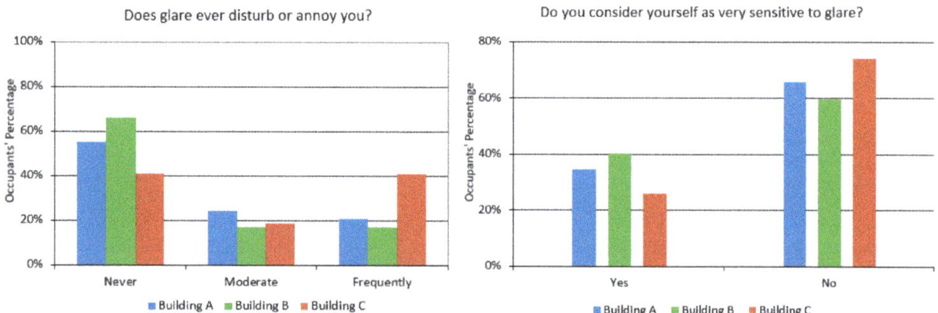

Figure 22. Results of how frequent the users are disturbed from glare (**left**) and if they consider themselves sensitive to glare (**right**).

4.4. Lighting Controls

Figures 13–15 show the different reactions of the occupant responses concerning the light controls. The relationship formed among the questions 14–16 (Table 2) and their corresponding variables was examined. This differs across the three buildings because of the different control algorithms (Table 1). For this reason, the pairwise Spearman correlation coefficients were estimated. The estimated ρ coefficients and their statistical significance are presented in Table 4. The estimated correlation coefficient between question 14 and 15 variables acquires statistical significance only when Building B is considered. This means that only in Building B it is both important, from the users, to have ability to control the lighting on their desk area and to be satisfied with the amount of control that they already have. As for the correlation of variables between question 14 and 16, this was found significant in buildings A and B, with the former estimation acquiring higher statistical significance (<0.01 to <0.10 level). For those buildings it seems that a positive relationship does exist between (a) the user's ability to control the lighting in their desk areas and (b) their satisfaction with the actual level of control. Finally, the third correlation test (between question 15 and 16 variables) yields the exact opposite results, as a strong and statistically significant correlation coefficient is found only for Building C. This finding denotes that the control of lighting and the satisfaction of employees regarding the actual control are correlated only in Building C. Finally, concerning the relationship between the importance on the ability to control lighting on their area, as opposed to the ability to do that separately from the adjacent desks, the correlation analysis returned a rather high rho coefficient (0.659) which was found statistically significant at the (<0.01) level, as expected.

Table 4. The results of the pairwise Spearman correlation analyses for the variables of question 14 (How important to you is that you are able to control the level of electric lighting over your desk?), question 15 (What degree of control do you have over the electric lighting above your work plane?) and question 16 (How satisfied are you with this level of control?) for the three examined buildings.

Rho	A	B	C
Question 14–Question 15	0.309	0.314 **	−0.026
Question 14–Question 16	−0.352 *	−0.428 ***	−0.292
Question 15–Question 16	0.048	−0.107	0.747 ***
N	29	64	23

Statistical significance: (***) at 0.01 level (**) at 0.05 level (*) at 0.10 level.

5. Conclusions

Buildings are established to be functional and satisfy the needs of their occupants. Hence, the goal of a building cannot be accomplished if its occupants are discontented by its performance. The goal of a post-occupancy evaluation study is to conduct a methodical and thorough assessment of the overall performance attributes of a facility after its occupation, to obtain feedback about a variety of features, such as energy consumption, quality of the indoor environment, occupant's satisfaction, productivity, etc. The results of a post-occupancy evaluation study are able to elucidate if a building responds to the expectations that were envisaged in the early stages of the design, by appraising the facility's performance with regards to human aspects such as the occupant's satisfactions or wellbeing, as well as nonhuman aspects such as metrics of their energy and thermal performance. In this study a questionnaire was generated so as to concentrate on the lighting of the examined buildings and especially on daylight utilization in buildings, where daylight controls with photosensors are installed.

A systematic documentation and analysis of the information gathered from the three examined buildings was conducted and some features that will be exploited in the future were identified. The majority of the occupants have a positive opinion about their lighting quality with the exception of the employees that occupy Building A. The discontentment of the occupants can be attributed to the external blinds that are installed in Building A, which obstruct their view. The users' strong preference for daylight illumination leads to the conclusion that daylight is more effective in increasing work place morale and providing visual comfort rather than artificial light. Another remark is that visual comfort, temperature and ventilation are evaluated as the most desirable characteristics of indoor environment quality in an office space. Additionally, low levels of noise and spaciousness received high, but not the highest, ranking. It should be noted that the only discontentment related to daylight penetration in buildings is mainly associated with glare, particularly on the computer monitors. Therefore, when designing with a daylight professional must implement all the appropriate measures so as to eradicate any visual discomfort caused by glare.

The lighting measurements along with a post-occupancy evaluation, can form a complete survey concerning the lighting system and can be part of a building energy audit and diagnosis. The users' satisfaction of their visual environment is in line with the lighting measurements and the corresponding target illuminance (500 ± 100lx, [86]). Two trends concerning their age and the lighting levels were documented. In Building C, the occupants were satisfied even when the lighting levels were lower than the target illuminance of 500lx for an office environment. This is due to the integral reset control algorithm (Building C) that had the poorest behavior from all three buildings since for 93% of the user positions were below 500lx. This of course is related to increased lighting energy savings in Building C, but with illuminances lower than the design illuminance. However, the acceptance rate of the users was high most probably due to their young age. The new edition of prEN12464-1 2019 [87] suggests a step increase of 250lx in office environments for aged users. However, the opposite action, a reduction of illuminance due to young users, does not occur as this reduction was documented in the current research. On the other hand, in Building B, the increased age group of people was satisfied with higher levels of illuminance verifying the recommendations the prEN12464-1 2019 [87]. These recommendations allow lighting professionals adopt an increased task illuminance for older occupants so as to counterbalance the decline of retinal illuminance due to age. This should be applied in specific task areas and note to whole area, as more energy consumption could weigh down the overall building energy equilibrium. Furthermore, the results show that in Building C increased energy savings can be expected since the users prefer lighting levels less than 500lx. A daylight control system can achieve up to 60% [89] energy savings in comparison to a manual control system provided that operates normally. Any erratic behavior could counterbalance the possible energy savings, as the users tend to override it and thus losing a substantial percentage of the energy savings. While the targeted POE for lighting control installations are limited, the erratic behavior of the integral reset control algorithm was verified according to the relative bibliography [36,39,45]. Examining the lighting levels with the satisfaction of the users the results were verified also with previous research [65]. The purpose of the study was mainly to examine the acceptance of control systems by building occupants recording all possible problems caused by their operation. This can be accomplished by a time limited survey. However, if the proposed POE method is performed at different seasons in a year can help to calibrate the daylight-harvesting system properly, maintaining the expected energy savings.

Most occupants praised the integration of automated daylight controls, although they revealed their strong preference towards individual control over the artificial lighting system. In other words, each occupant desires to be able to control his or her individual lighting conditions rather than having to accept lighting levels that are adjusted by others (e.g., lighting professionals) for them, even when the lighting conditions comply with international standards. For the full acceptance of a lighting control system, the occupants should become aware that the system does exist in their working environment. In this survey, 79% of the users knew that a daylight harvesting technique was implemented in their working space. In Building C, the aforementioned percentage dropped to 43%. In this building, lighting control was implemented only in the perimeter zone as a stand-alone solution for every luminaire. Although low illuminance levels were recorded, users did not bother to examine the cause of low illuminance and fix it. The age of the users was an important parameter which affected their tolerance for lower light levels. However, it is still important for the facilitators, to inform and educate the occupants regarding proper use of lighting control systems that are integrated to their building. Furthermore, the quick fluctuation of the dimming levels of the daylight-harvesting system was the most pivotal reason for the discontentment of the lighting system from occupants in all buildings. It is crucial that the commissioning of the system should be performed by a lighting expert [90]. A proper adjustment of the photosensor's time response could provide solution to this problem in the examined buildings. The proper commissioning of the installed photosensors could maximize energy savings.

The reality is that the occupants will never be satisfied by any daylight-harvesting system unless they have some benefit from it. The results showed that, for all the cases, (a) ease of use of lighting controls, (b) occupant awareness to lighting controls and (c) occupant training concerning lighting controls are of critical importance so as to establish desirable lighting conditions and simultaneously achieve energy savings.

Author Contributions: Conceptualization, L.T.D. and A.T.; Data curation, L.T.D., E.-N.M. and P.A.K.; Formal analysis, L.T.D., S.N., A.K., P.A.K., V.T.K., K.S., E.M. and M.S.; Investigation, L.T.D., E.-N.M., A.K., P.A.K., V.T.K. and K.S.; Methodology, L.T.D.; Supervision, L.T.D. and A.T.; Validation, A.T., K.S., M.S. and S.Z.; Writing—original draft, L.T.D., A.T., E.-N.M. and S.N.; Writing—review and editing, L.T.D., A.T., F.T. and S.Z. All authors have read and agreed to the published version of the manuscript.

Funding: This research received no external funding.

Conflicts of Interest: The authors declare no conflict of interest.

References

1. Wagiman, K.R.; Abdullah, M.N.; Hassan, M.Y.; Radzi, N.H.M.; Bakar, A.H.A.; Kwang, T.C. Lighting system control techniques in commercial buildings: Current trends and future directions. *J. Build. Eng.* **2020**, *31*, 101342. [CrossRef]
2. Zissis, G. Energy consumption and environmental and economic impact of lighting: The current situation. In *Handbook of Advanced Lighting Technology*; Karlicek, R., Sun, C., Zissis, G., Ma, R., Eds.; Springer International Publishing: Cham, Switzerland, 2016; pp. 1–13. [CrossRef]
3. Grigoropoulos, C.J.; Doulos, L.T.; Zerefos, S.C.; Tsangrassoulis, A.; Bhusal, P. Estimating the benefits of increasing the recycling rate of lamps from the domestic sector: Methodology, opportunities and case study. *Waste Manag.* **2020**, *101*, 188–199. [CrossRef]
4. Mavridou, T.; Doulos, L.T. Evaluation of different roof types concerning daylight in industrial buildings during the initial design phase: Methodology and case study. *Buildings* **2019**, *9*, 170. [CrossRef]
5. Tsangrassoulis, A.; Doulos, L.; Santamouris, M.; Fontoynont, M.; Maamari, F.; Wilson, M.B.; Jacobs, A.; Solomon, J.M.; Zimmerman, A.; Pohl, W.; et al. On the energy efficiency of a prototype hybrid daylighting system. *Solar Energy* **2005**, *79*, 56–64. [CrossRef]
6. Mantzourani, K.; Doulos, L.T.; Kontadakis, A.; Tsangrassoulis, A. The effect of the daylight zone on lighting energy savings. *IOP Conf. Ser. Earth Environ. Sci.* **2020**, *410*, 012099. [CrossRef]
7. Sibley, M.; Peña-García, A. Flat glass or crystal dome aperture? A year-long comparative analysis of the performance of light pipes in real residential settings and climatic conditions. *Sustainability* **2020**, *12*, 3858. [CrossRef]
8. Peña-García, A. Towards total lighting: Expanding the frontiers of sustainable development. *Sustainability* **2019**, *11*, 6943. [CrossRef]
9. Palarino, C. María beatriz piderit, optimisation of passive solar design strategies in Side-lit Offices: Maximising daylight penetration while reducing the risk of glare in different chilean climate contexts. *J. Daylighting* **2020**, *7*, 107–121. [CrossRef]
10. Tsangrassoulis, A.; Doulos, L.; Kontadakis, A.; Drakou, A. A Methodology to Model the Performance of a Dynamic Mirror Light-shelf Based on Solar Radiant Flux Impinging on the Window. In Proceedings of the Building Simulation 2019: 16th IBPSA International Conference and Exhibition (ISBN: 978-1-7750520-1-2), Rome, Italy, 2–4 September 2019; Available online: http://www.ibpsa.org/building-simulation-2019 (accessed on 30 June 2020).
11. Kontadakis, A.; Doulos, L.; Mantzourani, A.; Tsangrassoulis, A. Performance assessment of an active sunlight redirection system in areas with different climate: A comparison. *IOP Conf. Ser. Earth Environ. Sci.* **2020**, *410*, 012098. [CrossRef]
12. Kontadakis, A.; Tsangrassoulis, A.; Doulos, L.T.; Topalis, F. An active sunlight redirection system for daylight enhancement beyond the perimeter zone. *Build. Environ.* **2017**, *113*, 267–279. [CrossRef]
13. Eltaweel, A.; Mandour, M.A.; Lv, Q.; Su, Y. Daylight distribution improvement using automated prismatic louvre. *J. Daylighting* **2020**, *7*, 84–92. [CrossRef]
14. Kontadakis, A.; Tsangrassoulis, A.; Doulos, L.T.; Zerefos, S. A review of light shelf designs for daylit environments. *Sustainability* **2018**, *10*, 71. [CrossRef]
15. Arabatzis, G.; Kyriakopoulos, G.; Tsialis, P. Typology of regional units based on RES plants: The case of Greece. *Renew. Sustain. Energy Rev.* **2017**, *78*, 1424–1434. [CrossRef]
16. Kyriakopoulos, G.L.; Arabatzis, G.; Tsialis, P.; Ioannou, K. Electricity consumption and RES plants in Greece: Typologies of regional units. *Renew. Energy* **2018**, *127*, 134–144. [CrossRef]
17. Ntanos, S.; Skordoulis, M.; Kyriakopoulos, G.; Arabatzis, G.; Chalikias, M.; Galatsidas, S.; Batzios, A.; Katsarou, A. Renewable energy and economic growth: Evidence from european countries. *Sustainability* **2018**, *10*, 2626. [CrossRef]
18. Doulos, L.; Santamouris, M.; Livada, I. Passive cooling of outdoor urban spaces. The role of materials. *Solar Energy* **2004**, *77*, 231–249. [CrossRef]
19. Kyriakopoulos, G.L.; Arabatzis, G. Electrical energy storage systems in electricity generation: Energy policies, innovative technologies, and regulatory regimes. *Renew. Sustain. Energy Rev.* **2016**, *56*, 1044–1067. [CrossRef]

20. Doulos, L.T.; Sioutis, I.; Tsangrassoulis, A.; Canale, L.; Faidas, K. Revision of threshold luminance levels in tunnels aiming to minimize energy consumption at no cost: Methodology and case studies. *Energies* **2020**, *13*, 1707. [CrossRef]

21. Papalambrou, A.; Doulos, L.T. Identifying, examining, and planning areas protected from light pollution. The case study of planning the first national dark sky park in Greece. *Sustainability* **2019**, *11*, 5963. [CrossRef]

22. Ardavani, O.; Zerefos, S.; Doulos, L.T. Redesigning the exterior lighting as part of the urban landscape: The role of transgenic bioluminescent plants in Mediterranean urban and suburban lighting environments. *J. Clean. Prod.* **2020**, *242*, 118477. [CrossRef]

23. Doulos, L.T.; Sioutis, I.; Kontaxis, P.A.; Zissis, G.; Faidas, K. A decision support system for assessment of street lighting tenders based on energy performance indicators and environmental criteria: Overview, methodology and case study. *Sustain. Cities Soc.* **2019**, *51*, 101759. [CrossRef]

24. Anthopoulou, E.; Doulos, L.T. The Effect of the Continuous Energy Efficient Upgrading of LED Street Lighting Technology: The Case Study of Egnatia Odos, 2019 2nd ed. In Proceedings of the Balkan Junior Conference on Lighting, Balkan Light Junior 2019, Plovdiv, Bulgaria, 19 September 2019. Abstract number 8883662. [CrossRef]

25. Doulos, L.T.; Sioutis, I.; Tsangrassoulis, A.; Canale, L.; Faidas, K. Minimizing lighting consumption in existing tunnels using a no-cost fine-tuning method for switching lighting stages according revised luminance levels. In Proceedings of the 2019 IEEE International Conference on Environment and Electrical Engineering and 2019 IEEE Industrial and Commercial Power Systems Europe (EEEIC/I&CPS Europe), Genova, Italy, 11–14 June 2019; pp. 1–6. [CrossRef]

26. Peña-García, A.; Salata, F.; Golasi, I. Decrease of the maximum speed in highway tunnels as a measure to foster energy savings and sustainability. *Energies* **2019**, *12*, 685. [CrossRef]

27. Galatanu, C.D.; Canale, L.; Lucache, D.D.; Zissis, G. Reduction in light pollution by measurements according to EN 13201 standard. In Proceedings of the 2018 International Conference and Exposition on Electrical and Power Engineering (EPE), Iasi, Romania, 18–19 October 2018; pp. 1074–1079.

28. Peña-García, A.; Sędziwy, A. Optimizing lighting of rural roads and protected areas with white light: A compromise among light pollution, energy savings, and visibility. *LEUKOS* **2019**, *16*, 147–156. [CrossRef]

29. Madias, E.-N.D.; Kontaxis, P.A.; Topalis, F.V. Application of multi-objective genetic algorithms to interior lighting optimization. *Energy Build.* **2016**, *125*, 66–74. [CrossRef]

30. Madias, E.D.; Doulos, L.T.; Kontaxis, P.A.; Topalis, F.V. A decision support system for techno-economic evaluation of indoor lighting systems with LED luminaires. *Oper. Res. Int. J.* **2019**. [CrossRef]

31. Manolis, E.; Doulos, L.T.; Niavis, S.; Canale, L. The impact of energy efficiency indicators on the office lighting planning and its implications for office lighting market. In Proceedings of the 2019 IEEE International Conference on Environment and Electrical Engineering and 2019 IEEE Industrial and Commercial Power Systems Europe (EEEIC/I&CPS Europe), Genova, Italy, 11–14 June 2019; pp. 1–6. [CrossRef]

32. Doulos, L.T.; Tsangrassoulis, A.; Kontaxis, P.A.; Kontadakis, A.; Topalis, F.V. Harvesting daylight with LED or T5 fluorescent lamps? The role of dimming. *Energy Build.* **2017**, *140*, 336–347. [CrossRef]

33. Adam, G.K.; Kontaxis, P.A.; Doulos, L.T.; Madias, E.N.D.; Bouroussis, C.A.; Topalis, F.V. Embedded microcontroller with a CCD camera as a digital lighting control system. *Electronics* **2019**, *8*, 33. [CrossRef]

34. Bellia, L.; Fragliasso, F. Evaluating performance of daylight-linked building controls during preliminary design. *Autom. Constr.* **2018**, *93*, 293–314. [CrossRef]

35. Bellia, L.; Fragliasso, F. Automated daylight-linked control systems performance with illuminance sensors for side-lit offices in the Mediterranean area. *Autom. Constr.* **2019**, *100*, 145–162. [CrossRef]

36. Bellia, L.; Fragliasso, F.; Stefanizzi, E. Why are daylight-linked controls (DLCs) not so spread? A literature review. *Build Environ.* **2016**, *106*, 301–312. [CrossRef]

37. Beu, D.; Ciugudeanu, C.; Buzdugan, M. Circular economy aspects regarding LED lighting retrofit—From case studies to vision. *Sustainability* **2018**, *10*, 3674. [CrossRef]

38. Bonomolo, M.; Beccali, M.; Lo Brano, V.; Zizzo, G. A set of indices to assess the real performance of daylight-linked control systems. *Energy Build* **2017**, *149*, 235–245. [CrossRef]

39. Doulos, L.; Tsangrassoulis, A.; Topalis, F. Multi-criteria decision analysis to select the optimum position and proper field of view of a photosensor. *Energy Convers. Manag.* **2014**, *86*, 1069–1077. [CrossRef]

40. Topalis, F.V.; Doulos, L.T. Ambient Light Sensor Integration. In *Handbook of Advanced Lighting Technology*; Springer: Cham, Switzerland, 2017; pp. 607–634. [CrossRef]

41. Mistrick, R.; Casey, C.; Chen, L.; Subramaniam, S. Computer modeling of daylight-integrated photocontrol of electric lighting systems. *Buildings* **2015**, *5*, 449–466. [CrossRef]
42. Doulos, L.; Tsangrassoulis, A.; Topalis, F. The role of spectral response of photosensors in daylight responsive systems. *Energy Build.* **2008**, *40*, 588–599. [CrossRef]
43. Bertin, K.; Canale, L.; Ben Abdellah, O.; Méquignon, M.-A.; Zissis, G. Life cycle assessment of lighting systems and light loss factor: A case study for indoor workplaces in France. *Electronics* **2019**, *8*, 1278. [CrossRef]
44. Doulos, L.; Tsangrassoulis, A.; Topalis, F. Quantifying energy savings in daylight responsive systems: The role of dimming electronic ballasts. *Energy Build.* **2008**, *40*, 36–50. [CrossRef]
45. Madias, E.N.; Doulos, L.T.; Kontaxis, P.A.; Topalis, F.V. Multicriteria decision aid analysis for the optimum performance of an ambient light sensor: Methodology and case study. *Oper. Res. Int. J.* **2020**. [CrossRef]
46. Doulos, L.T.; Kontadakis, A.; Madias, E.N.; Sinou, M.; Tsangrassoulis, A. Minimizing energy consumption for artificial lighting in a typical classroom of a Hellenic public school aiming for near zero energy building using LED DC luminaires and daylight harvesting systems. *Energy Build.* **2019**, *194*, 201–217. [CrossRef]
47. Mistrick, R.G.; Thongtipaya, J. Analysis of daylight photocell placement and view in a small office. *J. Illum. Eng. Soc.* **1997**, *26*, 150–160. [CrossRef]
48. Choi, A.S.; Mistrick, R.G. Analysis of daylight responsive dimming system performance. *Building Environ.* **1999**, *34*, 231–243. [CrossRef]
49. Mistrick, R.; Chen, C.; Bierman, A.; Felts, D. A comparison of photosensor-controlled electronic dimming systems in a small office. *J. Illum. Eng. Soc.* **2000**, *29*, 66–80. [CrossRef]
50. Choi, A.S.; Sung, M.K. Development of a daylight responsive dimming system and preliminary evaluation of system performance. *Build. Environ.* **2000**, *35*, 663–676. [CrossRef]
51. Ranasinghe, S.; Mistrick, R. A study of photosensor configuration and performance in a daylighted classroom space. *J. Illum. Eng. Soc.* **2003**, *32*, 3–20. [CrossRef]
52. Mistrick, R.; Sarkar, A. A study of daylight-responsive photosensor control in five daylighted classrooms. *Leukos* **2005**, *3*, 51–74. [CrossRef]
53. Hadjri, K.; Crozier, C. Post-occupancy evaluation: Purpose, benefits and barriers. *Facilities* **2009**, *27*, 21–33. [CrossRef]
54. Nawawi, A.; Khalil, N. Post-occupancy evaluation correlated with building occupants' satisfaction: An approach to performance evaluation of government and public buildings. *J. Build.* **2008**, *4*, 59–69. [CrossRef]
55. Preiser, W.F.E. Post-occupancy evaluation: How to make buildings work better. *Facilities* **1995**, *13*, 19–28. [CrossRef]
56. Li, P.; Froese, T.M.; Brager, G. Post-occupancy evaluation: State-of-the art analysis and state-of-the-practice review. *Build. Environ.* **2018**. [CrossRef]
57. Zimmerman, A.; Martin, M. Post-occupancy evaluation: Benefits and barriers. *Build. Res. Inf.* **2001**, *29*, 168–174. [CrossRef]
58. Leaman, A.; Bordass, B. Assessing building performance in use 4: The Probe occupant surveys and their implications. *Build. Res. Inf.* **2001**, *29*, 129–143. [CrossRef]
59. Mustafa, F.A. Performance assessment of buildings via post-occupancy evaluation: A case study of the building of the architecture and software engineering departments in Salahaddin University-Erbil, Iraq. *Front. Archit. Res.* **2017**, *6*, 412–429. [CrossRef]
60. Vischer, J. Towards a user-centred theory of the built environment. *J. Build. Res. Inf.* **2008**, *36*. [CrossRef]
61. Thomas, L. Evaluating design strategies, performance and occupant satisfaction: A low carbon office refurbishment. *Build. Res. Inf.* **2010**, *38*. [CrossRef]
62. Agha-Hossein, M.M.; El-Jouzi, S.; Elmualim, A.A.; Ellis, J.; Williams, M. Post-occupancy studies of an office environment: Energy performance and occupants' satisfaction. *Build. Environ.* **2013**, *69*, 121–130. [CrossRef]
63. Ponterosso, P.; Gaterell, M.; Williams, J. Post occupancy evaluation and internal environmental monitoring of the new BREEAM "Excellent" Land Rover/Ben Ainslie Racing team headquarters offices. *Build. Environ.* **2018**. [CrossRef]
64. Veitch, J.A.; Charles, K.E.; Farley, K.M.J.; Newsham, G.R. A model of satisfaction with open-plan office conditions: COPE field findings. *J. Environ. Psychol.* **2007**, *27*, 177–189. [CrossRef]
65. Choi, J.H.; Loftness, V.; Aziz, A. Post-occupancy evaluation of 20 office buildings as basis for future IEQ standards and guidelines. *Energy Build.* **2011**, *46*, 167–175. [CrossRef]

66. Kim, J.; de Dear, R. Nonlinear relationships between individual IEQ factors and overall workspace satisfaction. *Build Environ.* **2012**, *49*, 33–40. [CrossRef]

67. Filippín, C.; Larsen, S.F.; Marek, L. Experimental monitoring and post-occupancy evaluation of a non-domestic solar building in the central region of Argentina. *Energy Build.* **2015**, *92*, 267–281. [CrossRef]

68. Park, J.; Loftness, V.; Aziz, A. Post-Occupancy evaluation and IEQ measurements from 64 office buildings: Critical factors and thresholds for user satisfaction on thermal quality. *Buildings* **2018**, *8*, 156. [CrossRef]

69. Candido, C.; Chakraborty, P.; Tjondronegoro, D. The Rise of office design in high-performance, open-plan environments. *Buildings* **2019**, *9*, 100. [CrossRef]

70. Göçer, Ö.; Candido, C.; Thomas, L.; Göçer, K. Differences in occupants' satisfaction and perceived productivity in high- and low-performance offices. *Buildings* **2019**, *9*, 199. [CrossRef]

71. Pastore, L.; Andersen, M. Building energy certification versus user satisfaction with the indoor environment: Findings from a multi-site post-occupancy evaluation (POE) in Switzerland. *Build. Environ.* **2019**, *150*, 60–74. [CrossRef]

72. Chastas, P.; Theodosiou, T.; Kontoleon, K.J.; Bikas, D. The Effect of embodied impact on the cost-optimal levels of nearly zero energy buildings: A case study of a residential building in Thessaloniki, Greece. *Energies* **2017**, *10*, 740. [CrossRef]

73. Touloupaki, E.; Theodosiou, T. Optimization of external envelope insulation thickness: A parametric study. *Energies* **2017**, *10*, 270. [CrossRef]

74. Balaras, C.A.; Droutsa, K.G.; Dascalaki, E.G.; Kontoyiannidis, S.; Moro, A.; Bazzan, E. Urban sustainability audits and ratings of the built environment. *Energies* **2019**, *12*, 4243. [CrossRef]

75. Droutsa, K.G.; Balaras, C.A.; Lykoudis, S.; Kontoyiannidis, S.; Dascalaki, E.G.; Argiriou, A.A. Baselines for energy use and carbon emission intensities in hellenic nonresidential buildings. *Energies* **2020**, *13*, 2100. [CrossRef]

76. Cholewa, T.; Balaras, C.A.; Nižetić, S.; Siuta-Olcha, A. On calculated and actual energy savings from thermal building renovations–Long term field evaluation of multifamily buildings. *Energy and Buildings.* **2020**, *223*, 110145. [CrossRef]

77. Cholewa, T.; Siuta-Olcha, A.; Balaras, C.A. Actual energy savings from the use of thermostatic radiator valves in residential buildings—Long term field evaluation. *Energy Build.* **2017**, *151*, 487–493. [CrossRef]

78. Pallis, P.; Gkonis, N.; Varvagiannis, E.; Braimakis, K.; Karellas, S.; Katsaros, M.; Vourliotis, P. Cost effectiveness assessment and beyond: A study on energy efficiency interventions in Greek residential building stock. *Energy Build.* **2019**, *182*, 1–18. [CrossRef]

79. Forouli, A.; Gkonis, N.; Nikas, A.; Siskos, E.; Doukas, H.; Tourkolias, C. Energy efficiency promotion in Greece in light of risk: Evaluating policies as portfolio assets. *Energy* **2019**, *170*, 818–831. [CrossRef]

80. Doukas, H.; Nikas, A. Decision support models in climate policy. *Eur. J. Oper. Res.* **2020**, *280*, 1–24. [CrossRef]

81. European Committee for Standardization. *European Norm 15193-1: Energy Performance of Buildings—Energy Requirements for Lighting—Part 1: Specifications, Module M9*; European Committee for Standardization: Brussels, Belgium, 2017.

82. European Committee for Standardization. *European Norm 12464-1: Light and Lighting—Lighting of Work Places Part 1: Indoor Work Places*; European Committee for Standardization: Brussels, Belgium, 2011.

83. Norusis, M. *SPSS 13.0 Statistical Procedures Companion*; Prentice Hall Publications: Upper Saddle River, NJ, USA, 2004.

84. Dodge, Y. *The Concise Encyclopedia of Statistics*; Springer Science & Business Media: New York, NY, USA, 2008.

85. Shier, R. *Statistics: 2.3 the Mann-Whitney U Test*; Mathematics Learning Support Centre: Loughborough University Leicestershire, UK, 2004.

86. Ministry of Environment and Energy of Greece. *Greek Energy efficiency Regulation of Buildings*; Greek Government: Athens, Greece, 2017.

87. European Committee for Standardization. *prEuropean Norm 12464-1: Light and Lighting—Lighting of Work Places Part 1: Indoor Work Places*; European Committee for Standardization: Brussels, Belgium, 2019.

88. Doulos, L.; Panagiotopoulou, E.K.; Taliantzis, S.; Edirneli, S.; Mehmet, A.; Fotiadis, I.; Gkika, M.; Konstantinidis, A.; Perente, I.; Dardabounis, D.; et al. Lighting needs of patients that underwent pseudophakic presbyopic corrections. A theoretical approach. In Proceedings of the 34th International Congress of the Hellenic Society of Intraocular Implant and Refractive Surgery, Athens, Greece, 7–10 July 2020.

89. European Committee for Standardization. *Technical Report 15193-2: Energy Performance of Buildings—Energy Requirements for Lighting—Part 2: Explanation and Justification of EN 15193-1, Module M9*; European Committee for Standardization: Brussels, Belgium, 2017.
90. What is the European Lighting Expert. Available online: https://europeanlightingexpert.org/en/what-is-the-ele/ (accessed on 30 June 2020).

Article

Measured and Simulated Energy Use in a Secondary School Building in Sweden—A Case Study of Validation, Airing, and Occupancy Behaviour

Jessika Steen Englund [1,*], Mathias Cehlin [1], Jan Akander [1] and Bahram Moshfegh [1,2]

[1]   Department of Building Engineering, Energy Systems and Sustainability Science, University of Gävle, 80176 Gävle, Sweden
[2]   Division of Energy Systems, Department of Management and Engineering, Linköping University, 58183 Linköping, Sweden
*   Correspondence: Jessika.Steen.Englund@hig.se; Tel.: +46-739920259

Received: 26 March 2020; Accepted: 25 April 2020; Published: 7 May 2020

Abstract: In this case study, the energy performance of a secondary school building from the 1960s in Gävle, Sweden, was modelled in the building energy simulation (BES) tool IDA ICE version 4.8 prior to major renovation planning. The objectives of the study were to validate the BES model during both occupied and unoccupied periods, investigate how to model airing and varying occupancy behaviour, and finally investigate energy use to identify potential energy-efficiency measures. The BES model was validated by using field measurements and evidence-based input. Thermal bridges, infiltration, mechanical ventilation, domestic hot water circulation losses, and space heating power were calculated and measured. A backcasting method was developed to model heat losses due to airing, opening windows and doors, and other occupancy behaviour through regression analysis between daily heat power and outdoor temperature. Validation results show good agreement: 3.4% discrepancy between space heating measurements and simulations during an unoccupied week. Corresponding monthly discrepancy varied between 5.5% and 10.6% during three months with occupants. Annual simulation indicates that the best potential renovation measures are changing to efficient windows, improved envelope airtightness, new controls of the HVAC system, and increased external wall thermal insulation.

Keywords: building energy simulation; school building; field measurements; validation; airing; windows and door opening; occupancy behaviour; energy efficiency measures

---

1. Introduction

Residential and public buildings account for around 40% of the annual energy use and 36% of greenhouse gas emissions in Europe [1,2]. In developed countries, the increase in energy use by heating, ventilation, and air conditioning systems is significant, with a share of about 50% of the buildings' total energy use and around 15% of a nation's energy use [2].

Reductions in energy demand and efficient energy use are seen as feasible ways for more sustainable energy use in the built environment. Many actions and methods are implemented for this purpose, such as stricter building codes, green building programs and certification systems such as green building certification systems LEED and BREEM, and energy-efficient renovation strategies for buildings. In the European Union (EU), policies for energy efficiency in buildings have been stricter according to the EU2030 goals to meet the EU's long-term 2050 greenhouse gas reductions target [3]. Targets for 2030 are a 40% reduction in greenhouse gas emissions compared to 1990 levels, at least a 27% share of renewable energy usage, and at least 27% energy savings compared with the business-as-usual scenario. The Energy Performance of Buildings Directive (EPBD) [4] and the Energy Efficiency Directive (EED) [5] are two pieces of legislation to reduce energy use and environmental impact.

Building energy simulation (BES) models are commonly used to predict and study building performance in terms of energy and indoor environment. BES modelling is often done in the design stage of a new building or when investigating the current status for an existing building to plan future renovation strategies. The main goal with BES modelling is to reach as accurate prediction as possible, and in this context, it is of high priority to validate the model.

Different ways to calibrate and validate BES models and tools have been widely discussed [6–8]. Three methods for validating numerical tools are analytical solutions, peer models, and empirical data [8]. Analytical solution validation means validation of the component models within the simulation tools and the analytical solution of the mathematical physics model. Peer model validation is validation by comparing simulation output with another similar simulation tool output, where the same input data have been used. Empirical validation means that the model output is compared to empirically collected data for the specific modelling case. Moreover, the validation can be divided into two main categories: idealised and realistic validation. In idealised validation, the mathematical laws, component models, and engineering assumptions within the model are tested, and these are most often tested against measured data in a test cell, which is unoccupied. In realistic validation, the model considers influence of occupants and occupant behaviour, and the BES output is compared to measured data from actual buildings [8].

The gap between the prediction of the BES model and the building's actual performance is known as the performance gap [9,10]. Occupancy behaviour is understood to be a major reason for this gap [7–9,11–13]. Today, there are many methods on how to model occupancy and behaviour, such as presence, window opening and shading, lighting, and HVAC systems [13–16]. However, there is a need for more research in the field of evaluation of occupancy models and how to more easily integrate these in BES programs [8,11,15]. Further challenges connected to modelling occupancy in institutional buildings are that they often have the characteristics of large scale and a high variation of occupancy number, variation of patterns and behaviour, and lack of awareness of energy use [13].

It is evident that occupants' window and door opening behaviour affects both indoor environment and energy use in buildings, and how to model this behaviour in BES is a growing topic of interest [14–20]. Airing activities vary due to different types of buildings such as residential, offices, and schools [17,19]. However, it is difficult to quantify the exact loss of heat due to airing activities by measurements and for office, residential, and institutional buildings. In Swedish conditions, default values from standards are often used in BES [21].

Many case studies including BES model predictions are not compared against measured data [22]. Coakley et al. [6] point out that the process of creating realistic, reliable, and accurate BES models is often a non-transparent and ad hoc approach where the model creator can "tune" and adjust input data to the BES model so that simulated output results will fit with measured or assumed data. They also highlight that there are often only requirements on accuracy in predicted energy use, with little focus on the accuracy of the input data or the simulated environment [6]. That is why it is important that the process of building a model is transparent and that input data to the model is built on evidence-based methodology [23].

The aim with this study is divided in four objectives: (1) to perform a detailed data collection process in order to obtain evidence-based input data and validation data; (2) to perform validation of the BES model during both occupied and unoccupied periods; (3) to investigate a method on how to handle the challenges connected to the modelling of airing and other varying occupancy behaviour of this building; and (4) to investigate the energy use in the school building and identify potential energy efficiency measures in renovation planning.

2. Case Study Description

The studied secondary school building is located in Gävle, Sweden, and it is owned by the municipality of Gävle. The entire facility consists of six buildings, four with mainly classrooms, a cafeteria, and a gymnasium building. The largest building is the case study; see Figure 1. The heated floor area of 4577 m^2 is divided in four stories, including a basement and an attic. It was built during 1961–1963 and there are about 100 rooms, encompassing classrooms, some offices, a coffeehouse, toilets, and smaller storage areas. According to the principal, there were about 140 pupils (ages 13–16) and about 20 personnel occupying the building during 2014–2015.

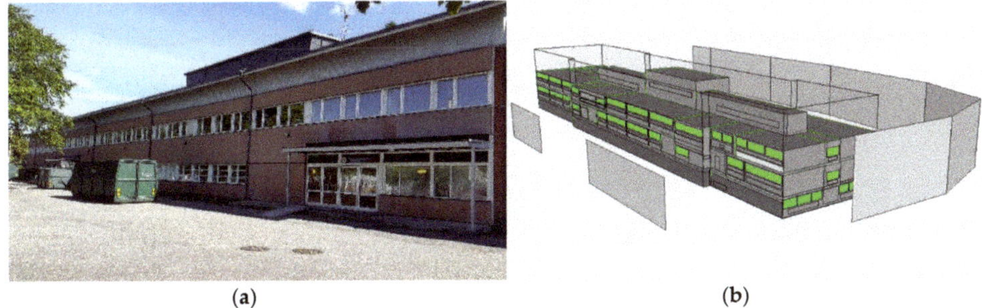

(a)   (b)

Figure 1. The school building, 97 m length, 18 m width, and 9 m height above the ground and the building energy simulation (BES) model, including shading objects such as trees shown in (**a**) and in (**b**) in form of screens.

A district heating system (DHS) supplies space and domestic hot water (DHW) heating. The space heating system is hydronic, consisting of three radiator circuits. Four air-handling units (AHU) supply ventilation air with a constant air flowrate through mechanical exhaust and supply systems with rotary heat exchangers. Heating coils are used to heat the supply air by district heating. The three smallest AHUs have separate control schemes, while the fourth and largest unit is controlled by both schemes and presence sensors. Any movement in the rooms with presence sensors will trigger forced ventilation and lighting for half an hour on a simple on/off basis.

U-values and areas of the building segments can be seen in Table 1. Construction drawings were available for all construction parts except for doors and windows. Based on site inspections and documentation, the materials and U-values were assessed for the whole building and inserted layer by layer in the BES tool IDA ICE construction component models. Additional assessment information on building construction and heating and ventilation units can be found in Section 3.2 and Table 5.

Table 1. Building construction description.

Building Segment	Area (m^2)	U-Value (W/(m$^2 \cdot °$C))
Roof	1700	0.15
Walls above ground	1579	0.32 [1]
Walls below ground	471	0.50 [1]
Floor towards ground	1700	0.22 [2]
Windows	465	2.7
Total	5956	0.51

[1] Average U-value of wall, consisting of different wall construction types. [2] Average U-value of floor construction towards ground and ground properties. The thermal resistance of the ground is simulated according to ISO 13,370 standard [24].

3. Methodology

The applied methods in this paper are field measurements, including logged data in Building Management System (BMS), and the utilisation of a BES tool. An overview of the methodology and research process to build and validate the BES model is shown in Figure 2, and it can be divided into five main steps. In the first step, a vast amount of information during both unoccupied and occupied time periods was collected. Input and validation data were collected from field measurements, BMS logging, and other sources, and the initial BES model was created. The only input data not included in the initial model was the occupants' airing behaviour. In the second step, the model was validated during an unoccupied time period when no occupants influenced the building. In the third step, a backcasting method on how to model airing and other varying occupancy behaviour was elaborated. The outcome of a function describing airing and occupancy behaviour variations was included as new input data in the model. In the fourth step, the model was validated during an occupied time period. Finally, this model was used to identify potential renovation measures.

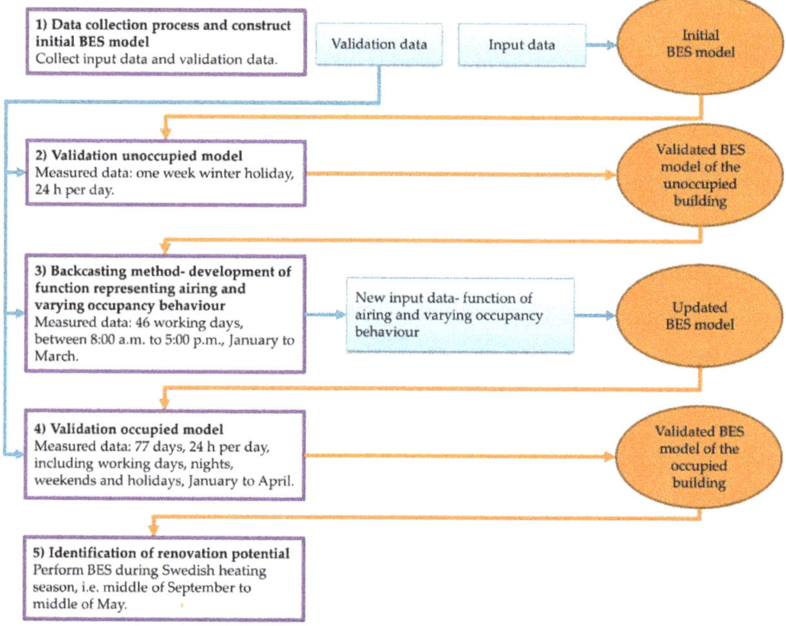

Figure 2. Overview of the research process.

Moreover, in steps two, three, and four in Figure 2, the measured power to the space heating system was compared with simulated data from the model during different time periods, and in step two and four, validation was performed. Validation criteria are set according to ASHRAE Guideline 14 [25] for discrepancies between measured and simulated hourly data. The Mean Bias Error (MBE) should not be higher than ±10% and Coefficient of Variation of Root Mean Square Error (CV(RMSE)) should not be higher than ±30%.

The strategy adopted to collect input and validation data was done by the evidence-based methodology suggested by Raftery et al. [23]. Data is categorised in a source hierarchy with the following order of priority: (1) long-term logged data from BMS; (2) field measurements; (3) direct observations and audits; (4) staff interviews; (5) operation documents; (6) construction documents; (7) benchmark studies and best practice guides; (8) standards, specifications and guidelines, and design information. Number one, logged data from BMS, has the highest source hierarchy and number eight (standards, specifications and guidelines, and design information) has the lowest source hierarchy. In Table 5, input and validation data to the model is summarised and the source hierarchy is denoted. Importantly, evidence-based input data with a high source hierarchy are desirable to obtain a reliable and valid model.

3.1. Field Measurements and Logged Data

Electricity and district heat usage provided by the local energy company was merged data for all six school buildings, and no separation between the different buildings was possible. Therefore, sub-metering through field measurements and logging in the BMS was necessary in order to collect input and validation data to the BES model. Table 2 summarises field measurements, and Table 3 summarises BMS logging. Both tables specify on which level the data was collected, room level or building level, and if the data were used as input and/or validation data. Table 4 presents equipment and measurement accuracy. The measurements were mainly performed from December 2014 until April 2015, except for the weather data, which was collected for one whole year.

Table 2. Overview of measurements.

Measurements	Measuring at Level		Use of Data	
	Room	**Building**	**Model Input**	**Model Validation**
Tracer gas for air leakage	X		X	
Room air temperature	X		X	X
Space heating power		X		X
Electric baseload power		X	X	
Weather data		X	X	
DHWC [1]		X	X	

[1] Domestic hot water circulation.

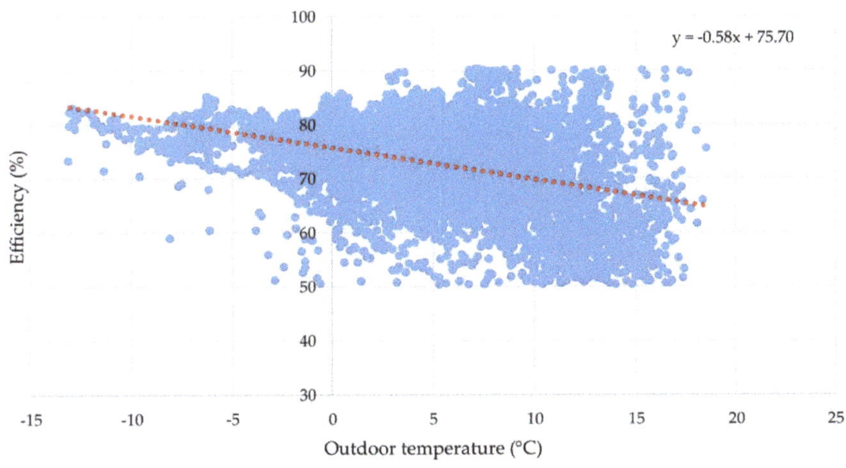

Figure 3. Linear correlation of rotary heat exchanger temperature efficiency in AHU and outdoor temperature; data set 1 September 2014 until 31 May 2015 (sampled every 10 min).

Table 3. Overview of Building Management System (BMS) logging.

BMS Logging	Logging at Level		Use of Data	
	Room	**Building**	**Model Input**	**Model Validation**
Damper positions [1]	X		X	
Air temperatures in AHU [2]		X	X	
AHU efficiency [3]		X	X	

[1] Data used to create schedules for internal loads and ventilation. [2] Data used as set point for supply temperatures in air-handling units (AHU). [3] Data used to create a function of AHU efficiency through linear regression, see Figure 3.

Table 4. Measurement equipment and accuracy.

Measurements	Equipment	Accuracy
Tracer gas for air leakage [1]	Passive sources and samplers	Total uncertainty of ±20% within 95% confidence interval [26].
Room air temperature	Mitec SatelLite-TH devices	Inaccuracy of ±0.4 °C for temperature [27].
Space heating power	TA Scope Premium	Inaccuracy of approx. ±5% of flow and <±0.2 °C of temperature [28].
Electric baseload	Tinytag Energy loggers	Inaccuracy of ± 2% [29]
Weather data	Vantage Pro2	Temperature inaccuracy of ±0.5 °C [30].
DHWC	Ultrasonic flowmeter Portaflow X	For pipe size ø13 to 50 mm, inaccuracy of 1.5% of flow rate (between 2 to 32 m/s) and 0.03 m/s (between 0 to 2 m/s) [31]. See also [32].

[1] Data used to calculate the mean age of air and average mean air change rate for the entire building [26].

An extraordinary measurement performed in this project was to quantify air infiltration by measuring the local mean age of air by tracer gas measurements during the non-occupied period [26]. This unintentional air flow is difficult to capture with other measurement methods and is often unknown input data in BES. The tracer gas measurements also gave proof of why the passive tracer gas method is not suitable in buildings with mechanical AHU systems that have large variations in air flow magnitudes [33].

Indoor temperatures and electrical equipment baseloads were measured during the non-occupied week with simultaneous inspections. Due to the installation setup, sub-metering of electricity for running HVAC systems and lighting and appliances in common spaces was difficult. Measurements of space heating power and energy in radiator systems were carried out at site with a manufacturer-calibrated TA Scope Premium balancing instrument with two-minute and 10-minute sampling intervals, respectively. A weather station was located on the roof. Temperature, wind speed, and direction from this unit were used to create a weather file in the simulation. The weather station only measured global radiation onto a horizontal surface, while IDA ICE requires diffuse and direct normal radiation onto a normal plane. Thus, complementary data to the weather file was created by using national meteorological weather station data, which were sourced from a station located about 7 km from the studied building and included solar radiation data through STRÅNG and the mesoscale analysis system called MESAN [34].

3.2. Building Energy Simulation Model Description and Collection of Data

A building model was created in the simulation tool IDA ICE version 4.8. In IDA ICE, thermal indoor climate and energy use can be studied by dynamic multi-zone simulations. IDA ICE has been certified and validated through idealised, empirical, analytical, and peer model validation in recent decades [35–39]. A typical simulation required approximately one core hour if performed on an Intel® Xeon® CPU E3-1505M v6 @ 3.00GHz computer. The BES simulations were mostly carried out at the University of Gävle's remote computer servers, Gävle, Sweden. In every simulation, the BES was simulated at least two weeks before the studied validation period in order to diminish initial value problems.

Approximately 100 rooms were modelled as 35 zones. The zoning strategy was done using the zone-typing approach according to Raftery et al. [23], which includes four major criteria: spatial location to the exterior, function of the space, conditioning methods (e.g., ventilation strategy and set points), and available measurements (including information about internal loads). Rooms were merged into zones in which additional internal mass was included to compensate for internal walls.

3.2.1. Description of Space Heating, Ventilation, and DHW Heating Systems

Space heating, ventilation, and DHW systems are all heated by heat exchangers connected to the local district heating net with an unlimited heat supply. The school is not equipped with cooling devices or air conditioning systems. Input data regarding DHW and DHWC systems are found in Table 5. Four AHUs with rotary heat exchangers were modelled with heat recovery efficiency based on linear regression on BMS logged data; see Figure 3. Values below 50% and above 90% were excluded, as these were assumed to be start up and shut down values. The linear function of temperature efficiency as a function of outdoor air temperature was implemented as a control curve into the AHU component models instead of using a fixed value for the efficiency of the rotary heat exchangers.

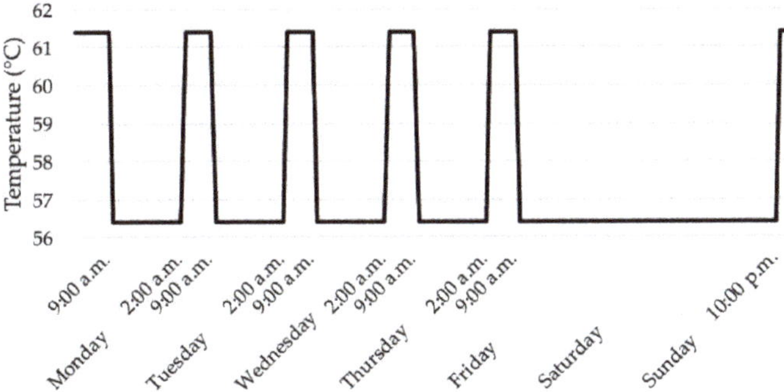

Figure 4. Hydronic radiator supply temperature schedule a week with −20 °C outdoor temperature.

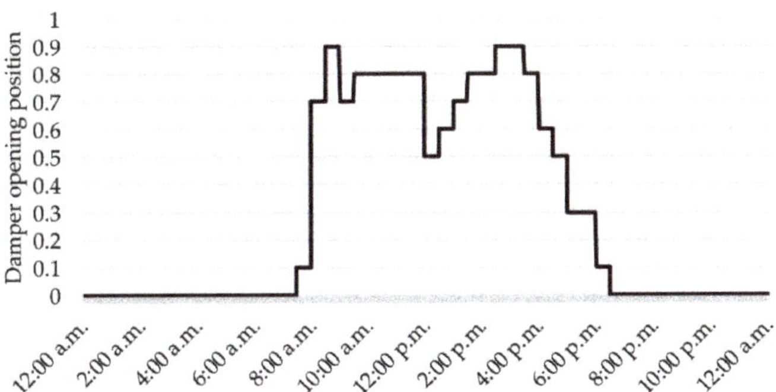

Figure 5. Schedule for Wednesday, classrooms in a normal school week, based on damper positions.

Table 5. Input and validation data overview. Values, source of collected data, and comments.

Parameter/Variable	Role	Value	Assessment Method [1]	Comment
Areas	Input	See Table 1	(3) (6)	Main measures checked with measurements.
Construction U-values	Input	See Table 1	(3) (6)	External walls had additional insulation, which was not updated in drawings but was estimated and measured at site. Material properties applied according to IDA ICE database.
Windows	Input	g-value 0.76	(3) (6) (7)	Window types with double glazing were observed and g-value assumed according to the BES tool ICA ICE database.
Solar shading	Input	Yes g-value 0.39	(3) (7)	Blind between window panes. Regulated by schedule and at insolation of 100 W/m^2 or more and multiplier for g-value according to the IDA ICE database.
Thermal bridges	Input	416 W/K	(6) and calculation	Total heat loss coefficient established by using Comsol Multiphysics 3.5. Typical thermal bridges were identified and modelled according to ISO 10211:2007 [40] and thermography.
Building air leakage	Input	0.12 ACH	(2)	Passive tracer gas method to measure average ACH for the whole building [26,33,41].
AHU flow rates	Input	4.25–0.55 l/s, m^2	(5)	From ventilation inspection protocols.
AHU supply air temp	Input	17.8–18.5 °C	(1)	Mean value of supply air temp. AHU1 = 18 °C, AHU2 = 17.8 °C, AHU3 = 18.5 °C, AHU4 = 18 °C. Assumed inaccuracy of ±0.5 °C for BMS logging.
AHU1 scheme	Input	1 schedule	(1)	Generalised schedule from three typical school weeks based on logged data of damper positions in room ventilation inlets. See Figure 5 and text Section 3.2.2 for description.
AHU2-4 schemes	Input	3 schedules	(5)	AHU2 = On 7:45 to 15:45 working days, AHU3 = On 6:30 to 17:30 working days, AHU4 = On 7:00–16:00 working days
AHU heat exchanger efficiency	Input	Control curve	(1)	Linear regression to model component performance in BES, based on logged data of AHU heat exchanger efficiency. See Figure 3 and text Section 3.2.1 for description.
Radiator system, design, and control	Input	Control curve	(5) (6)	Design heat power according to construction drawings and implemented as a component model within the BES model. Control curve according to BMS documentation. See Figure 4 and the text of Section 3.2.1 for a description.
Radiator system, space heating	Validation	See Section 3.1, Section 4.1.1, Section 4.2, and Section 4.3	(2)	Measurements of power and energy in radiator system. See Tables 2 and 4 and the text in Section 3.1 for a technical description. See Section 4.1.1, Section 4.2, and Section 4.3 for how measurements were used in the validation and backcasting method.
Indoor set-point temperatures	Input Indoor	17.1–22.3 °C	(2)	See Tables 2 and 4 and the text in Section 3.1 for technical description. See Section 4.1.2 for how the measurements have been used in validation.

Table 5. *Cont.*

Parameter/Variable	Role	Value	Assessment Method [1]	Comment
Electric baseload	Input	1.75 W/m^2	(2) (3)	Audits at site were done to map the location of the loads. See Tables 2 and 4 and the text in Section 3.1 for a technical description.
Lighting in zones	Input	8.4–15.6 W/m^2	(3)	Lighting power in different rooms is documented during the audits at site. See the text in Section 3.2.2 for a description of schedules.
Weather data	Input	Prn. File	(2)	Local weather prn file created from measurement data from the local weather station on roof and the national meteorological weather station. The weather file was used in both the validation and backcasting method. See Tables 2 and 4 and Section 3.1 for a technical description.
Domestic hot water use	Input	2 kWh/m^2, year	(8)	Swedish standard Sveby [21].
Domestic hot water circuit losses	Input	1 W/m^2	(2)	Ultrasonic flow measurements, see the Master's thesis for more details [32]. Most (75%) of the heat is assumed to be useful heat inside the building construction.
Occupant number	Input	0.038 occup./m^2	(4)	No. of occupants based on interviews and class schedules. In total, 160 occupants: 140 pupils and 20 staff.
Utilisation schedules	Input	8 varying schedules	(1) (5)	See Figure 5 and Section 3.2.2 for a description.
Airing	Input	See Section 3.4 and Section 4.2	(2)	Air leakage through windows, doors, and entrance openings. Backcasting method compiled; see Section 3.4 and Section 4.2 for a description.

[1] Evidence-based assessment method for data collection with the following source hierarchy: (1) logged data from BMS; (2) field measurements; (3) direct observation and audits; (4) staff interviews; (5) operation documents; (6) construction documents (7) benchmark studies and best practice guides: (8) standards, specification, and guidelines.

The AHUs' supply temperatures were set according to average temperatures logged in the BMS during daytime operation in March 2015; see the input values in Table 5. Three of the AHUs had fixed schemes; see Table 5. Presence sensors controlled the fourth AHU, and a generalised scheme was created as described in Section 3.2.2. The supply and exhaust air flow for each zone are set according to the ventilation inspection protocols for each room. At maximum, the total fan power is about 36 kW, and an air flow of about 10 m^3/s of supply and exhaust air are ventilating the building through all four AHUs.

Hydronic single plane radiators were modelled with emitting capacity from an IDA ICE component library and maximum heat output according to design calculations. Radiators were modelled with proportional thermostats with a dead band of 0.5 °C. Having the control strategy of outdoor compensated supply flow temperature, the heating system has a control curve with a maximum supply temperature of 61.4 °C at an outdoor temperature of −20 °C and a minimum temperature of 19.4 °C at an outdoor temperature of 18 °C. The corresponding curve implemented in the component model is based on the measured radiator circuit's temperature performance. Night setback decreases the supply temperature by 5 °C between 9:00 and 2:00 the next day Monday to Friday, 24 h during Saturday and until 20:00 during Sunday. Figure 4 show the supply temperatures control curve over a week with −20 °C outdoor temperature.

3.2.2. Schedules for Internal Loads and Ventilation

For electricity devices, a baseload was measured during the unoccupied time period and was allocated per m^2 area within the whole building and scheduled as "always on"; see the values in Table 5. For office zones and the coffeehouse, lighting and occupancy were scheduled according to the staffs' working hours, and the ventilation in these zones was controlled by AHU3. The rooms located at the basement level were ventilated by AHU2 and AHU4; see Table 5 for schedules for AHU2-4. Lighting in the corridor zones was controlled by a time schedule. The majority of the classrooms were controlled by presence sensors and ventilated by AHU1.

For all zones controlled by presence sensors, a scheme was created for every weekday based on BMS damper positions (10-minute sampling time) of ventilation inlets in 19 rooms. A generalised operation schedule for AHU1, occupants, and lighting was created for these zones. Figure 5 illustrates a Wednesday schedule, where 1.0 indicates presence control activation in all 19 rooms; 0.0 indicates that no room was used (closed dampers). A half-hourly mean value was created from three typical weeks, and the selection of these weeks was made by studying the measured electricity pattern for the whole building over a whole year.

In total, eight time schedules were created, used, and applied for lighting and occupancy depending on the zone type and functional use and data from BMS. Each of these eight schedules includes specific schemes for each weekday and are adjusted concerning weekend and holiday periods. These schedules can be assumed to capture the overall utilisation of the building in terms of occupants' presence, use of appliances and devices, and presence-activated ventilation rates. However, this scheme does not capture occupants' airing due to opening windows, doors, and entrances. Instead, energy losses due to airing, amount of persons, and other unknown energy usage are modelled with the backcasting method described in Section 3.3. In Table 5, the input and validation data to the model are summarised. The source hierarchy (described in the end of Section 2) of the input and validation data is specified in the Assessment method column in Table 5.

3.3. Validation of the Model—Unoccupied Building

Validation of the model during an unoccupied period was performed during a winter week, from 00:00 on 29 December 2014 to 00:00 on 5 January 2015. The building was fully heated, although occupants were on holiday and all AHUs were intentionally shut off. Validation included comparison between measurements against simulated results at the room and building level: indoor temperatures in three selected validation rooms at the room level, and power and energy for the heating system at the building level. This validation method captures the thermal performance of the building in a more valid and reliable way compared to validation with monthly purchased heat including the heating of space, ventilation, and DHW. Measured and simulated power to the heating system was compared on an hourly basis, and the variation of MBE and CV(RMSE) was calculated. The main purpose at this stage was to validate the building's thermal performance characteristics without the influence of occupants and occupant-dependent ventilation.

The three rooms that were chosen for validation at the room level were a classroom, a teachers' office (several teachers share an office), and the staff lunchroom. These rooms were modelled as separate zones and had the same areas in the model as in reality. Figure 6 illustrates the actual first floor plan. Figure 7 shows model zoning of the first and second floor and visualises the validation zones: staff lunchroom (a), teachers' office (b), and classroom (c).

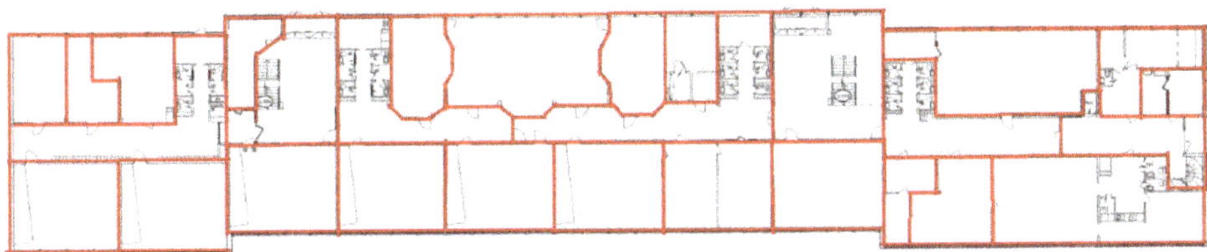

Figure 6. The actual rooms in the first floor plan.

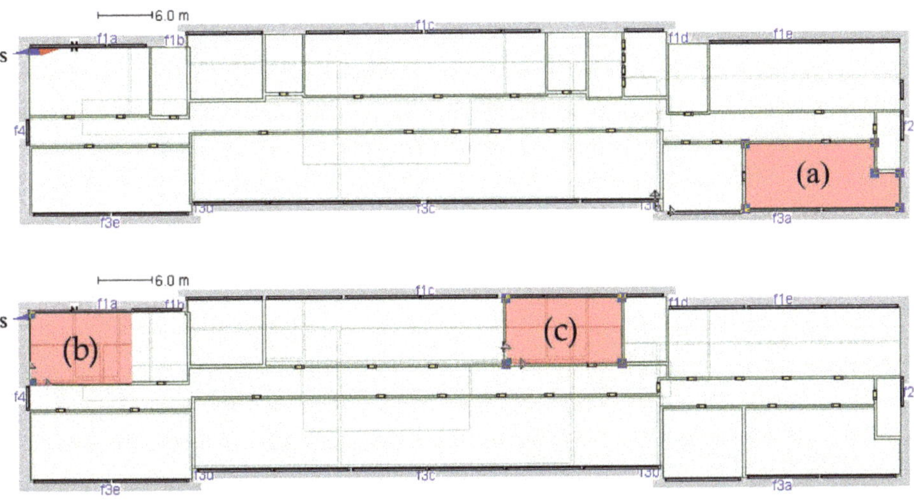

Figure 7. Top picture shows the zoning in the first floor in the model, including the validation zone staff lunchroom (**a**). Bottom picture shows the zoning in the second floor, including the validation zones teachers' office (**b**) and classroom (**c**).

3.4. Modelling Airing and Varying Occupancy Behaviour

Although the activities within the building are to a large extent scheduled and modelled on the basis of ventilation room damper positions as shown in Figure 5 and described in Section 3.2.2, there are irregularities which make it difficult to capture the true utilisation of the building in detail. These include the presence of pupils and staff, actual use of reserved rooms/time and activities therein, and equipment utilisation. Moreover, within modelling, a major uncertainty is how to schedule airing in terms of entrances, doors, and windows being opened. Instead of using best practice values for heat losses due to airing or creating schedules for window and door opening based on assumptions, a backcasting method has been developed. Due to occupancy and its consequences, this new method was developed for application to the occupied validation period. The daily mean power difference between measured and modelled heat power to hydronic radiators as a function of mean outdoor temperatures was calculated and a linear regression was applied, see the results in Section 4.2. The simulated and measured daily mean power was compared during 46 working days between 12 January and 27 March 2015, from 08:00 to 17:00 to study airing and variations in occupancy behaviour. Due to evidence-based input data in the model, it is possible to draw the conclusion that the difference in measured and modelled mean heating power can be attributed mostly to airing activities and other occupant-related energy use in the building, such as the influence of occupancy presence and traffic as well as the use of lighting, equipment, and ventilation. The linear correlation is used as input data and modelled as an extra heat load in the BES model during the validation of an occupied building and when performing a normalised annual BES.

3.5. Validation of the Model—Occupied Building

Validation for the period with the presence of occupants was performed during three months, from 12 January to 5 April 2015. This period represents a validation period including 77 days, 24 h per day of energy use comparison, both regular school weeks and holiday weeks. The linear correlation modelling airing and varying occupancy was included in this simulation. Measured and simulated power to the heating system was compared on an hourly basis, and the variation of MBE and CV(RMSE) was calculated.

3.6. Identification of Energy-Efficiency Measures

After the model was validated during occupied operation, an annual BES during the Swedish heating season, from the middle of September to the middle of May, was performed. A weather file compiled by the Sveby organisation [42] with typical weather data for the city of Gävle during the years 1981–2010 was applied in the simulation. In this climate file, diffuse and direct solar radiation is calculated through the altitude of the sun, solar angle, cloudiness, and some more parameters [43]. The heat losses were separated into different posts e.g., windows, external walls, air leakage through construction, etc., the existing U-values were compared to the Swedish Building Code U-values $(W/(m^2 \cdot {}^\circ C))$, and potential energy efficiency measures were identified and discussed.

4. Results

Results from unoccupied and occupied validation periods, how to handle airing and varying occupant behaviour, annual heat balance simulation, and potential energy efficiency measures in renovation planning are presented in this chapter.

4.1. Validation of the Model—Unoccupied Building

Model validation of the unoccupied building includes measurements of space heating at the building level and temperature measurements at the room level. Input data, which might be unknown in BES projects, are for example thermal bridges, infiltration, and other construction data. In this case study, much of the input data is evidence-based data with a high source hierarchy due to detailed data collection and the many field measurements where, for example, infiltration and domestic hot water circulation (DHWC) losses were measured. Validation was performed using both heating power demand and energy use during the above-mentioned week.

4.1.1. Building Level

Figure 8 illustrates measured and simulated heating power demand from the building's hydronic radiators. In the beginning of the validation period, the outdoor temperature was lower, and both measured and modelled heat power to the radiators were higher and vice versa during the middle of the validation period. An explanation of the higher and more fluctuating measured power during the beginning of the week can depend on air infiltration. In the model, infiltration is set to a fixed average value of 0.12 ACH, which is based on tracer gas measurement during the same period [26]. In reality, infiltration rates vary with wind conditions and outdoor air temperatures. Discrepancies are also due to initial value problems (since occupancy prior to the non-occupancy period is unknown), differences in control schedules and deficiencies in the building model concerning thermal mass within the building, and in the IDA ICE component models (for example, thermal bridges and the heating system lack thermal inertia).

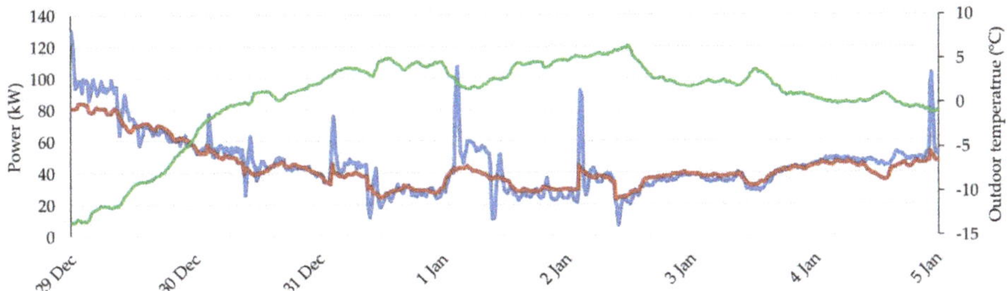

Figure 8. Measured (blue line) and simulated (red line) energy to the hydronic radiators and outdoor temperature (green line) during the unoccupied validation period.

The time resolution of 15 min mean power, based on measurements sampled every two minutes, shows the difference between measured and modelled heating power in terms of a more fluctuating pattern. This is due to a combination of the non-smooth control of the hydronic valves in the district heating substation, the inertia in the hydronic radiator system caused by the length of the piping, and the inertia in the operation of the radiator thermostats. The IDA ICE component models do not capture quick fluctuations in the heating system.

On 1 January, there was more measured heat power than simulated, which might be explained by the presence of office staff who perhaps were working during some of these holidays, even if the school is assumed to be empty. Since the ventilation systems were off during this week, the present staff might have opened windows due to deficient air quality. This clue was supported by the tracer gas measurements in one of the teachers' offices during the validation period that indicate windows have been opened in this room [26].

Figure 9 presents daily differences between measured and simulated energy use. The differences vary between 7.9% and −9.6% and 58 kWh and −139 kWh, where negative values represent higher measured than simulated values. The majority of days show that measured energy use is higher than the simulated use. The summarised measured energy use over the whole week is 7.3 MWh and the simulated energy use is 7.0 MWh. This results in a difference of −3.4%, meaning slightly higher measured energy use compared to simulated use. This difference is less than the accuracy of the measurement equipment. According to ASHRAE Guideline 14 [25], the difference between measured and simulated data should not be higher than ±10% for MBE and ±30% for CV(RMSE) when using a simulation time step of one hour. This criteria is achieved with a large margin for the unoccupied building; see Table 6. In this statistical analysis, the outliers that occur during nights, seen as peaks in measured values in Figure 8, are excluded. The peaks can be explained by the building's heating control causing an overshoot of heat every night at the time when the supply temperature increases by 5 °C; see the control curve in Figure 4. This type of overshoot is not modelled by IDA ICE components. In the statistical calculations, one hour of data are excluded every occasion this temperature increase occurs; in total, 5 of 168 values were removed.

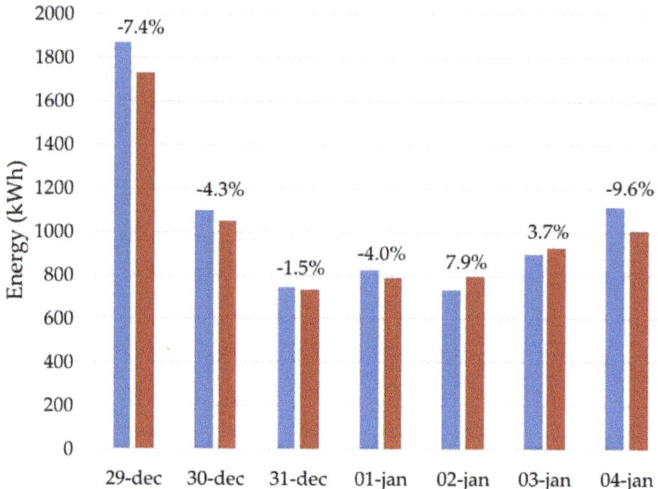

Figure 9. Measured (blue) and simulated (red) energy to the hydronic radiators per day including the difference percentage between measured and modelled energy during the unoccupied validation period.

Table 6. Differences between simulated and measured energy to the hydronic radiators. One week hourly data, 29 December 2014 to 5 January 2015.

Measured Energy Use (MWh)	Simulated Energy Use (MWh)	MBE Hourly Data (%)	CV(RMSE) Hourly Data (%)
7.3	7.0	3.3	14.2

In summary, the small differences in heat power demand and energy use might be due to insufficient settings for internal loads in the model such as occupants, use of lights and equipment, and differences in infiltration flows and insolation. Furthermore, in the basement, large heat losses due to partly uninsulated heating pipes exist, which are modelled as ideal heaters in two specific basement zones. There might be a few other parts of the heating energy system that should be covered by the zone radiators rather than by ideal basement heaters (ideal heaters respond perfectly to temperature changes).

Although the total DHWC losses have been measured [32], it is not known how much of these losses are gained in the building, since two separate buildings share this circulation system. In all simulations, 75% of the DHWC losses are assumed to be gained within the building. A sensitivity analysis concerning DHWC losses as gains varied between 50%, 75%, and 100%. The results from the sensitivity analysis showed that the measured energy use is higher than the simulated energy use. However, the sensitivity analysis shows that the percentage share of DHWC that is assumed to become useful heat in the BES model varies in the three cases between 1.7%, 3.4%, and 5.1%, which is not a critical assumption for the result output.

4.1.2. Room Level

Figure 10 shows indoor temperatures at the room level during the unoccupied validation period. The classroom reveals some peaks where the highest difference is 0.8 °C. The reasons for these variations might depend on insolation, also keeping in mind that IDA ICE zone air temperature is represented by one node (i.e., the room air is totally mixed).

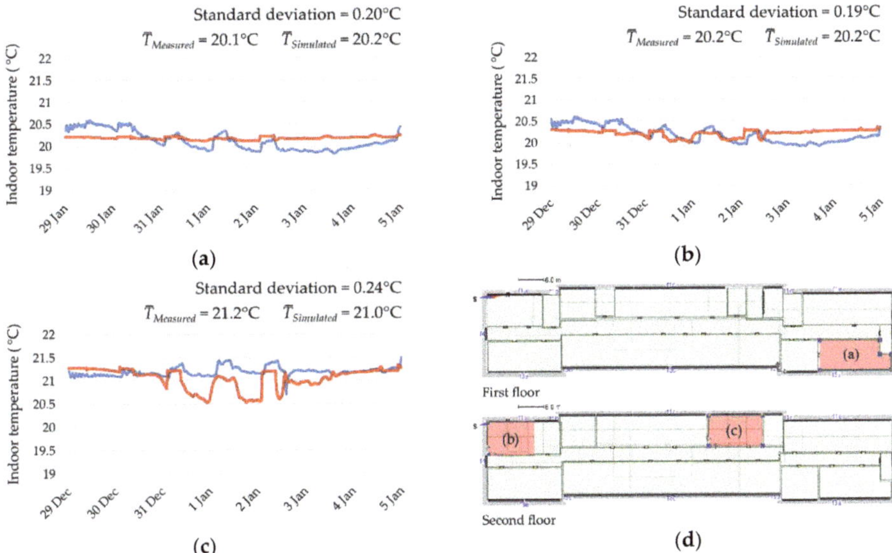

Figure 10. Measured (blue line) and simulated (red line) indoor temperatures in the three validation rooms, staff lunchroom (**a**), teachers' office (**b**), and classroom (**c**) during the unoccupied validation period and location of the validation zones at building floorplans (**d**).

The staff lunchroom located on the east side of the building is not affected by insolation to the same extent as the other rooms, which are located on the west side. However, the teachers' office is shaded by trees to a larger extent than the classroom. The standard deviation from measured indoor air temperatures is 0.20 °C for the staff lunchroom, 0.19 °C for the teachers' office, and 0.24 °C for the classroom. The low standard deviation values indicate good correlation between measured and simulated temperatures, and they are within measurement instrument errors.

4.2. Modelling Airing and Varying Occupancy Behaviour

At this stage, the building's technical characteristics have been validated and tested against measurements for the non-occupied period. The occupancy schemes described in Section 3.2.2 were implemented, and a three-month period with occupancy was simulated. In Figure 11, daily mean radiator power, measured (P_{measured}) and simulated ($P_{\text{simulated}}$) heat power to radiators are plotted against daily mean outdoor temperatures for 46 working days (data based on values from 08:00 to 17:00 p.m., 12 January until 27 March 2015). These show outdoor temperature dependency. Moreover, the differences between these two curves are illustrated. The plotted power differences can mostly be ascribed to airing activities by opening doors and windows and other possible behavioural patterns, as described in Section 3.4, keeping in mind that occupant presence influencing ventilation patterns and human heat emissions have been considered in BES in terms of schedules; see Section 3.2.2.

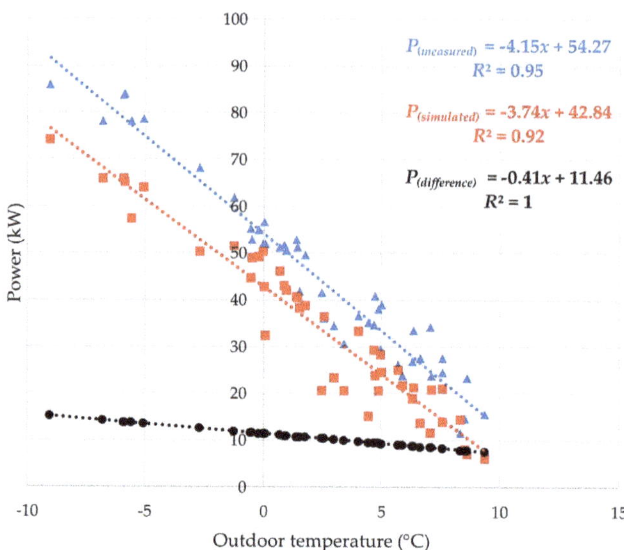

Figure 11. Mean power difference between measured and modelled heat power to hydronic radiators as a function of outdoor temperature daily values.

The linear correlation for the difference shows an overall power heat loss of 11.46 kW and an additional outdoor temperature dependency term of −0.41 kW/°C. The correlation curve indicates an increase in heating power difference between measured and simulated results as outdoor temperatures become colder. Outdoor temperatures between January and March vary in the same way as outdoor temperatures during autumn. This motivates why this linear correlation can be used to describe the mean power differences over the whole heating season in Sweden, which is often generalised to be from 15 September until 15 May. When performing the validation of an occupied building and during annual simulation with normalised weather data, the linear correlation of daily mean power difference, described in Figure 11, is added as an extra energy heat loss into the BES model.

4.3. Validation of the Model—Occupied Building

In Table 7, measured and simulated energy supplied to the radiators are compared during the time period from 12 January to 5 April by encompassing all hours during this period, excluding some days due to a lack of measured data. The monthly differences vary from 5.5% in January, 10.6% in February, and 9.7% in March/April as opposed to the unoccupied period, which resulted in 3.4%. The outliers, caused by the overshoot of heat during night control, are excluded in the same way as for the calculations of the unoccupied period. In the calculations presented in Tables 7 and 8, 54 hourly values are removed from the total values of 1847.

Table 7. Monthly measured and simulated energy to the hydronic radiators, including the difference in energy and percentage during the occupied validation period.

	Measured Energy Use (MWh)	Simulated Energy Use (MWh)	Difference (MWh)	Difference Percentage (MWh)
January	20.1	19.0	1.1	5.5
February	32.4	29.0	3.4	10.6
March/April	26.3	23.7	2.5	9.7

Table 8. Differences between simulated and measured energy to the hydronic radiators. Hourly values for three-month occupied time period, January to April 2015.

Measured Energy Use (MWh)	Simulated Energy Use (MWh)	MBE Hourly Data (%)	CV(RMSE) Hourly Data (%)
78.8	71.7	9.0	28.8

As can be seen in Table 7, the months of February and March/April might in reality be more divergent compared to the typical school week schedule and because of that, they show larger differences than January. The difference between measured and simulated energy to the space heating system is shown as statistical calculations in Table 8. The MBE and CV(RMSE) for the occupied period are met according to ASHRAE criteria [25]; however, the variations are larger for the occupied period compared to the unoccupied period. Some reasons for the observed differences might be the ventilation, light, and occupant schedules. These schedules are generalised and adjusted to represent typical school weeks over the whole year.

4.4. Annual Energy Use for a Typical Year and Identification of Energy Efficiency Measures

The simulated energy balance from the middle of September to the middle of May for heating and domestic hot water use for the school building results in 545 MWh, which represents supplied energy and heat losses, e.g., through the building envelope, ventilation, infiltration, and airing; see Figure 12. Internal gains contribute 155 MWh of heat, which includes heat from occupants (16 MWh), equipment (74 MWh), and lighting (65 MWh). Purchased district heating includes heating of the hydronic radiator system, the mechanical ventilation system, the DHWC losses, and other pipe heat losses. The heating of DHW is separated from the supplied heat and represents 8 MWh. The total amount of purchased heat is 73 kWh/m^2 for a normalised year.

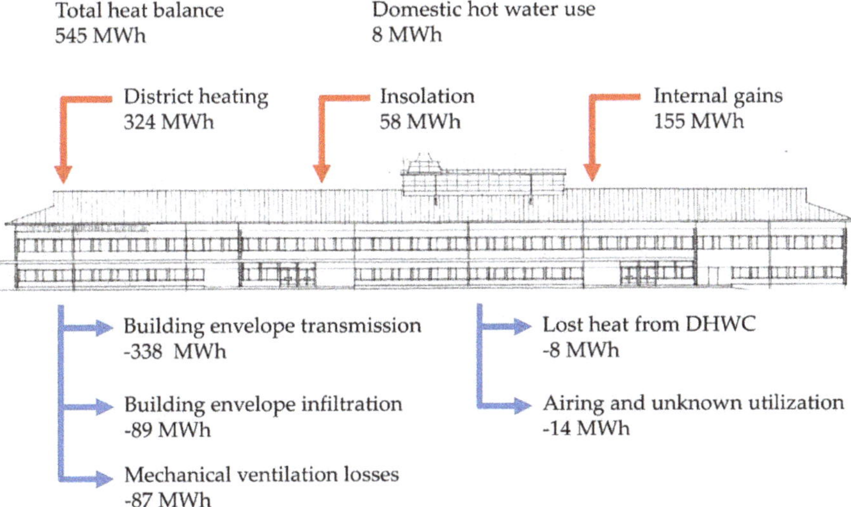

Figure 12. Heat balance of the school building, simulated during heating season, 1 September to 31 May, with normalised climate data.

The heat losses, which are arranged from highest to lowest contribution, are presented in Table 9. Window transmission losses account for 22% of the total. The second highest losses of 16% are due to infiltration. Future energy efficiency measures could be to replace windows with lower U values and to air-tighten the building envelope. When renovating a building in Sweden, attempts should be made to fulfil the requirements of the primary energy use, according to the current Swedish building code [44]. However, if this is not possible, minimum U-values for small residential houses can be seen as indicative values; see Table 9 [44]. It is clear that there is potential in future renovation to achieve lower U-values when renovating building segment parts. Furthermore, to build a more airtight building envelope will decrease the heat losses due to infiltration losses. The Swedish Forum of Energy Efficient Building (FEBY) set a maximal air leakage of 0.3 l/(s·m^2) through the building envelope area at a pressure difference of 50 Pa as a criterion in Swedish passive house certification [45]. The air leakage should be tested according to the standard SS-EN ISO 9972:2015. With a corresponding air tightness of 0.3 l/(s·m^2 at 50 Pa) pertaining to the enclosing envelope area, the BES results show a heat loss of about 25 MWh due to infiltration. This indicates an extensive reduction of heat loss compared to the current infiltration, which corresponds to 89 MWh during the heating season.

Table 9. Distribution of heat losses in heat balance, simulated during the heating season, 1 September to 31 May, with normalised climate data.

Title 1	Heat Loss (MWh)	Percentage	Existing Building U-Value (W/(m^2.$^\circ$C))	Swedish Building Code U-Value [44] (W/(m^2.$^\circ$C))
Windows	145	27%	2.7	1.2
Infiltration	89	16%	-	-
Mechanical ventilation	87	16%	-	-
Walls	81	15%	0.36 [1]	0.18
Thermal bridges	49	9%	416 [2]	-
Floor	35	6%	0.22	0.15
Roof	28	5%	0.15	0.13
Airing and unknown utilisation	14	3%	-	-

[1] Average U-value of wall consisting of different wall construction types, below and above ground. [2] W/K.

Another large heat loss post is the mechanical ventilation system; see Table 9. However, the rotary heat exchangers already have high temperature efficiency, as described in Section 3.2.1. Therefore, possible energy efficiency measures might be to minimise the specific fan power and adjust the air flows. The mechanical ventilation system is an on/off constant air volume (CAV) system controlled by presence sensors; however, the occupancy in the classrooms can vary significantly. Therefore, another control strategy can be a variable air volume (VAV) system controlled by CO_2 and temperature levels. This will reduce the risk of unnecessary air flows and thereby reduce the heat losses. Furthermore, a VAV and CO_2 controlled mechanical ventilation system might also improve the thermal comfort and indoor environment for pupils and staff.

The DHWC losses are simulated as 25% of the heat is lost, which accounts for 8 MWh; see Figure 12. Measurements on the DHWC together with observations at site visits and thermography pictures showed extensive heat losses through a partly uninsulated hot water piping system. This piping system supplies heat to five other school buildings. However, in the present study, it is not of interest to specify the exact amount of heat emitted from these partly uninsulated pipes, since this heat is assumed to be utilised within the building. However, if looking at the energy use for the whole school, including all six buildings, it is of high interest to decrease these heat losses due to partly uninsulated pipes, especially heat loss in distribution pipes between the buildings. High heating bills during summer holiday periods when the buildings are unoccupied confirms large pipe heat losses.

The heat losses due to airing and other unknown utilisation patterns were simulated to 14 MWh. This corresponds to a value of 3.1 kWh/m^2 per year and represents about 3% of the total amount of purchased district heat. A standard value commonly used for airing losses in BES in Sweden is 4 kWh/m^2 per year [21], which would result in 18 MWh for the studied building. The simulated energy lost due to airing is thereby somewhat lower than the standard Swedish value for BES simulations. At a glance, the power associated with airing and other variables is about 2.5 W/m^2.

5. Discussion and Conclusions

In this case study, an existing school building was modelled using the building energy simulation tool IDA ICE. The methodology developed in order to create the BES model can be summarised as follows: (1) collect evidence-based input and validation data with majority high source hierarchy; (2) separate validation of occupied and unoccupied time periods, where validation during the unoccupied period specifically concerns the buildings' technical characteristics; (3) applying a backcasting method in order to compile a heat load that represents airing and varying occupancy behaviour; and (4) perform BES simulation during the heating season to identify renovation potential. This methodology can entirely or partly be applied in other studies when creating BES models of existing buildings, and the ambition is to obtain reliable and validated BES models. The backcasting method can be especially useful to apply when buildings with irregular occupancy behaviour are modelled in BES.

The first objective was to obtain evidence-based input and validation data, and the study shows how the collection of evidence-based data with high source hierarchy, including both logged data in the building management system and detailed field measurements, constitutes a good base to perform validation of the building energy simulation model. This is of especially high importance in case studies including complex buildings such as school buildings. A strength of the study was the many field measurements providing input data, which often are unknown in BES case studies, such as for example infiltration leakage, which was measured through passive tracer gas measurements. Hot water circulation losses and space heating power were also measured and the performance of thermal bridges was calculated.

The second objective was to perform a separate validation of unoccupied and occupied time periods. The unoccupied period enables validation of the building thermal performance when uncertainties connected to occupancy behaviour input data are not present. The results from the unoccupied validation period show that the building's thermal performance is correctly modelled with only a discrepancy of 3.4% difference of simulated and measured heat to the hydronic radiators during one week. Furthermore, measurements of 15-min resolution capture the minor differences in simulated and measured operation of the heating system. Sensitivity analysis on the percentage of DHWC, which is gained as heat in the building, showed that this is not critical input data. At the room level, measured and simulated indoor temperatures in three validation rooms showed good agreement with a standard deviation of only 0.19–0.24 °C.

The third objective was fulfilled by the development of a backcasting method. As highlighted in the introduction, the airing and opening of windows and doors are often unknown input data that are difficult to model in BES. In this study, the energy loss due to airing and varying occupant behaviour was elaborated using a backcasting method. In this method, airing heat loss is modelled by creating a linear correlation including the daily mean power difference between measured and simulated heat to the hydronic radiators and outdoor temperature during 46 workdays, January to March, between 08:00 and 17:00. The result showed a mean power heat loss of 11.5 kW and additional outdoor temperature correlation of −0.41 kW/°C. This linear correlation was used as input data to represent varying occupancy behaviour and the majority of airing activities during the Swedish heating season from September to May. The linear correlation was implemented into the model and used in both validation of the occupied building and in the annual heat balance simulation. The annual heat loss due to airing was predicted at 14 MWh, corresponding to 3.1 kWh/m^2 and year and implying an overall power loss of 2.1 W/m^2 during working hours/use of the school. Validation of the model during occupied time periods resulted in monthly discrepancies between measured and simulated energy use for space heating varying between 5.5% and 10.6% during three months. This can be seen as an acceptable discrepancy for a valid BES model during an occupied period. It was more important that this model contains schedules of occupancy behaviour and mechanical ventilation operation that represent typical behaviour during the whole year than to calibrate or "tune" the model to achieve a perfect match between measurement and simulations during a specific time period. This backcasting method is a suitable complement to the strategy of generating schedules for room ventilation and internal loads based on the presence of controlled ventilation damper positions.

Validating the BES model and handling modelling uncertainties are preconditions to fulfil the final goal with the modelling: to perform as accurate and realistic a prediction of the real system modelled as possible. In this case study, the final BES model predicts a total need of purchased heat at 73 kWh/m^2, 332 MWh in the school building. The fourth objective was to identify energy-efficiency measures, and the ones that have the greatest impact on energy use (in order of highest first) were changing to energy-efficient windows, improved airtightness of the building envelope (which corresponds to 0.12 ACH), new controls of the HVAC system, and increased wall insulation. In future research, the validated model will be used for investigations on life-cycle cost optimisation of energy-efficient measures in renovation planning.

Author Contributions: Conceptualisation, J.S.E., M.C., J.A. and B.M.; Data curation, J.S.E.; Formal analysis, J.S.E., M.C., J.A. and B.M.; Funding acquisition, B.M.; Investigation, J.S.E. and J.A.; Methodology, J.S.E., M.C., J.A. and B.M.; Project administration, J.S.E.; Software, J.S.E. and J.A.; Supervision, M.C., J.A. and B.M.; Validation, J.S.E., M.C., J.A. and B.M.; Visualisation, J.S.E.; Writing—original draft, J.S.E.; Writing—review and editing, J.S.E., M.C., J.A. and B.M. All authors have read and agreed to the published version of the manuscript.

Funding: This research was funded by Gavlefastigheter AB, and the Knowledge Foundation (KK-stiftelsen) grant number 20120273.

Acknowledgments: The work has been carried out under the auspices of the industrial post-graduate school Reesbe, which is financed by the Knowledge Foundation (KK-stiftelsen).

Conflicts of Interest: The authors declare no conflict of interest.

References

1. International Energy Agency. *Energy Technology Perspectives—Scenarios and Strategies to 2050*; International Energy Agency: Paris, France, 2010.
2. Pérez-Lombard, L.; Ortiz, J.; Pout, C. A review on buildings energy consumption information. *Energy Build.* **2008**, *40*, 394–398. [CrossRef]
3. European Council. *EUCO 169/14—Conclusions 23/24 October 2014*; European Council: Brussels, Belgium, 2014.
4. European Council. Directive 2010/31/EU of the European Parliament and of the Council of 19th May on the energy performance of buildings. In *Directive 2010/31/EU, DOUE 153*; European Council: Brussels, Belgium, 2010.
5. European Council. EED Directive 2012/27/EU of the European Parliament and of the Council of 25 October 2012 on energy efficiency. In *2012/27/EU*; European Council: Brussels, Belgium, 2012.
6. Coakley, D.; Raftery, P.; Keane, M. A review of methods to match building energy simulation models to measured data. *Renew. Sustain. Energy Rev.* **2014**, *37*, 123–141. [CrossRef]
7. Fabrizio, E.; Monetti, V. Methodologies and Advancements in the Calibration of Building Energy Models. *Energies* **2015**, *8*, 2548–2574. [CrossRef]
8. Ryan, E.M.; Sanquist, T.F. Validation of building energy modeling tools under idealized and realistic conditions. *Energy Build.* **2012**, *47*, 375–382. [CrossRef]
9. Sunikka-Blank, M.; Galvin, R. Introducing the prebound effect: The gap between performance and actual energy consumption. *Build. Res. Inf.* **2012**, *40*, 260–273. [CrossRef]
10. De Wilde, P. The gap between predicted and measured energy performance of buildings: A framework for investigation. *Autom. Constr.* **2014**, *41*, 40–49. [CrossRef]
11. Gaetani, I.; Hoes, P.-J.; Hensen, J.L. Occupant behavior in building energy simulation: Towards a fit-for-purpose modeling strategy. *Energy Build.* **2016**, *121*, 188–204. [CrossRef]
12. Hoes, P.; Hensen, J.; Loomans, M.; De Vries, B.; Bourgeois, D. User behavior in whole building simulation. *Energy Build.* **2009**, *41*, 295–302. [CrossRef]

13. Yang, J.; Santamouris, M.; Lee, S.E. Review of occupancy sensing systems and occupancy modeling methodologies for the application in institutional buildings. *Energy Build.* **2016**, *121*, 344–349. [CrossRef]

14. Dong, B.; Yan, D.; Li, Z.X.; Jin, Y.; Feng, X.H.; Fontenot, H. Modeling occupancy and behavior for better building design and operation-A critical review. *Build. Simul.* **2018**, *11*, 899–921. [CrossRef]

15. Yan, D.; O'Brien, W.; Hong, T.; Feng, X.; Gunay, H.B.; Tahmasebi, F.; Mahdavi, A. Occupant behavior modeling for building performance simulation: Current state and future challenges. *Energy Build.* **2015**, *107*, 264–278. [CrossRef]

16. Haldi, F.; Robinson, D. Interactions with window openings by office occupants. *Build. Environ.* **2009**, *44*, 2378–2395. [CrossRef]

17. Fabi, V.; Andersen, R.V.; Corgnati, S.; Olesen, B.W. Occupants' window opening behaviour: A literature review of factors influencing occupant behaviour and models. *Build. Environ.* **2012**, *58*, 188–198. [CrossRef]

18. Pisello, A.L.; Castaldo, V.L.; Piselli, C.; Fabiani, C.; Cotana, F. How peers' personal attitudes affect indoor microclimate and energy need in an institutional building: Results from a continuous monitoring campaign in summer and winter conditions. *Energy Build.* **2016**, *126*, 485–497. [CrossRef]

19. Dutton, S.; Shao, L. Window opening behaviour in a naturally ventilated school. In Proceedings of the Simbuilt 2010, New York, NY, USA, 11–13 August 2010; pp. 260–268.

20. Zhang, Y.; Bai, X.; Mills, F.P.; Pezzey, J.C. Rethinking the role of occupant behavior in building energy performance: A review. *Energy Build.* **2018**. [CrossRef]

21. Sveby. *Brukarindata Undervisningsbyggnader version 1.0*; Svebyprogramet: Stockholm, Sweden, 2016; Available online: http://www.sveby.org (accessed on 14 November 2019).

22. Moran, P.; Hajdukiewicz, M.; Goggins, J. Achieving Nearly Zero Energy Buildings-A Lifecycle Assessment Approach to Retrofitting Buildings. In Proceedings of the Advanced Building Skins, At Graz, Austria, 23–24 April 2015.

23. Raftery, P.; Keane, M.; O'Donnell, J. Calibrating whole building energy models: An evidence-based methodology. *Energy Build.* **2011**, *43*, 2356–2364. [CrossRef]

24. International Standard ISO 13370:2017. Thermal performance of buildings—Heat transfer via the ground—Calculation methods. International Organization for Standardization: Geneva, Switzerland, 2017.

25. Ashrae, G. Guideline 14-2002. In *Measurement of Energy and Demand Savings*; American Society of Heating, Ventilating, and Air Conditioning Engineers: Atlanta, Georgia, 2002.

26. Steen Englund, J.; Akander, J.; Björling, M.; Moshfegh, B. Assessment of Airflows in a School Building with Mechanical Ventilation Using Passive Tracer Gas Method. In *Mediterranean Green Buildings & Renewable Energy*; Sayigh, A., Ed.; Springer: Cham, Switzerland, 2017; pp. 619–631. [CrossRef]

27. Mitech Instrument. Mitec SatelLite-TH. Available online: http://www.mitec.se/sv/produkter_dataloggrar_kompaktloggrar.html (accessed on 11 November 2019).

28. IMI Hydronic. TA-SCOPE Instrument. Available online: https://www.imi-hydronic.com/sites/EN/international/products/balancing-control-actuators/measuring-tools/instruments/TA-SCOPE/170c70d5-5229-4e20-8058-1501fdfbca15 (accessed on 10 November 2019).

29. Gemini Data Loggers. Tinytag Energy logger. Available online: https://www.geminidataloggers.com/data-loggers/tinytag-energy-data-logger (accessed on 7 November 2019).

30. Davis Instruments. Vantage Pro. Available online: https://www.davisinstruments.com/weather-monitoring/ (accessed on 11 November 2019).

31. Fuji Electric. Portable Flowmeter Fuji Portaflow X. Available online: https://www.coulton.com (accessed on 13 November 2019).

32. Salazar Navalón, P. Evaluation of Heat Losses from a Domestic Hot Water Circulation System. MSC Thesis, University of Gävle, Gävle, Sweden, 2015.

33. Björling, M.; Akander, J.; Steen Englund, J. On Measuring Air Infiltration Rates Using Tracer Gases in Buildings with Presence Controlled Mechanical Ventilation Systems. In Proceedings of the Indoor Air 2016, 14th International Conference of Indoor Air Quality and Climate, Ghent, Belgium, 3–8 July 2016.

34. Lundström, L. Shiny weather data. Available online: https://rokka.shinyapps.io/shinyweatherdata/ (accessed on 5 May 2019).

35. Equa Simulation AB. *Validation of IDA Indoor Climate and Energy 4.0 build 4 with respect to ANSI/ASHRAE Standard 140-2004*; EQUA Simulation Technology Group: Stockholm, Sweden, 2010.

36. Equa Simulation AB. *Validation of IDA Indoor Climate and Energy 4.0 with respect to CEN Standard EN 15255-2007 and EN 15265-2007*; EQUA Simulation Technology Group: Stockholm, Sweden, 2010.

37. Kropf, S.; Zweifel, G. *Validation of the Building Simulation Program IDA-ICE According to CEN 13791 "Thermal Performance of Buildings–Calculation of Internal Temperatures of a Room in Summer Without Mechanical Cooling–General Criteria and Validation Procedures"*; Hochschule fur Technik+ Architektur Luzern: Horw, Switzerland, 2001.

38. Loutzenhiser, P.; Manz, H.; Maxwell, G. *Empirical validations of shading/daylighting/load interactions in building energy simulation tools*; Technical Report; Swiss Federal Laboratories for Materials Testing and Research: Duebendorf, Switzerland; Iowa State University: Ames, IA, USA, 2007.

39. Moosberger, S. *IDA-ICE CIBSE-validation, Test of IDA Indoor Climate and Energy version 4.0 according to CIBSE TM33, issue 3*; Hochschule fur Technik+ Architektur Luzern: Horw, Switzerland, 2007.

40. International Standard ISO 10211:2007. *Thermal Bridges in Building Construction–Heat Flows and Surface Temperatures–Detailed Calculations*; International Organization for Standardization: Geneva, Switzerland, 2007.

41. International Standard ISO 16000-8:2007. *Indoor air—Part 8: Determination of local mean ages of air in buildings for characterizing ventilation conditions*; International Organization for Standardization: Geneva, Switzerland, 2007.

42. Levin, P. Klimatdatafiler Sveby SMHI 1981-2010. Svebyprogrammet 2016. Available online: www.sveby.org (accessed on 22 August 2019).

43. Taesler, R.; Andersson, C.J.E. Method for solar radiation computations using routine meteorological observations. *Energy Build.* **1984**, *7*, 341–352. [CrossRef]

44. National Board of Housing Building and Planning. Boverkets Byggregler—föreskrifter och allmänna råd, BBR 26—BFS 2018:4. Available online: https://www.boverket.se/ (accessed on 4 December 2019).

45. Forum för Energieffektivt byggande. FEBY 18—Kravspecifikation för energieffektiva byggnader- bostäder och lokaler. Available online: https://www.feby.se/ (accessed on 4 December 2019).

Review

Evaluation of School Building Energy Performance and Classroom Indoor Environment

Jitka Mohelníková [1], Miloslav Novotný [1] and Pavla Mocová [2,*]

[1] Faculty of Civil Engineering, Brno University of Technology, 602 00 Brno, Czech Republic; mohelnikova.j@fce.vutbr.cz (J.M.); novotny.m@fce.vutbr.cz (M.N.)
[2] Faculty of Forestry and Wood Technology, Mendel University, 613 00 Brno, Czech Republic
* Correspondence: pavla.mocova@mendelu.cz

Received: 21 April 2020; Accepted: 12 May 2020; Published: 15 May 2020

Abstract: Existing building stock represents potential for energy saving renovations. Energy savings and indoor climate comfort are key demands for sustainable building refurbishment. Especially in schools, indoor comfort is an extremely important issue. A case study of energy consumption in selected school buildings in temperate climatic conditions of Central Europe region was performed. The studied buildings are representatives of various school premises constructed throughout the last century. The evaluation was based on data analysis of energy audits. The goal was aimed at assessment of the school building envelopes and their influence on energy consumption. One of the studied schools was selected for detailed evaluation. The school classroom was monitored for indoor thermal and visual environments. The monitoring was performed to compare the current state and renovation scenarios. Results of the evaluation show that the school buildings are highly inefficient even if renovated. Indoor climate in classrooms is largely influenced by windows. Solar gains affect interior thermal stability and daylighting. Thermal insulation quality of building envelopes and efficient solar shading systems appear to be fundamental tasks of school renovation strategies.

Keywords: energy performance; energy audits; school buildings; indoor climate

---

1. Introduction

Current trends are aimed at the improvement in building energy performance and sustainability [1,2]. Modern construction methods facilitate more energy conscious buildings compared to old building stocks [2–4]. Newly designed buildings comply with top-level energy efficiency [5,6]. Nevertheless, building refurbishment represents a rational approach towards sustainable development [7,8]. Energy conscious renovation became one of the promising ways to reduce carbon dioxide emissions from buildings [9,10]. Existing buildings in the EU region represent massive potentials for energy savings [1,2]. Energy renovation has utmost importance for the residential and public buildings sectors. These buildings exert high energy consumption for heating and ventilation or cooling systems as well as artificial lighting.

Especially school buildings have been objects of energy saving interests. Considerable numbers of programs are focused on school buildings performance [11–13]. Zero emissions and high quality of indoor environment were topical tasks of the project "School of the Future", within the frame of the EU programme [14]. The aim of the project was to communicate examples of efficient buildings under different European climates.

Strategies for energy consumption [15–21] and thermal comfort assessments [22–26] were developed for educational buildings.

Energy efficiency, together with indoor climate comfort, are key features of good practice school performances in accordance with sustainable architecture principles [27,28]. Building envelopes play a substantial role in the built environment [29,30]. Particularly windows influence indoor thermal and visual comfort [31–33].

The abovementioned overview shows the importance of energy efficiency and indoor climate conscious attitudes [34] towards educational buildings. A study focused on the assessment of selected representatives of school buildings in temperate climatic conditions was performed to evaluate their energy saving potentials in accordance with recent trends of highly energy efficient schools in Central Europe [35].

2. Materials and Methods

A case study aimed at an analysis of energy consumption in selected school buildings in the Central Europe region was performed. The studied buildings are representatives of various school premises. The evaluation was based on the analysis of data from energy audits for specification of current state of building envelopes and their influence on energy consumption. One of the studied schools was selected for detailed evaluation. The school classroom was monitored for indoor thermal and visual climate.

2.1. Analysis of Energy Audits

The selected schools [36] were studied for their energy consumption. They are representatives of an existing school building stock of construction styles throughout the last century. Many schools are very old and some of them are listed as historically protected premises. Most of them are energy inefficient. Thermal insulation improvements and windows retrofit are required for upgraded efficiency. The case study is focused on analysis of the buildings' energy performance and renovation potentials.

The school buildings were selected in accordance with the building site locality and climatic conditions, (Figure 1 and Table 1). The buildings' construction is commonly represented by solid brick (70%) and ceramic block (18%) masonry systems or reinforced concrete prefabricated technology (12%). Central gas heating is widely used in the schools, while electric heating has minor application. The school buildings' energy consumption is spent on heating and hot water services (80%), artificial lighting and electric appliances (12%) and auxiliary energy demands (8%).

Figure 1. Map with localities of the studied school buildings 1 to 18, (map: https://geology.com/world/czech-republic-satellite-image.shtml).

Table 1. Buildings location, geometry and heat losses.

| Building | Year of Construction (Renovation) | Locality | | | Building Geometry | | Building Envelope | |
		Altitude [m]	Latitude Longitude [°]	Aver. Winter Temperature [°C]	Building Volume [m³]	Gross Floor Area/Heated Volume [m⁻¹]	Heat Loss [kW]	Average U Value [W m⁻² K⁻¹]
1	1900 (2000)	248	49.763969 N 17.180405 E	4.1	3264	0.63	90	0.907
2	1950	210	49.456479 N 17.450230 E	3.9	23,278	0.40	373.1	1.27
3	1994	272	49.038646 N 17.814872 E	3.9	19,255	0.4	303.6	1.136
4	1931	272	49.038646 N 17.814872 E	3.9	3448	0.47	61.1	0.954
5	1890	334	49.458565 N 18.056868 E	3.8	1938	0.574	50.1	0.97
6	1984	304	49.6819311 N 18.3673219 E	3.7	23,955	0.45	434	1.087
7	1949	336	49.712716 N 13.204605 E	3.6	3491	0.43	88.7	1.23
8	1978	387	49.443259 N 13.248114 E	3.7	1471	0.85	61.1	0.95
9	1937	520	49.7161561 N 13.9473069 E	3.5	7163	0.47	102	0.90
10	1930	440	49.4248869 N 13.8817589 E	3.7	11,427	0.503	276.5	1.05
11	1988	225	50.289161 N 14.824512 E	3.8	5288	0.40	99.4	0.82
12	1960	188	48.9.07468 N 16.775371 E	4.5	4140	0.48	82.6	1.18
13	1929	450	50.129276 N 16.499965 E	3.6	6670	0.40	160.6	1.21
14	1967	179	49.059797 N 17.495850 E	3.6	3326	0.46	78	1.33
15	1980	675	49.908449 N 17.211115 E	3.1	4131	0.68	130	1.02
16	1887 (1962)	378	49.303454 N 14.158029 E	3.7	21,777	0.37	380.6	1.1
17	1980	334	49.820923 N 18.262524 E	3.6	22,423	0.25	424.5	0.76
18	1894 (2014)	280	50.655668 N, 14.724856 E	2.9	17,244	0.24	396	0.72

2.2. School Classroom Thermal and Daylight Evaluation

One of the studied buildings was selected for detail evaluations—building 18 (Table 1). The main goal was thermal and daylight assessment of the school classrooms. The school building constructed in 1894 is listed as a historically protected premise (Figure 2) [37]. The three-storey building has a solid brick masonry and roof truss load-bearing structures. The ground floor is dedicated to the main entrance and school facilities as well as management and administrative departments. Educational rooms are on the first and second floor.

Figure 2. Historical building of primary school, locality Mimoň, CZ [37].

The school building maintenance has been limited by design obstructions like many other historical premises. The building was renovated in 2014 under an architectural preservation review. The renovation was mainly aimed at the building envelope. Window and facade retrofit was completed. It was not allowed to change the historical style of the facade and for this reason, external walls were thermally insulated from the interior side.

The school has spacious side lit classrooms. One of them, located on the second floor, was selected for the evaluation. Firstly, the classroom thermal assessment was performed for the current state and compared with renovation scenarios. Secondly, the classroom daylight evaluation was carried out. Daylight illuminance and luminance measurements were taken and completed with a daylight simulation study.

2.2.1. Thermal Evaluation

The thermal study was focused on the building facade and its influence on indoor climate. The external wall thermal transmittance and condensation risks were studied. The evaluation was carried out for three variations:

- current state: wall with 8 cm of thermal insulation on the interior side;
- renovation scenario I: wall with 15 cm of thermal insulation on the interior side;
- renovation scenario II: wall with 15 cm of thermal insulation on the exterior surface.

The software Teplo [38] was used for the evaluation. The software is intended for fundamental analyses of building constructions like thermal resistance R (m^2KW^{-1}) and U-value (W m^{-2} K^{-1}) calculations, temperature profiles, interstitial condensation analysis and specification of annual balance of condensed and evaporated amount within building constructions in accordance with standard methodology of ISO 6946 and ISO 13788 [39,40] and other standard requirements [41].

Boundary conditions for the thermal evaluation according to [41] are following:

- design outdoor temperature −13.0 °C, locality—GPS: 50.655668 N, 14.724856 E;
- design indoor air temperature 21.0 °C (classroom);
- design relative humidity of outdoor air 84.0%;
- design relative humidity of indoor air 55%.

Two-dimensional temperature distribution of the wall details at window jambs in the current state and design variations with internal and external thermal insulation were simulated using the software Area [38]. The software is dedicated to complex thermal analyses of building construction details for specification of potential thermal bridges in two dimensional stationary heat transfer and water vapour diffusion simulations in accordance with standard methodology of ISO 10211 [42].

A study of the influence of the renovated facade on indoor thermal stability was performed [41,43]. The thermal stability was selected as an indicator of indoor thermal climate. Thermal stability of the classroom was simulated for winter and summer season conditions using the software Stabilita [38]. The internal temperature drop during a heating lapse in winter and summer indoor temperature rise were calculated. The goal of the evaluation is to compare influence of windows on the indoor thermal environment.

2.2.2. Daylight Evaluation

Natural light influences indoor climate and wellbeing. Especially in educational buildings, visual comfort is crucial. Daylight positively affects students' alertness and health [44]. Reviews of the importance of daylighting in schools [45,46] show relationships between the occupants' responses and natural lighting. The positive impact of daylight in classrooms was proven [47,48]. Daylighting in educational buildings has been a topical task of professional projects [49–51]. Extensive surveys of US schools in different climate conditions were performed [52]. The EU programme [53] promotes daylight integration for high performance indoor environment in schools. These activities are in agreement with the main principles of sustainable development [54,55] and architectural design strategies [56–59] as well as standard recommendations [60–62].

Daylight Measurements

Daylight illuminance was monitored in the classroom. The intention was to study the classroom visual environment under the most characteristic daylight conditions throughout the year. The measurement time was limited due to the accessibility schedule into the classroom. Measurements were performed without pupils' occupancy over weekends in March 2017.

The classroom is a spacious place of floor area 11.9 × 7.63 m and clearance height 4 m. It has three big windows, of width about 2 m and height 2.5 m (Figures 3 and 4). The daylight measurements were taken for a set of sixty points on a working plane. The plane is located 0.85 m over the classroom floor level (Figure 4). The set of points is positioned in distance of 1 m around the room perimeter. Spacing of the points is 1.10 m by 1.126 m. Simultaneously with the interior measurements, the external illuminance was monitored on unshaded horizontal plane outdoors. Sky luminance was also studied.

Figure 3. View into the classroom [37].

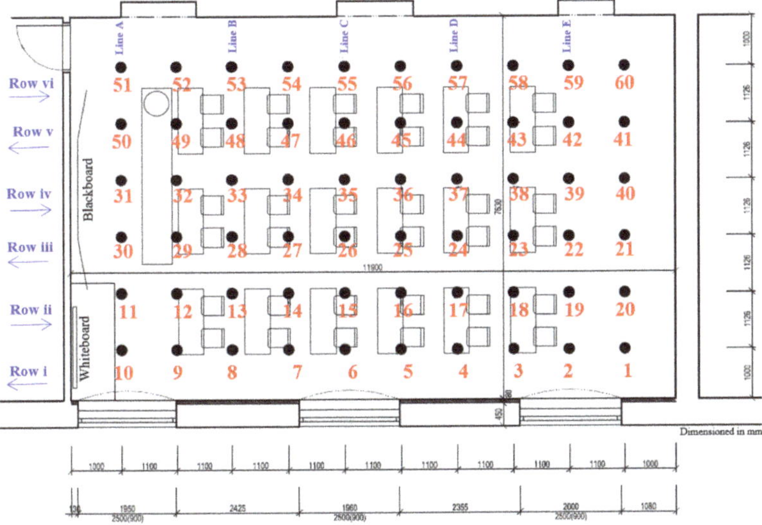

Figure 4. Plan of the classroom with points 1 to 60 on the working plane (0.85 m over the floor).

The daylight illuminance was measured using two calibrated illuminance meters Testo 545 (calibration 2017). One illuminance meter was taken for interior measurements and the second device measured external horizontal illuminance. Laser rangefinder Bosch GLM 50 C Professional was used to set points on the working plane. Tripods were also used to stabilize the illuminance meter at the required height. The external illuminance sensor was positioned on the roof. The measuring instruments were synchronized. Outdoor and indoor illuminance values were measured at the same time. The illuminance data processing was carried out in MS Excel and Statistica software [63].

Furthermore, monitoring of the surface luminance was taken in the classroom visual field. Luminance Meter LS-100 Konica Minolta is used for measurements.

Daylight Simulations

The classroom visual environment was also studied on the basis of daylight simulations. The current state daylighting is compared with designed variations. The design state is represented by renovation scenario II with two variations of window glazing (double or triple glass units). Daylight simulations were run in software Daylight Visualizer [64] for the following parameters:

- light reflectance ρ [-] of the classroom surfaces in current state (resp. designed state): floor finishing 0.35 (resp. 0.5), wall surfaces 0.7 (resp. 0.9), ceiling 0.84 (resp. 0.9).
- window glass light transmittance τ [-]: double glazed units 0.81, triple glazed units 0.73.
- the south-east orientation of the classroom windows (Figure 5).

Daylight simulations were run for an annual balance of internal horizontal illuminance under two sky models [65]:

- CIE clear sky model to simulate sunlight conditions.
- CIE overcast sky model for consideration of the most unfavourable daylight situation.

The balance was simulated for the 21st day of every month and daytime 12:00.

Finally, the classroom daylighting was also simulated for designed variations and compared with standard requirements according to EN 17037 [61] as follows:

- Daylight illuminance simulated for the clear sky model on 21st June, at 12:00 was compared with target illuminance 300 lux.
- Daylight factor simulation for the overcast sky model was compared with target daylight factor $D_T = 2\%$.

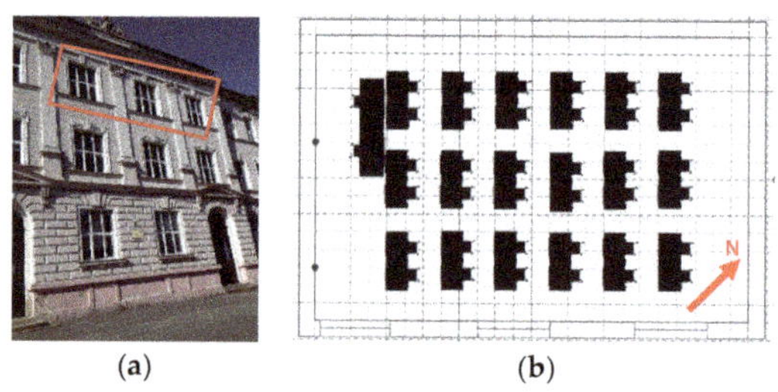

(a)   (b)

Figure 5. Orientation of windows in the daylight model. (**a**) Photograph of a part of the school building facade (the classroom windows are on the 2nd floor). (**b**) Geometric model of the classroom (Daylight Visualizer).

3. Results

The above mentioned energy auditing evaluation achieved notable results for the analysis of the school buildings efficiency. Outputs of indoor thermal and daylight evaluations in the selected representative of school classrooms show potential problems in the current state and give an overview about some renovation scenarios.

3.1. Results of the Energy Audits Analysis

The analysis of selected school buildings energy audits gives an overview about their envelopes and their influence on heat losses. Total heat transmission and ventilation losses of buildings vary from 61.1 kW to 424.5 kW. Total annual energy consumption in the schools is between 265 and 3305 GJ per year. Percentage of heat transmission losses of their building envelopes are following:

- 23.6% to 57.0% of external walls (U = 0.57 to 1.83 W m^{-2} K^{-1}).
- 18.3% to 36.0% of roofs (U = 0.36 to 1.50 W m^{-2} K^{-1}).
- 17.4% to 55.3% of windows (U = 2.30 to 3.50 W m^{-2} K^{-1}), external doors (U = 3.50 to 6.50 W m^{-2} K^{-1}).

The annual consumption of energy in dependence of the building volume is between 23.7 and 64.7 kWh.m^{-3} per year and consumption of energy for heating and domestic hot water vary from 19.8 to 61.7 kWh.m^{-3} per year. The heating energy consumption is quite high in the school buildings compared to demands for low energy buildings (less than 50 kWh/m^2 per year) and passive houses (less than 15 kWh/m^2 per year) [5] (Figure 6).

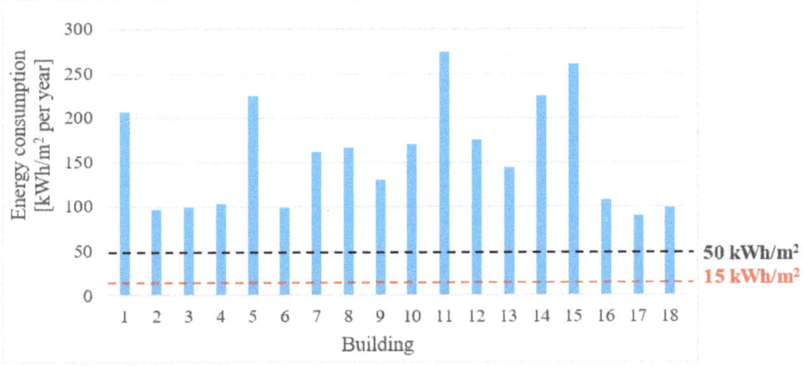

Figure 6. Comparison of heating energy consumption in studied school buildings 1 to 18.

3.2. Results of the Thermal Evaluation

Results of the external wall hygro-thermal evaluation are presented in Figure 7. The current wall (U = 0.43 W m^{-2} K^{-1}) and renovation scenarios (U = 0.29 W m^{-2} K^{-1}) with internal and external thermal insulation were studied. The figure shows schemes of vapour pressure distribution within the wall and specification of potential condensation regions. An annual balance of condensed/evaporated amount inside of the wall is summarized in graphs. It is clear that the interstitial condensation is fully eliminated in the wall with external thermal insulation (renovation scenario II).

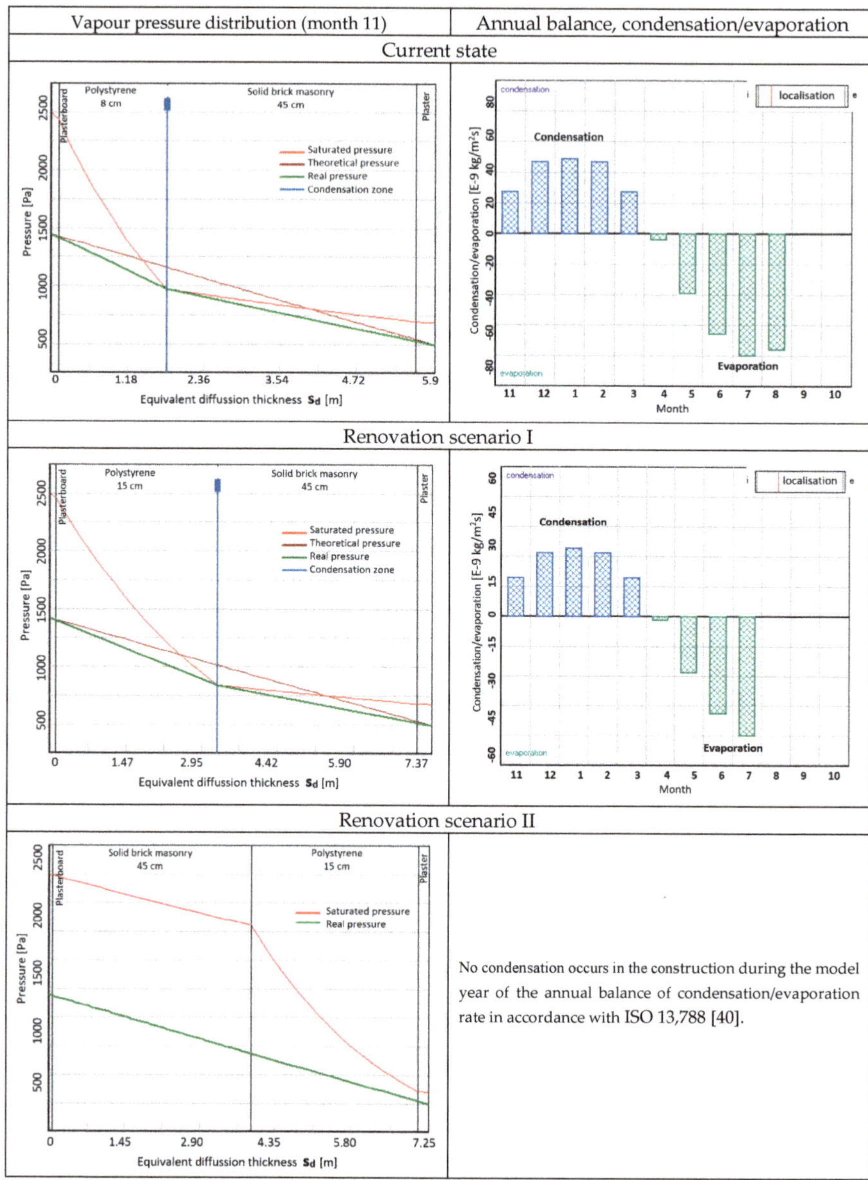

Figure 7. Vapour pressure distribution within the external wall and annual balance of condensation/evaporation rate according to ISO 13788 [40].

Simulation outputs of two-dimensional temperature distribution of the window jamb details in the current state and in the two renovation scenarios are shown in Figure 8. It is obvious that the variation of the wall with external thermal insulation represents better temperature distribution and more convenient design solution.

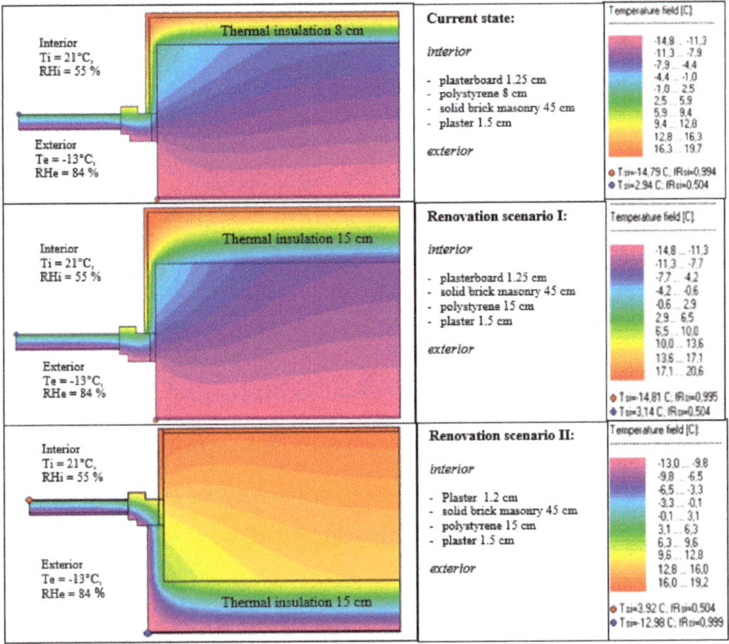

Figure 8. Temperature distribution of the external wall detail at the window jamb.

Thermal stability of the classroom was evaluated for winter and summer season conditions. The indoor temperature drop during a 24 h heating lapse in winter does not vary significantly when comparing the existing state and two renovation scenarios (Figure 9). It is because of the relatively small facade area compared to the big volume of the classroom. Three large windows of southeast orientation represent massive solar gains and indoor temperature rise in summer seasons (Table 2). It could bring about overheating problems during intensive solar shining periods. Solar gains can affect indoor visual discomfort. For this reason, daylighting in the classroom was also evaluated for clear sky conditions.

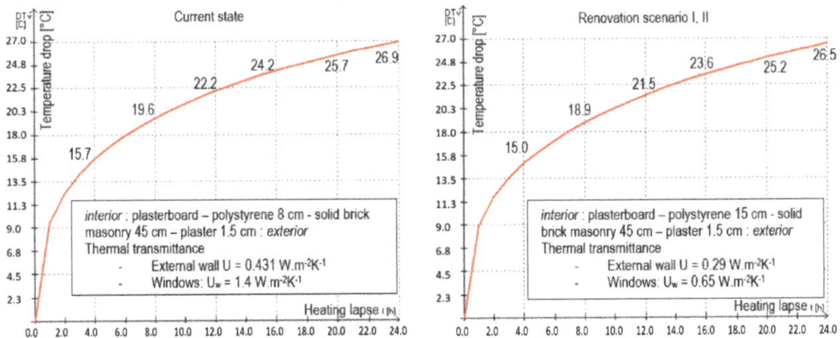

Figure 9. Results of assessment of the classroom thermal stability in winter.

Table 2. Results of assessment of the classroom thermal stability in summer.

Current State		Renovated Scenario I, II	
Total solar transmittance of window	g = 0.65	Total solar transmittance of window	g = 0.53
Solar gains	2141.64 W	Solar gains	1746.26 W
Heat gain through facade	4857.30 W	Heat gain through façade	3961.7
Heat ventilation loss		Heat ventilation loss	
(for ventilation rate 0.5 h^{-1})	−32.78 W	(for ventilation rate 0.5 h^{-1})	−32.78 W
Total heat gain	**6966.16 W**	**Total heat gain**	**5675.24 W**
Max. indoor temperature rise per day	**16.5 °C**	**Max. indoor temperature rise per day**	**14.7 °C**

3.3. Daylight Study Resultss

The classroom daylighting was analysed for illuminance and luminance measured data and daylight simulation outputs.

3.3.1. Measured Data Analysis

Data from illuminance measurements on the horizontal working plane in the classroom (from Figure 4) are summarized in Figures 10 and 11. The daylight level is reduced with the distance from windows. It is obvious that big differences in illuminance are in Row i close to the window line. The illuminance differences are minimized in the room depth. Higher illuminance values in positions of Line A and B at Row i are influenced due to the glossy whiteboard surface located near the big window (Figure 3). It is also obvious from Figure 11, where illuminance is intensified at points 8, 9 and 10 of Row i. Illuminance at points 2, 6 and 10 is increased because of windows positions. Illuminance difference between window/wall positions is reduced in the room depth—Rows iii to vi.

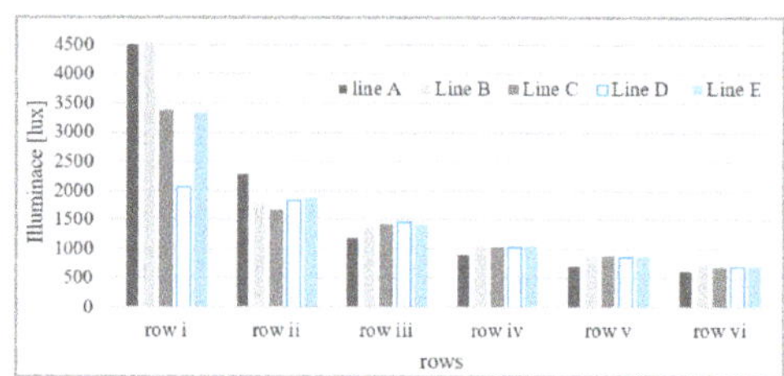

Figure 10. Classroom daylight illuminance—dependence on distance from windows.

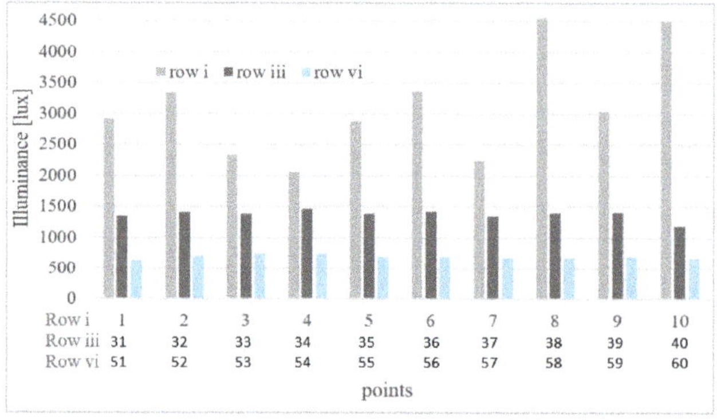

Figure 11. Daylight illuminance: window line (row i), middle (row iii), room depth (row vi).

The internal daylight illuminance measurement was carried out for simultaneous external horizontal illuminance monitoring (Figure 12). Sky luminance was also controlled during the illuminance measurement time. The luminance data recorded for four different directions (1 to 4) and elevation angles 0°, 15°, 45° and 90° are summarized in Table 3.

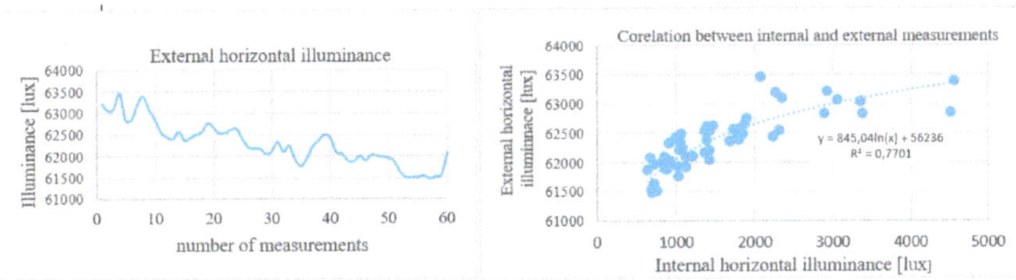

Figure 12. External horizontal illuminance measurements.

Table 3. Sky luminance data (4th March 2017).

Luminance [cd/m²]		$L_{horizon}$	$L_{15°}$	$L_{45°}$	L_{zenith}
time	**Elevation Angle**	**0°**	**15°**	**45°**	**90°**
	direction 1	1923	890	8337	1716
	direction 2	961	4906	3393	1727
11:45	direction 3	1897	1634	1143	2065
	direction 4	5437	2978	2148	1880
	direction 1	4883	7204	1422	1514
	direction 2	1081	2066	1070	1333
12:50	direction 3	824	2309	1752	1244
	direction 4	1187	257	811	1299

Results of the luminance distribution monitored in the classroom are shown in Figure 13. Sunny sky conditions were selected for the monitoring. The classroom windows are often affected by intensive solar radiation. It is clear that surfaces with significantly different luminance in the visual field of the classroom desks cause visual discomfort. Luminance distribution varies from 50 cd/m² of the blackboard to window glass luminance 9534 cd/m². The ratio of minimal/maximal luminance values is then 1:191, which is not in agreement with visual comfort recommendation for max 1:10 ratio [62]. Extremely high luminance ratio results in glare problems. Shading blinds must be activated on the classroom windows (Figure 3). However full shading activation could minimise indoor daylighting. An installation of efficient shading systems is recommended for dynamic reduction in solar transmittance during intensive solar shining periods.

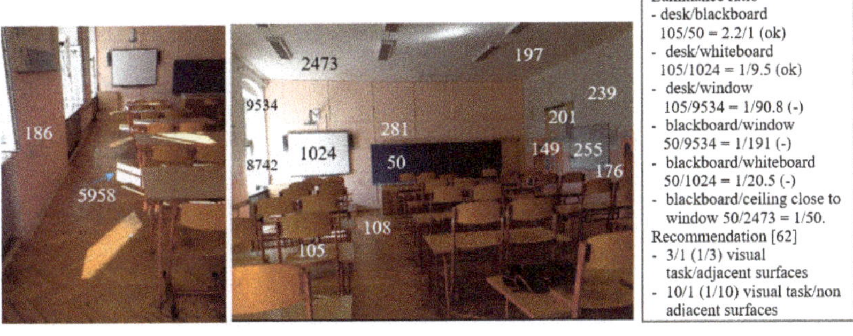

Luminance ratio
- desk/blackboard
  105/50 = 2.2/1 (ok)
- desk/whiteboard
  105/1024 = 1/9.5 (ok)
- desk/window
  105/9534 = 1/90.8 (-)
- blackboard/window
  50/9534 = 1/191 (-)
- blackboard/whiteboard
  50/1024 = 1/20.5 (-)
- blackboard/ceiling close to
  window 50/2473 = 1/50.
Recommendation [62]
- 3/1 (1/3) visual
  task/adjacent surfaces
- 10/1 (1/10) visual task/non
  adjacent surfaces

Figure 13. Classroom surfaces luminance [cdm⁻²].

3.3.2. Daylight Simulation Outputs

The annual balance of the classroom illuminance on the working plane shows monthly mean, median, minimum and maximum values calculated for 21st day of every month and daytime 12:00 under clear and overcast sky conditions (Figure 14). Illuminance can be increased more than 300 lux in the case of clear sky compared to cloudy sky conditions.

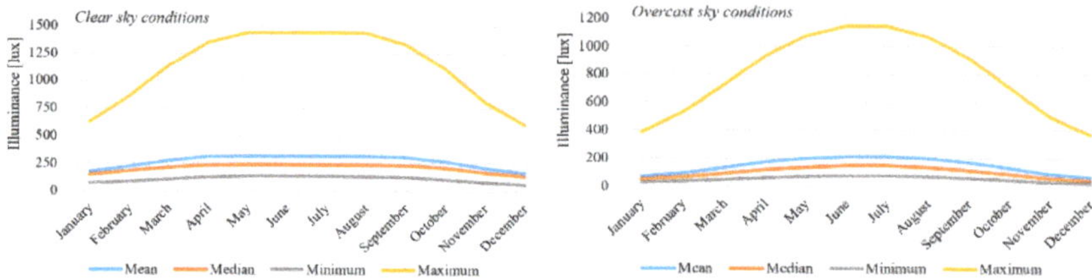

Figure 14. Annual balance of the classroom illuminance for clear and overcast sky conditions.

Illuminance [lux] level on the horizontal working plane is shown in colour iso-lines in Figure 15 for the clear sky simulation on 21st June at 12:00. Mean, median, minimum and maximum illuminance and uniformity values in the current and designed state of the classroom (renovated scenario II with two variations of window glazing transmittance) are compared with standard target values [61,62] for desks min 500 lux and desk surroundings min 300 lux. Despite high illuminance levels, a part of the classroom does not comply with target illuminances.

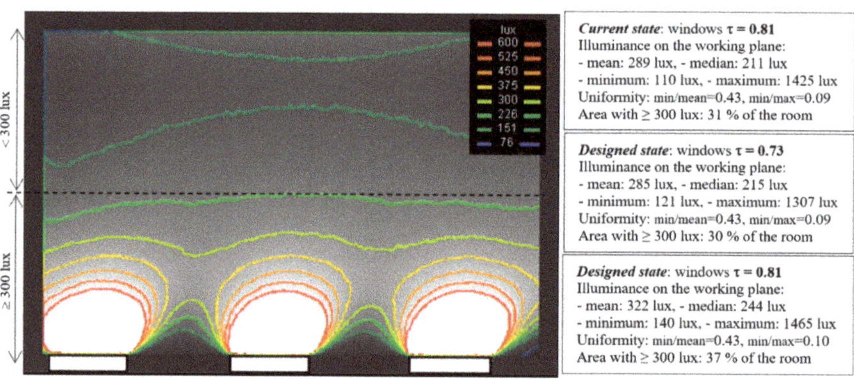

Figure 15. Working plane daylight illuminance, CIE clear sky (21st June, 12:00).

Daylight factor DF [%] distribution on the working plane was also simulated for the CIE overcast sky conditions. The daylight factor was controlled with a target daylight factor in accordance with EN 17037 [61]. The target daylight factor $D_T = 2\%$ is specified for target indoor illuminance 300 lux and external diffuse horizontal illuminance 14,900 lux (locality Prague, Czech Republic). Minimum 50% of the classroom working plane should have daylight level at least adequate to the target daylight factor [61]. However, the daylight factor simulations show that the side lit classroom is too large to comply with the mentioned standard requirements (Figure 16).

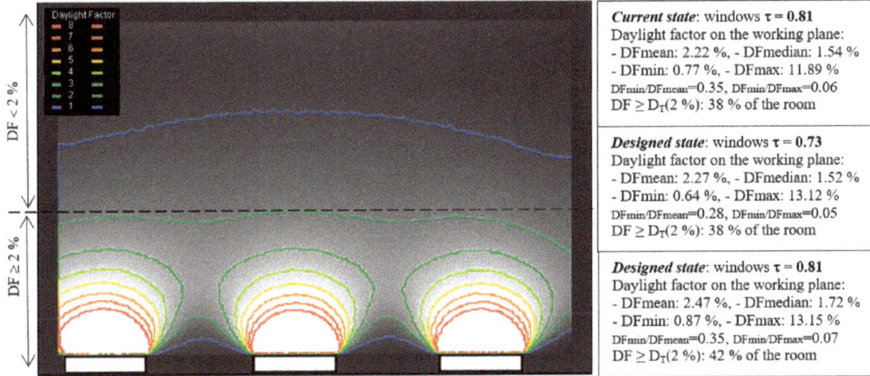

Current state: windows $\tau = 0.81$
Daylight factor on the working plane:
- DFmean: 2.22 %, - DFmedian: 1.54 %
- DFmin: 0.77 %, - DFmax: 11.89 %
DFmin/DFmean=0.35, DFmin/DFmax=0.06
DF $\geq$ D$_T$(2 %): 38 % of the room

Designed state: windows $\tau = 0.73$
Daylight factor on the working plane:
- DFmean: 2.27 %, - DFmedian: 1.52 %
- DFmin: 0.64 %, - DFmax: 13.12 %
DFmin/DFmean=0.28, DFmin/DFmax=0.05
DF $\geq$ D$_T$(2 %): 38 % of the room

Designed state: windows $\tau = 0.81$
Daylight factor on the working plane:
- DFmean: 2.47 %, - DFmedian: 1.72 %
- DFmin: 0.87 %, - DFmax: 13.15 %
DFmin/DFmean=0.35, DFmin/DFmax=0.07
DF $\geq$ D$_T$(2 %): 42 % of the room

Figure 16. Daylight factor simulation outputs for the CIE overcast sky model.

4. Discussion

The analysis of heating energy demands in existing old school buildings shows the importance of renovations for their energy efficiency. Indoor comfort demands are fundamental for schools and they have even priority over energy saving requirements. The thermal and daylight evaluation results show that renovation improvements could be sometimes counter-productive from the indoor comfort point of view. The facade renovation with additional thermal insulation system and new windows reduces heat transmission losses but also solar gains and daylighting. Building envelopes can be affected by interstitial condensation, particularly constructions that are insulated from the interior side (Figure 6). The renovation scenario of the brick wall with external thermal insulation appears to be the most convenient design solution (Figures 7 and 8).

Simulation outputs of the temperature drop during the winter heating lapse show that the part of the insulated facade has not desired effect on indoor thermal stability of the such voluminous classroom (Figure 9). Results display that better thermal insulation quality of windows have not significant influence on the winter thermal stability. Solar gains through big glazed areas cause potential overheating problems during intensive solar shining periods, especially in summer seasons (Table 2).

Windows are key elements influencing indoor climate comfort. The simulations show that the classroom windows are not of great importance for heating energy reduction. However, they are relevant to cooling needs, which is in agreement with design principles [32,33]. It indicates priority of design optimisation of window glazing/shading parameters in dependence on their size, position and orientation with regards to the indoor environment [59].

Window solar gains also influence indoor visual comfort. Daylight illuminance level was controlled in the classroom (Figure 4). Daylight measurements display that the big southeast-oriented glazed areas inflict uneven daylight distribution (Figures 10–12). It is also obvious from increased luminance rates in the classroom visual field (Figure 13). Noticeable variations of interior luminance cause visual discomfort. For example, surfaces in the visual field between the desk and the blackboard. In addition, the positioning of the white glossy board near the window and next to the blackboard is not convenient (Figures 3 and 13). Application of grey boards would be a more reasonable solution in this case.

Daylight simulations show that despite big windows, the voluminous classroom is not properly daylit. The design variation, with additional facade insulation and triple glazed windows, represents daylight level reduction compared to the current room daylighting (Figures 15 and 16) (designed state, $\tau = 0.73$). Activation of shading blinds could also cause discomfort (Figure 3). Special shadings that transmit diffusive skylight and reflect direct solar radiation [66] can be recommended for schools.

Application of light guiding systems and anidolic ceilings [67] could bring better daylight uniformity in large classrooms.

5. Conclusions

Existing school premises represent massive potentials for energy savings. The studied school buildings were renovated in several stages and time periods. However, in many cases, they do not perform the desired energy efficiency. Renovations of old buildings are often limited to partial improvements. Nevertheless, the holistic approach towards a comfortable and energy efficient indoor environment appears to be the most convenient strategy, especially for school buildings renovation. This strategy is in agreement with temporary trends of sustainable buildings development. Complex retrofits focused on building envelopes and their window/shading systems together with the installation of efficient technical systems for heating, ventilation, cooling and artificial lighting could bring desirable effects.

Author Contributions: Conceptualization and data analysis, P.M., J.M.; supervision, M.N. All authors have read and agreed to the published version of the manuscript.

Funding: This research received no external funding. The APC was funded by the Institute of Building Structures, Faculty of Civil Engineering, Brno University of Technology.

Acknowledgments: This article has been worked out under the project No. LO1408 "AdMaS UP—Advanced Materials, Structures and Technologies", supported by Ministry of Education, Youth and Sports of the Czech Republic under the "National Sustainability Programme I". School energy audits were used with permission of the State Environmental Fund of the Czech Republic.

Conflicts of Interest: Authors declare no conflict of interest.

References

1. Directive (EU) 2018/844 of the European Parliament and of the Council of 30 May 2018. Amending Directive 2010/31/EU on the Energy Performance of Buildings and Directive 2012/27/EU on Energy Efficiency. Available online: https://eur-lex.europa.eu/legal-content/EN/TXT/?toc=OJ%3AL%3A2018%3A156%3ATOC& uri=uriserv%3AOJ.L_.2018.156.01.0075.01.ENG (accessed on 15 January 2020).

2. Energy Transition of the EU Building Stock. Available online: https://www.openexp.eu/sites/default/ files/publication/files/Reports/energy_transition_of_the_eu_building_stock_full_report.pdf (accessed on 15 January 2020).

3. Pérez-Lombard, L.; Ortiz, J.; Pout, C. A review on buildings energy consumption information. *Energy Build.* **2008**, *40*, 394–398. [CrossRef]

4. Butala, V.; Novak, P. Energy consumption and potential energy savings in old school buildings. *Energy Build.* **1999**, *29*, 241–246. [CrossRef]

5. Mahdavi, A.; Doppelbauer, E. A performance comparison of passive and low-energy buildings. *Energy Build.* **2010**, *42*, 1314–1319. [CrossRef]

6. Mihai, M.; Tanasiev, V.; Dinca, C.; Badea, A.; Vidu, R. Passive house analysis in terms of energy performance. *Energy Build.* **2017**, *144*, 74–86. [CrossRef]

7. Gohardani, N.; Björk, F. Sustainable refurbishment in building technology. *Smart Sustain. Built Environ.* **2012**, *1*, 241–252. [CrossRef]

8. Leung, B.C.-M. Greening existing buildings [GEB] strategies. *Energy Rep.* **2018**, *4*, 159–206. [CrossRef]

9. Zimmermann, M. School Building Renovation for Sustainable Second Life, in IEA ECBCS Annex 50 Prefabricated Systems for Low Energy Renovation of Residential Buildings. Building Renovation Case Studies, Empa, Duebendorf. 2011. Available online: https://annex53.iea-ebc.org/Data/publications/EBC_ Annex_50_Documented_CaseStudies.pdf (accessed on 14 January 2020).

10. Erhorn-Kluttig, H.; Mørck, O. Energy-Efficient Renovation of Educational Buildings, PEB Exchange, Programme on Educational Building 2005/9. OECD Publishing. Available online: https://ideas.repec. org/p/oec/eduaaa/2005-9-en.html (accessed on 17 January 2020).

11. Burman, B.; Kimpian, J.; Mumovic, J. Building Schools for the Future: Lessons Learned from Performance Evaluations of Five Secondary Schools and Academies in England. *Front. Built Environ.* 2018. Available online: https://www.frontiersin.org/articles/10.3389/fbuil.2018.00022/full (accessed on 17 January 2020).

12. Advanced Energy Design Guide for K-12 School Buildings. Achieving 50% Energy Savings toward a Net Zero Energy Building. American Society of Heating, Refrigerating and Air-Conditioning Engineers. 2011. Available online: file:///D:/Downloads/Energy_Design_Guide_for_Schools_k_12_LEE.pdf (accessed on 14 January 2020).

13. Schools. Carbon Trust. 2012. Available online: https://www.carbontrust.com/media/39232/ctv019_schools.pdf (accessed on 14 January 2020).

14. School of the Future. Towards Zero Emission with High Performance Indoor Environment. EU 7th Framework Programme—EeB-ENERGY, Project Number: 260102. The Project Final Report. Available online: http://www.ectp.org/fileadmin/user_upload/documents/E2B/School_of_the_future/1_School_of_the_Future_Final_public_report.pdf (accessed on 7 May 2020).

15. Raatikainen, M.; Skön, J.; Leiviskä, K.; Kolehmainen, M. Intelligent analysis of energy consumption in school buildings. *Appl. Energy* **2016**, *165*, 416–429. [CrossRef]

16. Erhorn, H.; Mroz, T.; Mørck, O.; Schmidt, F.; Schoff, L.; Thomsen, K.L. The Energy Concept Adviser—A tool to improve energy efficiency in educational buildings. *Energy Build.* **2008**, *40*, 419–428. [CrossRef]

17. Hong, S.; Paterson, G.; Mumovic, D.; Steadman, P. Improved benchmarking comparability for energy consumption in schools. *Build. Res. Inf.* **2014**, *42*, 47–61. [CrossRef]

18. Agdas, D.; Srinivasan, R.S.; Frost, K.; Masters, K.J. Energy use assessment of educational buildings: Toward a campus-wide sustainable energy policy. *Sustain. Cities Soc.* **2015**, *17*, 15–21. [CrossRef]

19. Najjar, M.K. Integrating parametric analysis with building information modelling to improve energy performance of construction projects. *Energies* **2019**, *12*, 1515. [CrossRef]

20. Becker, R.; Goldberger, I.; Paciuk, M. Improving energy performance of school buildings while ensuring indoor air quality ventilation. *Build. Environ.* **2007**, *42*, 3261–3276. [CrossRef]

21. Ascione, F.; Bianco, N.; De Masi, R.F.; Mauro, G.M.; Vanoli, G.P. Energy retrofit of educational buildings: Transient energy simulations, model calibration and multi-objective optimization towards nearly zero-energy performance. *Energy Build.* **2017**, *144*, 303–319. [CrossRef]

22. Zomorodian, Z.S.; Tahsildoost, M.; Hafezi, M. Thermal comfort in educational buildings: A review article. *Renew. Sustain. Energy Rev.* **2016**, *59*, 895–906. [CrossRef]

23. Barbhuiya, S.; Barbhuiya, S. Thermal comfort and energy consumption in a UK educational building. *Build. Environ.* **2013**, *68*, 1–11. [CrossRef]

24. Wang, Y.; Kuckelkorn, J.; Zhao, F.; Liu, D.; Kirschbauma, A.; Zhange, J. Evaluation on classroom thermal comfort and energy performance of passive school building by optimizing HVAC control systems. *Build. Environ.* **2015**, *89*, 86–106. [CrossRef]

25. Chatzidiakou, E.; Mumovic, D.; Summerfield, A.J.; Altamirano, H.M. Indoor air quality in London schools. Part 1: Performance in use. Intelligent Buildings International, Taylor & Francis, Special Issue: Designing Intelligent School Buildings: What do We Know. *Intell. Bulid. Int.* **2015**, *7*, 101–129. Available online: http://www.tandfonline.com/doi/abs/10.1080/17508975.2014.918870#.VSlx-txfWHg (accessed on 12 May 2020).

26. Manfren, M.; Nastasi, B.; Piana, E.; Tronchin, L. On the link between energy performance of building and thermal comfort: An example. *AIP Conf. Proc.* **2019**, *2123*, 020066. [CrossRef]

27. Gelfand, L.; Freed, E.C. *Sustainable School Architecture: Design for Elementary and Secondary Schools*; John Wiley & Sons: Hoboken, NJ, USA, 2010.

28. Gonzalo, R.; Habermann, K.J. *Energy-Efficient Architecture: Basics for Planning and Construction*; Birkhäuser: Basel, Switzerland, 2002.

29. Jaffe, S.B.; Fleming, R.; Karlen, M.; Roberts, S.H. *Sustainable Design Basics*; John Willey and Sons: Hoboken, NJ, USA, 2020.

30. Dubey, K.; Dodonov, A. *Mapping of Existing Technologies to Enhance Energy Efficiency in Buildings in the UNECE Region, Joint Task Force on Energy Efficiency Standards in Buildings*; United Nations Economic Commission for Europe: Geneva, Switzerland, 2018; Available online: https://www.unece.org/fileadmin/DAM/hlm/Meetings/2018/10_03_Geneva/1_Study_on_Mapping_of_EE_technologies_v2.pdf (accessed on 29 April 2020).

31. Windows—Making It Clear: Energy, Daylight and Thermal Comfort (NF78). National House Building Council (NHBC) Foundation, 2017. Available online: https://www.buildup.eu/en/practices/publications/windows-making-it-clear-energy-daylight-and-thermal-comfort-nf78-0 (accessed on 29 April 2020).

32. Persson, M.L.; Roos, A.; Wall, M. Influence of window size on the energy balance of low energy houses. *Energy Build.* **2006**, *38*, 181–188. [CrossRef]

33. Ochoa, C.E.; Aries, M.B.C.; Loenen, E.J.; Hensen, J.L.M. Considerations on design optimization criteria for windows providing low energy consumption and high visual comfort. *Appl. Energy* **2012**, *95*, 238–245. [CrossRef]

34. Wierzbicka, A.; Pedersen, E.; Persson, R.; Nordquist, B.; Stålne, K.; Gao, C.; Harderup, L.E.; Borell, J.; Caltenco, H.; Ness, B.; et al. Healthy indoor environments: The need for a holistic approach. *Int. J. Environ. Res. Public Health* **2018**, *15*, 1874. [CrossRef] [PubMed]

35. Wang, Y.; Du, J.; Kuckelhorn, J.M.; Kirschbaum, A.; Gu, X.; Li, D. Identifying the feasibility of establishing a passive house school in central Europe: An energy performance and carbon emissions monitoring study in Germany. *Renew. Sustain. Energy Rev.* **2019**, *113*, 109256. [CrossRef]

36. Energy Audits of School Buildings, Archive of the State Environmental Fund of the Czech Republic, Prague. Available online: https://www.sfzp.cz/en/ (accessed on 21 April 2020).

37. Primary School Mirova, Mimon, 2015, Mimon. Available online: http://zsamsmirova.cz/ (accessed on 16 January 2020).

38. Svoboda, Z. Thermal Assessment Software, CVUT Prague. Available online: http://kcad.cz/cz/stavebni-fyzika/tepelna-technika/ (accessed on 12 January 2020).

39. *ISO 6946:2017 Building Components and Building Elements—Thermal Resistance and Thermal Transmittance—Calculation Methods*; ISO: Geneva, Switzerland, 2017.

40. *ISO 13788:2012 Hygrothermal Performance of Building Components and Building Elements—Internal Surface Temperature to Avoid Critical Surface Humidity and Interstitial Condensation—Calculation Methods*; ISO: Geneva, Switzerland, 2012.

41. ČSN 730540 Thermal Protection of Buildings (Tepelná Ochrana Budov). Available online: https://www.buildup.eu/en/practices/publications/czech-republic-standard-csn-73-0540-thermal-protection-buildings (accessed on 12 May 2020).

42. *ISO 10211:2017 Thermal Bridges in Building Construction—Heat Flows and Surface Temperatures—Detailed Calculations*; ISO: Geneva, Switzerland, 2017.

43. Šála, J. Interpretation of Provisions of the Czech Standard ČSN 73 0540 Thermal Protection of Buildings for Residential Wooden Houses and Design Recommended Practice. Prague, CZ. 2007. Available online: https://www.competitionline.com/upload/downloads/16xx/1673_3028045_Interpretation_of_thermal_standards.pdf (accessed on 29 April 2020).

44. Shishegar, N.; Boubekri, M. Natural light and productivity: Analyzing the impacts of daylighting on students' and workers' health and allertness. *Int. J. Adv. Chem. Eng. Biol. Sci.* **2016**, *3*, 72–77. Available online: https://iicbe.org/upload/4635AE0416104.pdf (accessed on 12 May 2020).

45. Wu, W.; Ng, E. A review of the development of daylighting in schools. *Light. Res. Technol.* **2003**, *35*, 111–125. [CrossRef]

46. Nocera, F.; Faro, A.L.; Constanzo, V.; Raciti, C. Daylight Performance of Classrooms in a Mediterranean School Heritage Building. *Sustainability* **2018**, *10*, 3705. [CrossRef]

47. Küller, R.; Ballal, S.; Laike, T.; Mikellides, B.; Tonello, G. The impact of light and colour on psychological mood: A cross-cultural study of indoor work environments. *Ergon. Des.* **2006**, *49*, 1496–1507. [CrossRef]

48. Küller, R.; Lindsten, C. Health and behaviour of children in classrooms with and without windows. *J. Environ. Psychol.* **1992**, *12*, 305–317. [CrossRef]

49. Pellegrino, A.; Cammarano, S.; Savio, V. Daylighting for Green schools: A resource for indoor quality and energy efficiency in educational environments. *Energy Procedia* **2015**, *78*, 3162–3167. [CrossRef]

50. Maesano, C.; Annesi-Maesano, I. Impact of lighting on school performance in European classrooms. In Proceedings of the CLIMA 2013—the 12th REHVA World Congress, Aalborg, Denmark, 22–25 May 2016; pp. 1–14.

51. Costanzo, V.; Evola, G.; Marletta, L. A review of daylighting strategies in schools: State of the art and expected future trends. *Buildings* **2017**, *7*, 41. [CrossRef]

52. Heschong, L.; Wright, R.L.; Okura, S. Daylighting impacts on human performance in school. *J. Illum. Eng. Soc.* **2013**, *31*, 101–114. [CrossRef]
53. Renew School. Available online: https://www.renew-school.eu/en/about/ (accessed on 15 January 2020).
54. Energy Efficiency. IEA, 2018. Available online: https://www.iea.org/efficiency2018/ (accessed on 16 January 2020).
55. Guzowski, M. *Daylighting for Sustainable Design*; McGraw-Hill: Hoboken, NJ, USA, 2000.
56. Houck, F. A novel approach on assessing daylight access in schools. *Procedia Econ. Financ.* **2015**, *21*, 40–47. [CrossRef]
57. Piderit, M.B.; Labarca, C.Y. Methodology for assessing daylighting design strategies in classroom with a climate-based method. *Sustainability* **2015**, *7*, 880–897. [CrossRef]
58. Bellia, L.; Pedace, A.; Barbato, G. Lighting in educational environments: An example of a complete analysis of the effects of daylight and electric light on occupants. *Build. Environ.* **2013**, *68*, 50–65. [CrossRef]
59. Zomorodian, Z.S.; Korsavi, S.S.; Tahsildoost, M. The effect of window configuration on daylight performance in classrooms: A field and simulation study. *Int. J. Arch. Urban Plan.* **2016**, *26*, 15–24. Available online: https://www.researchgate.net/publication/305026369_The_effect_of_window_configuration_on_daylight_performance_in_classrooms_A_field_and_simulation_study#fullTextFileContent (accessed on 12 May 2020).
60. IESNA Lighting Handbook. *Illuminating Engineering Society of North America*, 9th ed.; IESNA: New York, NY, USA, 2000.
61. *BS EN 17037:2018 Daylight in Buildings*; BSI: London, UK, 2018.
62. EN 12464-1. *Light and Lighting—Lighting of Work Places—Part 1: Indoor Work Places*; CEN (European Committee for Standardization: Brussels, Belgium, 2011.
63. Statistica Software. Available online: http://www.statsoft.com/Products/STATISTICA-Features (accessed on 10 October 2019).
64. Daylight Visualizer, Velux. Available online: https://www.velux.com/article/2016/daylight-visualizer (accessed on 10 October 2019).
65. *ISO 15469:2004 Spatial Distribution of Daylight—CIE Standard General Sky*; ISO: Geneva, Switzerland, 2004.
66. Mandalaki, M.; Tsoutsos, T. *Solar Shading Systems: Design, Performance and Integrated Photovoltaics*; Springer Briefs in Energy; Springer: Berlin, Germany, 2019. [CrossRef]
67. Scartezzini, J.-L.; Courret, G. Anidolic daylighting systems. *Sol. Energy* **2002**, *73*, 123–135. [CrossRef]

Article

Energy Effects of Retrofitting the Educational Facilities Located in South-Eastern Poland

Anna Życzyńska [1], Zbigniew Suchorab [2,*], Jan Kočí [3] and Robert Černý [3]

[1] Faculty of Civil Engineering and Architecture, Lublin University of Technology, 40 Nadbystrzycka Str., 20-618 Lublin, Poland; a.zyczynska@pollub.pl

[2] Faculty of Environmental Engineering, Lublin University of Technology, 40B Nadbystrzycka Str., 20-618 Lublin, Poland

[3] Faculty of Civil Engineering, Czech Technical University in Prague, Thákurova 7, 166 29 Praha 6, Czech Republic; jan.koci@fsv.cvut.cz (J.K.); cernyr@fsv.cvut.cz (R.Č.)

* Correspondence: z.suchorab@pollub.pl; Tel.: +48-81-538-4756

Received: 7 April 2020; Accepted: 9 May 2020; Published: 13 May 2020

Abstract: One way to decrease the greenhouse gas emissions in the building sector is to improve the building energy performance, which can be mainly achieved by the reduction of energy consumption. In the case of the existing objects, this goal could be achieved by the thermo-modernization of the building partitions and equipment. This article concerns the issue of heat consumption for heating purposes after a comprehensive retrofitting of nine educational buildings (two kindergartens and seven schools) located in south-eastern Poland where both the total efficiency of the heating installation and the thermal insulation of building partitions were improved. The evaluation of the real energy effects was made on the basis of the measurements performed over the 8 year period of operation for each building. The obtained values were compared with the boundary values of the factors that were in force in Poland during the period when all of the buildings were retrofitted. Additionally, they were compared with the results of theoretical calculations included in the energy audits of the example of three objects and an attempt to describe the reasons for the discrepancies was made. All obtained results were discussed with the available literature sources and summarized with the suitable conclusions.

Keywords: retrofitting; thermo-modernization; final energy; primary energy; energy consumption

---

1. Introduction

The treaty on climate change from the conference in Kyoto, which was held in 1997, was ratified by 141 countries and entered into force on February 16, 2005. The countries that ratified it, including Poland, were obliged to reduce their own greenhouse gas emissions, and thus to achieve the overarching goal of environmental protection. This goal can be achieved, among others, by reducing the energy consumption in the building sector, which is one of the most energy-consuming branches of the economy [1–4]. Therefore, energy savings have been sought in this area for a long time, setting increasingly high requirements in terms of the thermal insulation of building partitions and energy efficiency for modernized and newly constructed buildings as well as the technical systems that constitute their equipment [5,6]. The activities to reduce the energy consumption in buildings have been conducted for over 20 years in Poland, and even longer in some other European countries (Germany, Denmark) which have particularly intensified after the introduction of the European Parliament Directive [7]. In the case of the existing buildings, the reduction of energy consumption is achieved through their comprehensive retrofitting, which gives better results than individual improvements [8–10]. In order to achieve the intended goal, work usually begins with the legislative process and implementation of national legal acts in the form of directives, laws and regulations [7,11].

In Poland, one of the first acts that contributed to the promotion of the activities leading to a reduction of heat consumption in the existing buildings was the Act on supporting thermo-modernization projects of 1998 and its executive regulations of 1999 regarding the principles of preparing an energy audit and its verification [12,13]. For the first time, the procedures were formalized to obtain financial support from the state budget for comprehensive thermal modernization of buildings, which resulted in the dynamic development of deep thermal modernization of primarily residential and public buildings [14,15]. Additional financial support for the measures to reduce the energy consumption in the existing buildings was received by Poland after joining the European Union, which resulted in the launch of many thermo-modernization investments, especially in the buildings belonging to the local government units.

In the abovementioned Act [12], limiting the issue to the building and its equipment with technical systems, a thermo-modernization project is defined as an improvement or a set of improvements based on the technical activities as a result of which the energy demand of a building for heating and ventilation as well as hot water preparation is reduced. The most common technical actions taken in buildings concern two areas; the first one is related to the parameters of the building balance shield, i.e., the improvement of thermal insulation of opaque building partitions (most often external walls, roofs, flat roofs, internal partitions between heated and unheated spaces), replacement of windows and doors with tighter ones, characterized by much lower heat transfer coefficient values [16–20]. The second one, covers the improvement of the overall efficiency of the building heating system and the system used to prepare hot water in the building [21–24]. The modernization activities are carried out to improve the technical systems, but often, especially in the buildings where the technical systems are characterized by a significant degree of exploitation and wear, new heating and hot water preparation systems are implemented. Then, the devices using renewable energy sources are often employed. A necessary condition for a properly conducted retrofitting is to adapt the heating system to technical requirements in such a way that the system automatically adapts to the energy demand of a building, depending on the changing parameters of the external and internal environment in the building.

Due to the fact that the financial support for thermo-modernization investments most often comes from the national or EU funds, it is necessary to properly prepare the investment process. One of the elements of this preparation is the performance of an energy audit, which aims to comprehensively analyze the condition of the existing building and determine the optimal, in terms of scope and costs, technical measures to reduce the energy demand of a building and improve the energy efficiency. An audit is often referred to as a technical and economic analysis of a building thermo-modernization; it is a common study required when conducting this type of investment [15,25,26]. This study, among other things, aims to provide the forecast of energy savings resulting from the thermo-modernization measures assuming standard boundary conditions for the indoor and outdoor environment of the building. The energy saving level calculated in the audit is the forecasted or expected effect of measures to improve the energy efficiency. The requirements in terms of heat transfer coefficients of building partitions, as well as the efficiency of technical systems set for buildings after retrofitting are usually the same as for the newly constructed buildings. Retrofitting contributes to achieving measurable energy effects, and thus economic and ecological benefits, as well as improves the thermal comfort in the heated rooms. Often, barriers and restrictions of a legal, technical or financial nature arise when planning a deep thermo-modernization. The assessment of the thermo-modernization energy effects can be made by comparing the measured energy consumption before and after the investment is carried out over a period of several years or heating seasons, taking into account at least the variability of the external environment conditions. For this purpose, it is reasonable to calculate the unit indicators related to $1~m^2$ of the heated building surface. The indicators can also be used to compare and evaluate buildings in terms of the energy quality. A comparative analysis of the level of annual energy consumption in individual years and buildings are the key issues related to the monitoring and rational energy management in the building.

The aim of the research presented in this work was to determine the actual energy effects obtained as a result of comprehensive thermo-modernization in educational buildings. The energy saving levels obtained from the measurements under operating conditions were compared with the predicted results of the theoretical calculations contained in audits available for some of the considered buildings. The analysis was based on measuring the data from long-term exploitation of the buildings before and after retrofitting. Additionally, in three cases theoretical investigation was conducted based on the Polish regulations [26].

2. Materials and Methods

In this work, the subject of analyses involves only the thermal needs of the buildings related solely to heating. All considered energy factors in the presented investigation refer to a year, that is why the annual heat consumption was measured for each building. Using these data, the decrease in heat consumption for heating due to comprehensive retrofitting was calculated, real factors of annual final and non-renewable primary energy consumption were determined, and then they were compared with the values included in the national technical and construction regulations required during the thermo-modernization investment period. At that time, the limit values in these provisions were determined as a function of the building shape factor [27].

The analysis is a case study of 9 educational buildings (hereinafter referred to as SUCs–"system use cases"): two pre-school buildings serving as kindergartens (SUC1 and SUC2) and seven primary school buildings (SUC3–SUC9). The number of students in the examined objects was relatively constant within the considered period of time. In the case of three buildings, the results of energy audits carried out before proceeding with the thermo-modernization investment were available. All buildings are located in a large city of eastern Poland and are supplied with heat from the same provider. None of the objects are equipped with a cooling system. Five buildings are supplied with the heat from a centralized heating system covering the thermal needs for heating and hot water preparation, also four from a centralized system, but only covering the needs for heating purposes (see the Table 1). The heat source for the distribution system is a municipal combined heat and power (CHP) plant operating in a cogeneration system (electricity and heat, coal and gas combustion). For each building, the value of the shape factor was determined, the heated usable area was given, and the year in which the thermo-modernization was carried out was determined. The analysis used the results of annual measurements of heat consumption from an eight-year period, several years before and several years after retrofitting. The measured values were corrected with a factor taking into account the variability of the number of degree-days, which is characteristic for a given year in relation to the number of degree-days specified under standard conditions for a given location using a correction coefficient φ calculated according to the following Formulas (4) and (5). The measurements from heat meters and data for determining the correction factor were obtained from the heat supplier, who manages the centralized heat distribution system in the city and supplies heat to the considered buildings.

In the case of the buildings that drew heat from the heating network, also to cover the needs related to hot water preparation, the heat consumption measurements obtained from the supplier were available in the summer months, in which the heat consumption for heating purposes was zero. In the considered location, deep water was used and seasonal water temperature fluctuations could be neglected, which allowed estimating the annual heat consumption for hot water preparation and to extract the heat consumption for heating purposes according to Formulas (1)–(3).

In all buildings, the retrofitting projects included: insulation of external walls (U_{wall}), roofs (U_{roof}) or flat roofs ($U_{flat\ roof}$), replacement of windows (U_{window}) and external doors (U_{door}), modernization or replacement of central heating installations. The calculated heat transfer coefficients of building partitions after insulation are presented in the Table 1. The assumed values of the average efficiencies of the building heating systems were the following: efficiency of heat generation $\eta_{H,g} = 1.0$ (heat source behind the balance cover of the building), efficiency of heat transfer within the balance cover $\eta_{H,d} = 0.95$, efficiency of heat accumulation $\eta_{H,s} = 1.0$ (systems without heat accumulators), efficiency of heat adjustment and utilization $\eta_{H,e} = 0.90$. These values are assumed as constant coefficients depending on the type and technical condition of the installations, according to the Polish regulations [28]. The data are the same for all buildings due to the identical range of heating system modernization.

Table 1. Basic parameters of the buildings examined for energy consumption.

Object	Construction Year	Technology of Construction	A/V	Heated, Usable Area [m^2]	Retrofitting Year	Energy Needs
SUC1	1968	traditional	0.90	721	2006	heating
SUC2	1981	industrialized	0.80	625	2007	heating + hot water
SUC3	1970	industrialized	0.44	3624	2006	heating
SUC4	1963	traditional	0.43	3458	2007	heating + hot water
SUC5	1985	industrialized	0.50	11,654	2007	heating + hot water
SUC6	1961	traditional	0.40	2855	2008	heating
SUC7	1974	industrialized	0.58	4000	2008	heating
SUC8	1982	industrialized	0.42	9216.2	2004	heating + hot water
SUC9	1983	industrialized	0.42	9216.2	2004	heating + hot water

	Before Retrofitting			After Retrofitting		
Object	U_{walls} [W/m^{2}K]	$U_{roof}/U_{flat\ roof}$ [W/m^{2}K]	U_{window}/U_{door} [W/m^{2}K]	U_{walls} [W/m^{2}K]	$U_{roof}/U_{flat\ roof}$ [W/m^{2}K]	U_{window}/U_{door} [W/m^{2}K]
SUC1	0.95 **	1.28 **	2.6/2.5 **	0.24 **	0.22 **	1.8/1.8 **
SUC2	1.16 ***	0.70 ***	3.0/5.6	0.25 *	0.22 *	1.8/1.8 *
SUC3	1.12	0.85	5.1/3.5	0.25 *	0.22 *	1.8/1.8 *
SUC4	1.16 ***	0.70 ***	5.6/5.6	0.25 *	0.22 *	1.8/1.8 *
SUC5	1.13	0.85	2.6/2.5	0.25*	0.22 *	1.8/1.8 *
SUC6	1.16 ***	0.70 ***	3.0/5.6	0.25 *	0.22 *	1.8/1.8 *
SUC7	1.16 ***	0.70 ***	3.0/5.6	0.25*	0.22 *	1.8/1.8 *
SUC8	1.12 **	1.54/2.37 **	2.8/5.6 **	0.24 **	0.21/0.22 **	1.3/1.3 **
SUC9	1.12 **	1.54/2.37 **	2.8/5.6 **	0.24 **	0.21/0.22 **	1.3/1.3 **

* based on requirements [13]. ** based on energy audits. *** based on requirements during erecting the building. without asterisks—based on archival documentation or manager's information.

Additionally, the authors of the investigation had received the information from the object managers and directors that before retrofitting the buildings were unheated and after retrofitting the parameters of thermal comfort were respected in all institutions.

All buildings were monitored for heat consumption for the period of 8 years (since 2003 until 2010). During this period, thermo-modernization was conducted, which significantly influenced the heat-meter readouts. The years of thermo-modernization are presented in the Table 1 together with other basic technical data of the buildings.

In order to determine the energy effects due to thermo-modernization of buildings, the following algorithm was used (also presented as a block diagram in Figure 1):

1. Collecting the data from the measurements of legalized heat meters under real conditions for eight years (several years before and several after retrofitting).

    - Measurement of the heat consumption for heating purposes in the buildings not equipped with a central hot water installation by means of legalized heat meters installed on the main pipelines of the installation, before the distributors (Q_p, GJ/a);
    - Measurement of the heat consumption in total for the purposes of heating and hot water preparation in the buildings equipped with a central hot water installation supplied from the heating network (Q_p*, GJ/a);
    - Measurement of the heat consumption only for the purposes of hot water preparation in June, July, August, (outside the heating season) in the buildings equipped with a central hot water system supplied from the heating network, used to estimate the heat consumption for hot water;
    - Estimated heat consumption for hot water (Q_W, GJ/a) using one of the two methods:
        - for the objects without summer break (SUC2) according to:

$$Q_W = q_{w,j} \cdot d \tag{1}$$

where: $q_{w,j}$—unit daily heat consumption for hot water calculated from the measurements during three summer months (June, July, August) in the particular year, GJ/day; d—number of days during the year, 365 days/a.

○   for the objects with summer break (July, August) according to:

$$Q_W = q_{w,j}{}^* \cdot d^* + QW^* \tag{2}$$

where: $q_{w,j}{}^*$—unit daily heat consumption for hot water calculated from the measurements in June in the particular year, GJ/day; d^*—number of days during the year without July and August, 303 days/a (10 months); $Q_W{}^*$—measured heat consumption for hot water preparation in July and August of the particular year (during 32 days of the year), GJ.

- evaluation of the heat consumption for heating of the building equipped with the hot water system powered from the heating network (Q'_p, GJ/a):

$$Q'_p = Q_p{}^* - Q_W \tag{3}$$

2. Collecting the data concerning the duration of the heating period and month average temperatures of the outside air.

3. Calculation the number of degree-days for each year covered the with analysis according to the following formula:

$$Sd = \sum (\theta_{int,H} - \theta_{e,m}) \cdot Ld_m \tag{4}$$

where: Sd—number of degree-days calculated for each year, day·K/a; $\theta_{e,m}$—average monthly temperature of outdoor air for the particular year, °C; $\theta_{int,,H}$—temperature of indoor air in the heating zone, established 20 °C; Ld_m—number of heating days in the particular month for each year, day.

4. Calculation of correction factor resulting from the variation of degree-day according to the following dependence:

$$\varphi = \frac{Sd_0}{S_d} \tag{5}$$

where: φ—correction coefficient; Sd_0—number of degree-days in standard year, calculated for the standard year using average month outdoor air temperatures from multi-year measurement and theoretical duration of the heating period (222 days), which equals 3825.2 day·K/a for the location of the analyzed buildings (constant value for the considered location). Table 2 presents values of correction factor calculated using Equation (5).

5. Correction of the measured values to the standard year according to the following formulas:

$$Q_0 = Q_p \cdot \varphi \tag{6}$$

$$Q'_0 = Q'_p \cdot \varphi \tag{7}$$

where: Q_0–Adjusted Annual Energy Consumption (under standard conditions), GJ/a; Q'_0–Adjusted Annual Energy Consumption (under standard conditions) in the buildings with the hot water system powered from the heating network, GJ/a; Q_p–measured annual energy consumption, GJ/a; Q'_p–estimated annual energy consumption in the buildings with the hot water system powered from the heating network, GJ/a.

6. Determination of the Annual Final Energy Factor for Heating (FEF_H) according to the following relation:

$$FEF_H = \frac{1000 \cdot Q_0}{3.6 \cdot A_f} \tag{8}$$

or

$$FEF_H = \frac{1000 \cdot Q'_0}{3.6 \cdot A_f} \tag{9}$$

where: FEF_H—Final Energy Factor for Heating (FEF_H), kWh/(m²·a); A_f—usable heating area of the building, m².

7. Determination of the Annual Primary Energy Factor for Heating (PEF_H) according to the following relation:

$$PEF_H = w_H \, FEF_H \tag{10}$$

where: w_H—coefficient of non-renewable primary energy input assumed for cogeneration as 0.8 according to Polish regulations [27,28].

8. Calculation of the boundary value of the Annual Primary Energy Factor for Heating ($PEF_{H,0}$) in relation to the Building Shape Factor, according to the national (Polish) requirements from the period when thermo-modernization was carried out, according to the relation [27]:

$$PEF_{H,0} = 1.15 \cdot [55 + 90 \cdot (A/V)] \tag{11}$$

where: $PEF_{H,0}$—maximal value of the Annual Primary Energy Factor for Heating (PEF_H), kWh/(m²·a); A/V—Building Shape Factor–ratio between the sums of the areas of building boundaries serving the balance cover and heated volume of the building measured in outer contour, 1/m.

9. Determination of the energy consumption savings according to the following dependences:

$$\Delta Q_{\%,avg} = (Q_{01,avg} - Q_{02,avg})/Q_{01,avg} \cdot 100 \tag{12}$$

$$\Delta Q_{\%,min} = (Q_{01,avg} - Q_{02,min})/Q_{01,avg} \cdot 100 \tag{13}$$

$$\Delta Q_{\%,max} = (Q_{01,avg} - Q_{02,max})/Q_{01,avg} \cdot 100 \tag{14}$$

where: $\Delta Q_{\%,avg}$, $\Delta Q_{\%,min}$, $\Delta Q_{\%,max}$—average, minimal and maximal (respectively) obtained decrease of Annual Energy consumption after thermo-modernization related to the annual value of the average Annual Energy Consumption before thermo-modernization, %; $Q_{01,avg}$—average Annual Energy Consumption before thermo-modernization reduced to the standard conditions, GJ/a; $Q_{02,avg}$, $Q_{02,min}$, $Q_{02,max}$—average, minimal and maximal (respectively) Annual Energy Consumption after thermo-modernization reduced to the standard conditions, GJ/a.

Table 2. Coefficients reflecting the harshness of winter in the particular year of the examination period.

Year	Sd_0	Sd	φ
2003		3938.3	0.971
2004		3714.7	1.030
2005		3844.5	0.995
2006	3825.2	3788.8	1.010
2007		3677.4	1.040
2008		3542.5	1.080
2009		3669.2	1.043
2010		4263.9	0.897

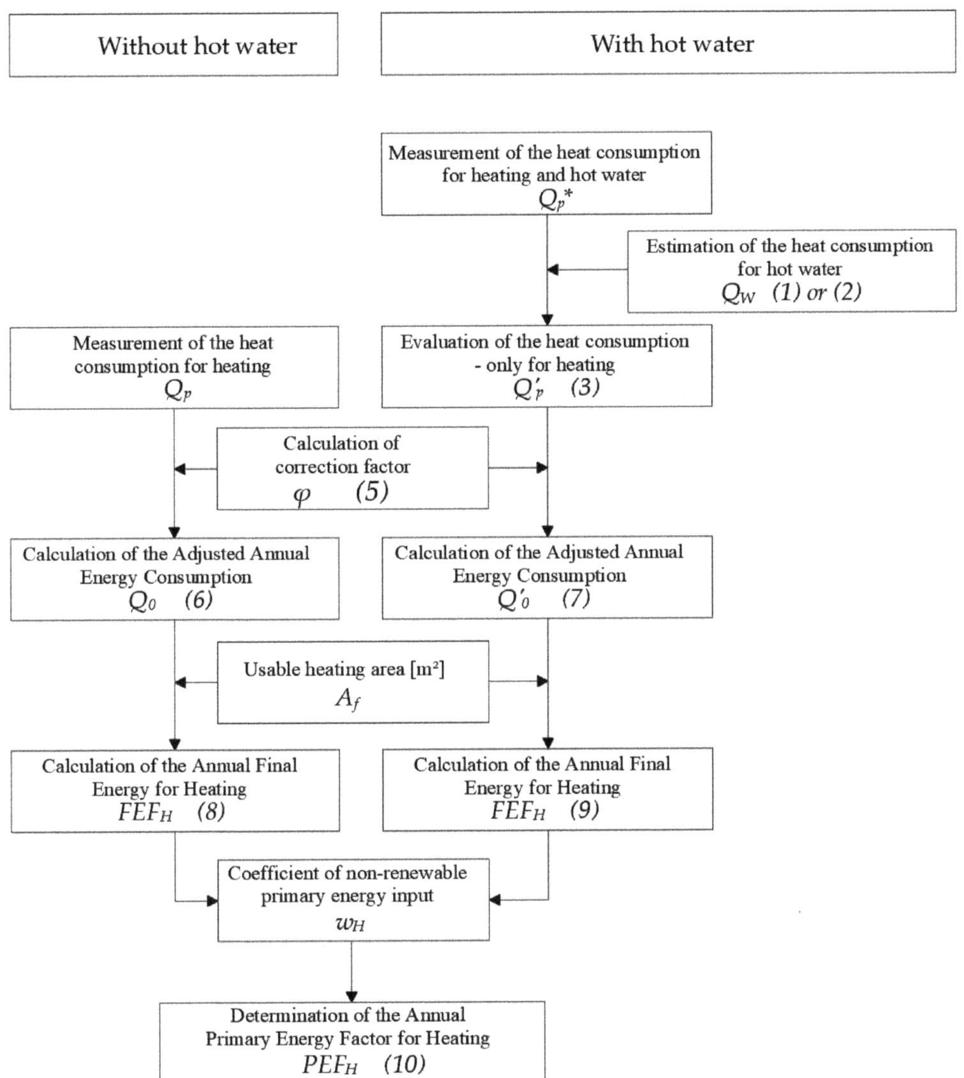

Figure 1. Block diagram of the calculation algorithm (numbers of formulas and explanation of the symbols presented below).

3. Results

3.1. Readouts of Energy Consumption

The readouts of energy consumption in the considered period of time (8 years) were presented in Table 3 for the particular examined objects. The columns present the values of Annual Energy Consumption for heating purposes (Q_p or Q'_p) [GJ/a]. The years when retrofitting was conducted (transition years), are indicated using asterisks.

Table 3. Annual Energy Consumption (Q_p or Q'_p) [GJ/a] for heating purposes in the particular year of the conducted examination read by the heat-meters.

Year	SUC1	SUC2	SUC3	SUC4	SUC5	SUC6	SUC7	SUC8	SUC9
2003	1062	662.98	2153	2592	7550.34	1761.8	2245.9	6273	6531.9
2004	1031.2	650.36	2148.8	2421.4	7323.28	1736	1924.6	4651.5 *	4274.8 *
2005	1032.4	643.6	2189.2	2505.6	7556.95	1768.3	1901.3	4143.9	2536
2006	794.5 *	663	1716.7 *	2347.9	7425.35	1675.5	1968.1	4058.4	3201.4
2007	434.4	546.9 *	1010.8	1683.9 *	4636.12 *	1544.3	1816.2	4056.2	3101.9
2008	441	501.4	1042.25	1146.9	2970.3	1038.3 *	1701.6 *	4064.9	2910.1
2009	434.2	473.8	1071.4	1118.4	3153.72	727.4	1251.1	3742.2	2641.8
2010	500	329.5	1235.8	1295.4	3575.89	993.7	1460.1	3485.5	2313.8

* transition year.

3.2. Calculation of the Thermo-Modernization Efficiency

3.2.1. Evaluation of the Boundary Value of the Annual Primary Energy Factor for Heating

The first stage of the theoretical investigation was to evaluate the boundary value of the Annual Primary Energy Factor for Heating ($PEF_{H,0}$) which was assessed in relation to the Building Shape Factor, serving as a reference value to verify the quality of the thermo-modernization procedure. All data were calculated from the A/V ratio values presented in Table 1, and recalculated using Equation (11). Those data are also presented in Table 4.

Table 4. Evaluation of energy savings and Primary Energy Factor for Heating after retrofitting.

	SUC1	SUC2	SUC3	SUC4	SUC5	SUC6	SUC7	SUC8	SUC9
$PEF_{H,0}$ [kWh/(m^2·a)]	156.4	146.1	108.8	107.6	115.0	104.7	123.3	106.7	106.7

3.2.2. Evaluation of Annual Energy Consumption and Energy Factors for Heating

With the data presented in Table 3, the value of the Adjusted Annual Energy Consumption (Q_0 or Q'_0) [GJ/a] was calculated using Equations (6) and (7), Final Energy Factor for Heating (FEF_H) [kWh/m^{2}a], calculated using Formulas (8) or (9), and Primary Energy Factor for Heating (PEF_H) [kWh/m^{2}a] calculated using Equation (10). All calculated data are presented in Table 5, with the transition years marked using asterisks, similarly to Table 3.

The data presented in Table 2 describing the harshness of winter period by the φ coefficient prove that the conditions during winter period do not influence the heat consumption significantly. In the considered region and period of time, these values varied between 0.897 and 1.08. This means that the difference between both values differs only by 20%, but the average value of φ coefficient for the period of 8 years equals 1.008. It should be also noticed that through most of the examined period (between 2004 and 2009) these differences were even smaller (winters of 2003 and 2010 were more severe); the range of φ varies between 0.095 and 1.08 which means 8.5% in the difference between those values. Nevertheless, the Adjusted Annual Energy Consumption (Q_0) being a measure of standard heating period independent of weather conditions, was used for the comparative purposes. This value is more objective than the Annual Energy Consumption (Q_0), not being influenced by the hardness of the winter period

Table 5. Adjusted Annual Energy Consumption (AECH), Final Energy Factor (FEF) and Primary Energy Factor (PEFH) for heating achieved from the heat meter readouts in the particular year of the research.

	SUC1			SUC2		
Year	Q_0 (Q'_0) [GJ/a]	FEF_H [kWh/m²a]	PEF_H [kWh/m²a]	Q_0 (Q'_0) [GJ/a]	FEF_H [kWh/m²a]	PEF_H [kWh/m²a]
2003	1031.2	397.29	317.83	643.8	286.11	228.89
2004	1062.1	409.21	327.37	669.9	297.72	238.18
2005	1027.2	395.76	316.61	640.4	284.61	227.69
2006	802.4 *	309.16 *	247.32 *	669.6	297.61	238.09
2007	451.8	174.05	139.24	568.8 *	252.79 *	202.23 *
2008	476.3	183.50	146.80	541.5	240.67	192.54
2009	452.9	174.48	139.58	494.2	219.63	175.71
2010	448.5	172.79	138.23	295.6	131.36	105.09

	SUC3			SUC4		
Year	Q_0 (Q'_0) [GJ/a]	FEF_H [kWh/m²a]	PEF_H [kWh/m²a]	Q_0 (Q'_0) [GJ/a]	FEF_H [kWh/m²a]	PEF_H [kWh/m²a]
2003	2090.6	160.24	128.19	2516.8	202.17	161.74
2004	2213.3	169.65	135.72	2494.0	200.34	160.28
2005	2178.3	166.96	133.57	2493.1	200.27	160.21
2006	1733.9 *	132.90 *	106.32 *	2371.4	190.49	152.39
2007	1051.2	80.58	64.46	1751.3 *	140.68 *	112.54 *
2008	1125.6	86.28	69.02	1238.7	99.50	79.60
2009	1117.5	85.65	68.52	1166.5	93.70	74.96
2010	1108.5	84.97	67.97	1162.0	93.34	74.67

	SUC5			SUC6		
Year	Q_0 (Q'_0) [GJ/a]	FEF_H [kWh/m²a]	PEF_H [kWh/m²a]	Q_0 (Q'_0) [GJ/a]	FEF_H [kWh/m²a]	PEF_H [kWh/m²a]
2003	7331.4	174.75	139.80	1710.7	166.44	133.15
2004	7543.0	179.79	143.83	1788.1	173.97	139.18
2005	7519.2	179.22	143.38	1759.5	171.19	136.95
2006	7499.6	178.76	143.00	1692.3	164.65	131.72
2007	4821.6 *	114.92 *	91.94 *	1606.1	156.26	125.01
2008	3207.9	76.46	61.17	1121.4 *	109.10 *	87.28 *
2009	3289.3	78.40	62.72	758.7	73.82	59.05
2010	3207.6	76.45	61.16	891.3	86.72	69.38

	SUC7			SUC8		
Year	Q_0 (Q'_0) [GJ/a]	FEF_H [kWh/m²a]	PEF_H [kWh/m²a]	Q_0 (Q'_0) [GJ/a]	FEF_H [kWh/m²a]	PEF_H [kWh/m²a]
2003	2180.8	151.44	121.15	6091.1	183.59	146.87
2004	1982.3	137.66	110.13	4791.0 *	144.40 *	115.52 *
2005	1891.8	131.37	105.10	4123.2	124.27	99.42
2006	1987.8	138.04	110.43	4099.0	123.54	98.84
2007	1888.8	131.17	104.94	4218.4	127.14	101.72
2008	1837.7 *	127.62 *	102.10 *	4390.1	132.32	105.85
2009	1304.9	90.62	72.49	3903.1	117.64	94.11
2010	1309.7	90.95	72.76	3126.5	94.23	75.39

	SUC9		
Year	Q_0 (Q'_0) [GJ/a]	FEF_H [kWh/m²a]	PEF_H [kWh/m²a]
2003	6342.5	191.16	152.93
2004	4403.0 *	132.71 *	106.17 *
2005	2523.3	76.05	60.84
2006	3233.4	97.46	77.96
2007	3226.0	97.23	77.79
2008	3142.9	94.73	75.78
2009	2755.4	83.05	66.44
2010	2075.5	62.56	50.04

* transition year.

Additionally, the data contained in Table 5 are visualized in Figure 2. The adjusted energy consumption is set with Primary and Final Energy Factors for Heating. The diagrams were extended with the graphical presentation of the boundary value of the Annual Primary Energy Factor for Heating ($PEF_{H,0}$) in relation to the Building Shape Factor serving as a critical value to evaluate the efficiency of retrofitting efforts. The presented data correspond to the values of energy consumption in the Polish educational buildings available in the scientific literature, where the annual heating energy consumption in Polish schools was evaluated to be at the level of 120 kWh/(m²·a) [29,30].

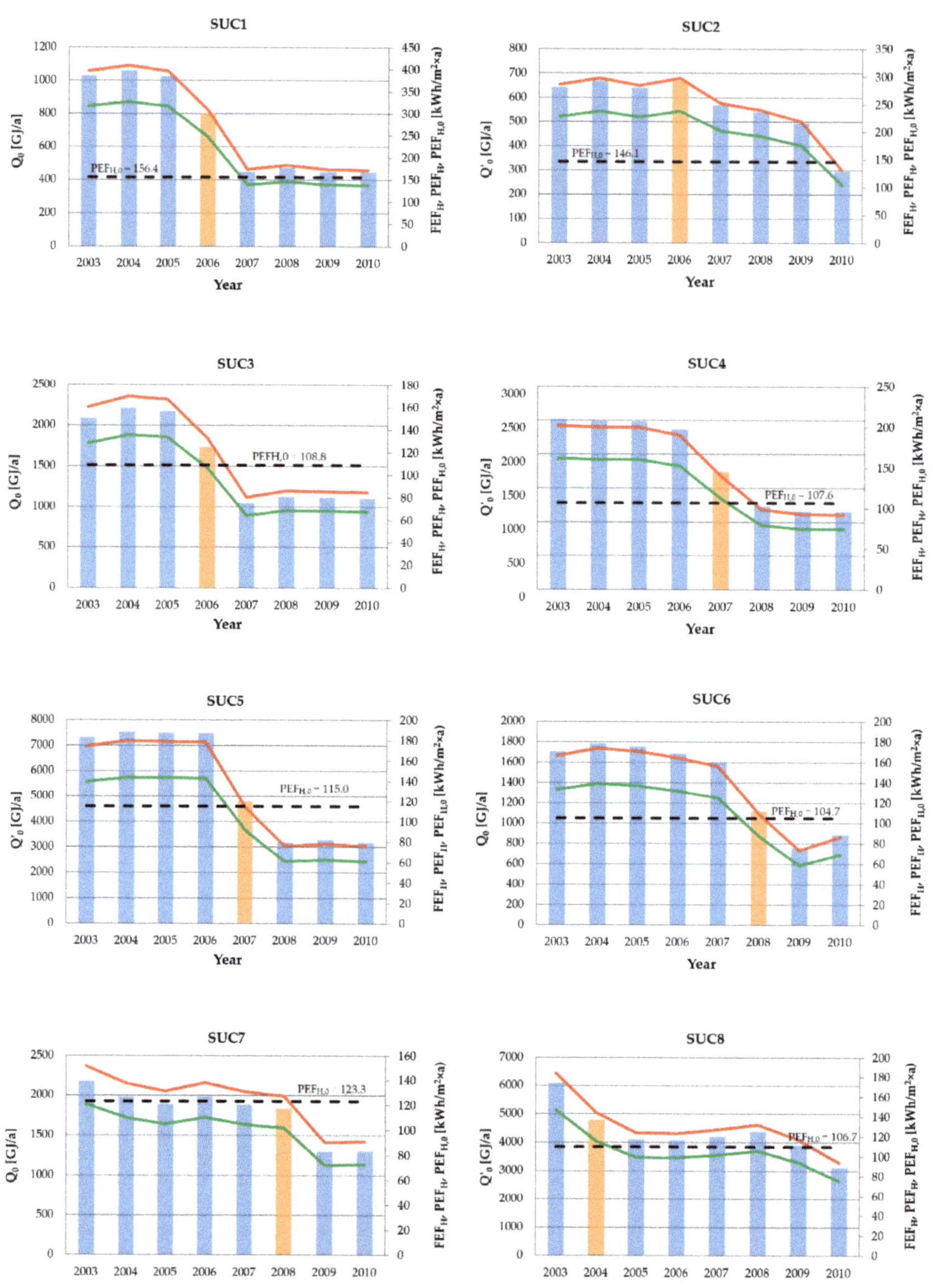

Figure 2. *Cont.*

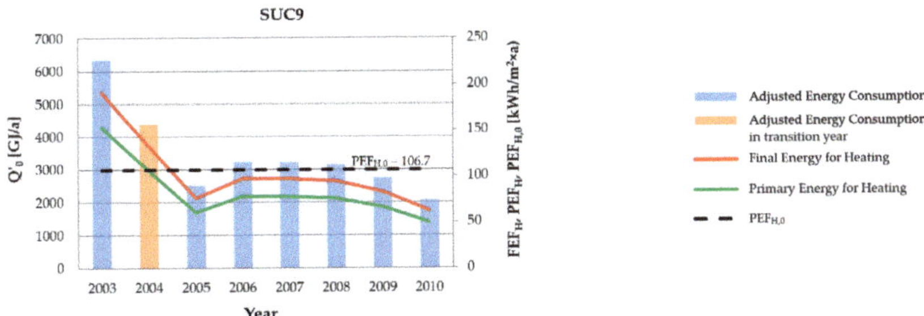

Figure 2. Presentation of changes of the Annual Adjusted Energy Consumption within the investigated period together with Final and Primary Energy Factor for Heating.

4. Discussion

With the data presented in Table 5 and Figure 2, it is clearly visible that, due to the thermo-modernization processes, both the Energy Consumption (Q_p and Q_0) and values of Final and Primary Energy Factors decreased. This confirms the efficiency of this process. In the case of SUC1 with transition year 2006, the decrease of heat consumption after 2006 is clearly visible. Even during the transition year, the heat consumption bar shows the value of 802 GJ/a, which is about 23% lower than the previous building status. Following the completion of the modernization process, the Q_0 value equals about 450 GJ/a, which proves the efficiency of the modernization process. The Final and Primary Energy Factors confirm this trend. From the diagram it can be also noted that the PEF$_H$ value is smaller than the reference PEF$_{H,0}$ value, which means that after this procedure the building fulfills heat standards that became effective in 2009 for that type of object.

In the case of SUC2 (Kindergarden), the efficiency of thermo-modernization is not that visible. It is caused by the prolonged period of the retrofitting procedures, which was conducted for the period of 3 years. This is visible on the diagram as a gradual decrease of the annual heat consumption. The final result of the thermo-modernization is visible in 2010, when the Primary Energy Factor value is below the reference value PEF$_{H,0}$ according to the standards from the discussed period.

In the case of the other buildings that served as Primary Schools (SUC3-9), the energy performance was comparable to SUC1 (Kindergarten), where a partial consumption decrease was visible during the transition year and became more significant afterwards.

For better visualization, the averaged values of Q_0 before and after modernization determined in all tested buildings are summarized in Table 6.

Table 6. Averaged values of the Adjusted Energy Consumption before, during and after the transition year together with comparison of the energy savings.

Object	Q_0 before [GJ/a]	Q_0 Transition Year [GJ/a]	Q_0 after [GJ/a]	$\Delta Q_{\%,avg}$ [%]	$\Delta Q_{\%,min}$ [%]	$\Delta Q_{\%,max}$ [%]
SUC1	1040.2	802.0	457.4	56.0	54.2	56.9
SUC2	651.3	568.5	295.6	54.6	-	-
SUC3	2160.7	1733.9	1100.7	49.1	47.9	51.3
SUC4	2468.8	1751.3	1189.0	51.8	49.8	52.9
SUC5	7473.3	4821.6	3234.9	56.7	56.0	57.1
SUC6	1711.3	1121.4	825.0	51.8	47.9	55.7
SUC7	1986.3	1837.7	1307.3	34.2	34.1	34.3
SUC8	6091.1	4791.0	3976.7	34.7 *	27.9 *	48.7 *
SUC9	6342.5	4403.0	2826.1	55.4 *	49.0 *	67.3 *

* related to only the first year of the monitored period before thermo-modernization (2003).

The averaged values of the Adjusted Heat Consumption for Heating in each case decrease after the thermo-modernization process. During the transition period, the energy consumption was in all cases between the values before and after that period. This can be explained by the fact that the thermo-modernization process was conducted partially, mainly during the non-heating period, and the advantages of the procedure were observed only in December.

The thermo-modernization procedure provided the owners of the buildings with energy savings at a very high level. The average percentage of energy savings from all buildings was equal to 46.8%, which meant that for all considered objects, almost 50% of the measured energy was saved. It must be noticed that the real (measured) energy savings were smaller than those projected by the auditing methodology. This can be explained by the fact that, before thermo-modernization, those buildings were unheated, and after the renovation, the heat comfort of those buildings increased, which required higher energy consumption. Three of the examined objects were additionally investigated theoretically using the auditing methodology. The calculated energy savings expressed in heat consumption would reach the level 70.2%. Averaged heat consumption after thermo-modernization read by heat meters reached in this case 56%. This value is very promising, but significantly below the theoretical estimation. In the case of the SUC8, this difference is more significant: 82.3% of savings from theoretical evaluation but only 34.7% of real readouts. In the case of SUC9, these differences were smaller but still visible—82.4% for theoretical evaluation and 55.4% from heat-meter readouts—after adjusting.

The theoretical energy consumption calculated in audits (Table 7) is significantly higher than the measured adjusted energy consumption (Table 6). After thermo-modernization, the relations between the abovementioned values are reversed. For example, in the case of SUC1, the measured value of heat consumption before thermo-modernization Q_0 was smaller and equal to 1040.2 GJ/a, while the calculated value was higher and equaled 1324.9 GJ/a. On the other hand, after the renovation procedure, the measured value was higher from the calculated one (457.4 GJ/a and 395.0 GJ/a respectively). Similar relations were observed for the primary schools (SUC8 and SUC9). It is supposed that the reason for these differences is the insufficient heating of the rooms and improper ventilation (real ventilation flux was lower than the calculated one). After the thermo-modernization procedure, higher temperatures could occur but also the assumed weakening in heating may be omitted within the theoretical calculations. Irrational energy management in the buildings could also have taken place, if there was no person responsible for monitoring heat consumption or no central heating system control. The abovementioned reasons cause the difference in the measured heat consumption between schools SUC8 and SUC9 (with the same design and shape factor), and the differences in the calculations are due to various assumed weakening factors in the heating system performance. It should be emphasized that a more thorough analysis should be conducted where all objects would be investigated under the operating conditions and (theoretically) energy audits would be calculated.

Table 7. Heat consumption values before and after thermo-modernization according to energy audits (theoretical calculations).

Object	Q before [GJ/a]	Q after [GJ/a]	Energy Savings [%]
SUC1	1324.9	395.0	70.2
SUC8	9373.7	1662.6	82.3
SUC9	8472.4	1487.5	82.4

Additionally, the influence of building shape ratio on Primary Energy Factor for Heating was checked. Figure 3 presents the relation between these two factors characterizing the tested buildings before and after the retrofitting processes. From the diagrams it can be noticed that before the modernization, the values of PEF_H were significantly higher, independent of the building shape. It was observed that in the case of low A/V ratio values before the thermo-modernization, the influence of building shape was not significant and was even negative. Accordingly, it can be interpreted that those buildings were operated differently and maybe some of them could be unheated. In the case of the buildings with a worse (higher) value of A/V ratio, higher energy consumption expressed as Primary Energy Factor for Heating was higher and rising. The discussed relation can be described by the polynomial model presented in Figure 3 with the determination coefficient equal to 0.96.

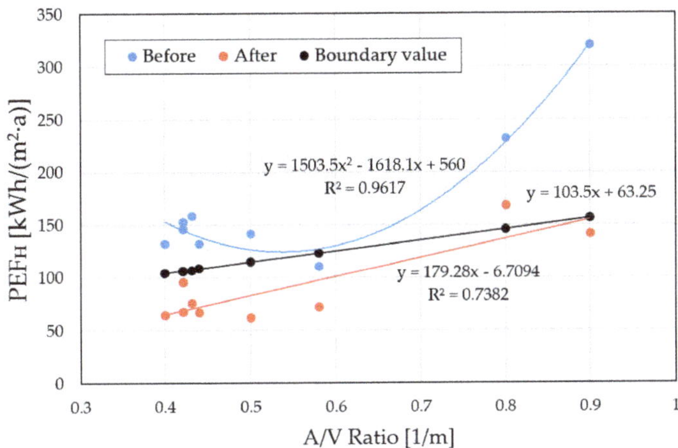

Figure 3. Relation between Primary Energy Factor for Heating evaluated by measurements and A/V ratio of all tested buildings, before and after retrofitting.

On the other hand, the relation between the building shape factor and Primary Energy Factor for Heating after retrofitting is clearer and could be also described by the linear regression model presented in Figure 3. Coefficient of determination equals 0.74 in that case, but there is no visible significant decrease of PEF_H value in the case of the lower values of the A/V ratio.

Additionally, the boundary value of the Annual Primary Energy Factor for Heating ($PEF_{H,0}$) in relation to the Building Shape Factor was presented (black line) to compare the energy consumption before and after the retrofitting procedure. It can be observed that, before thermo-modernization, averaged consumption did not exceed the boundary value (only in the case of the primary school SUC7). In the case of the buildings with a higher value of A/V ratio, the energy consumption was significantly higher. After thermo-modernization, nearly all the buildings had lower value of energy consumption compared to the $PEF_{H,0}$ value. Both relations are represented by the linear regression models and importantly, the y-intercept values differ in value–boundary values of PEF_H in the case of small A/V ratio values and are higher compared to the real ones. On the other hand, the slope is smaller. This can be interpreted that all the examined buildings with small A/V ratio had better energy properties compared to the boundary values coming from the local regulations [27]. In the case of the buildings with more complicated shapes, the energy savings were not that significant and in the case of Kindergarten SUC2, $PEF_{H,0}$ was exceeded even after thermo-modernization.

The analyzed objects indicate that different energy effects can be obtained in the same building under operating conditions, as shown in the Table 6—e.g., the SUC6 building where the minimum value of energy saving is 47.9% and the maximum value 55.7%.

Comparing the obtained results to the literature reports, it can be noticed that similar dependences can be found in the scientific sources. Marrone et al. [30] compared the energy savings resulting from thermo-modernization evaluated by theoretical analysis and from real readouts. The savings from theoretical calculations varied up to 40%, but the savings from real data reached 33%. Comparing the cited results with those presented in Table 6, it should be noticed that average energy savings achieved in seven Polish educational objects were higher and in two cases (SUC7 and SUC8) they were comparable. On the other hand, Mazzola et al. [31] elaborated a method of auditing the energy renovation of the historical buildings based on energy simulations. The differences between the readouts of the real objects and the simulations varied between 1% and 5%.

In their theoretical investigation presented in the article [32], Biserni et al. found that retrofitting a building by changing the existing windows caused the energy savings at the level of 13%, additional thermal insulation of the external walls provided the energy savings at the level of 50.1% and finally, complex retrofitting of the external building envelope covering the exchange of the windows, walls and roof additional thermal insulation, provided energy savings of 70.3%. Those values were similar or smaller by about 10% compared to those calculated in our research and presented in Table 7.

Mancini and Nastasi [33] conducted the simulations of the energy consumption in the residential buildings based on surveys and readouts from the bills for energy use. The investigated objects were modernized for many scenarios, from the elements of the building envelopes through the heating systems. After retrofitting of the envelopes, they achieved average primary energy savings between 8.5% and 30.8%. Retrofitting of the heating systems decreased the energy consumption by about 7%. Combined modernization of the envelopes and heating systems gave energy savings at the level of 30%–35%, depending on the situation. This was still less compared to the results achieved in our investigation (Table 6).

Under the Polish conditions, Zender-Swiecz and Telejko [34] investigated the residential buildings that were retrofitted—renovation relied on the insulation of the building external envelope which sealed the building and had a significant negative influence on the amount of the air delivered to the rooms. That worsened the indoor microclimate causing negative effects on the human health. Similar conclusions were drawn by Gładyszewska-Fiedoruk [35] who investigated kindergartens located in eastern Poland. After retrofitting a building by replacing the old windows with the new tight ones, the ventilation of the building was hindered and the indoor air conditions deteriorated.

Woroniak and Piotrowska-Woroniak [15] examined small church buildings in eastern Poland, using a similar methodology of energy considerations. They achieved between 31% and 66% decrease in the seasonal demand. Życzyńska, in the article [17] described thermo-renovation energy effects of the group of the buildings built in the traditional technology in the 1960s and 1970s. The thermo-renovation concerned only the external walls, roofs and the replacement of some windows. The decrease of energy consumption equaled between 16.3% and 21.5% for the whole group of the objects. The technical requirements expressed as $PEF_{H,0}$ reference factor were not met.

5. Conclusions

A thorough analysis of the obtained results from the heat consumption measurement and theoretical calculations enables us to draw the following conclusions:

- Under the operating conditions, the state calculated theoretically is not obtained. The actual energy consumption decreases are much lower than predicted, which mainly results from the improvement of thermal comfort in the rooms of the building.
- In the cases analyzed, the energy consumption under operating conditions decreased in the range of 34%–56%, which gave an average decrease in heat consumption of 46.8%.
- For better building energy management, the heat consumption for heating and hot water production should be monitored separately.
- In the buildings with identical shape factors and a similar manner of use, different energy effects are obtained under operating conditions.
- The efficiency of thermo-modernization in the case of the buildings with a simple shape was better compared to the objects with a higher A/V ratio.

Author Contributions: Conceptualization, A.Ż. and Z.S.; Formal analysis, A.Ż. and Z.S.; Methodology, A.Ż.; Project administration, A.Ż.; Supervision, J.K. and R.Č.; Validation, J.K. and R.Č.; Writing—original draft, A.Ż. and Z.S.; Writing—review and editing, J.K. and R.Č. All authors have read and agreed to the published version of the manuscript.

Funding: This research was financially supported by Ministry of Science and Higher Education in Poland within the statutory research of scientific units under subvention for science program (FN-61/ILT/2019, FN-70/IŚ/2019) and by the Grant Agency of the Czech Technical University in Prague under project No. SGS19/143/OHK1/3T/11.

Conflicts of Interest: The authors declare no conflict of interest.

References

1. Moretti, E.; Barbanera, M.; Foschini, D.; Buratti, C.; Cotana, F. Energy and environmental performance analysis of biomass-fuelled combined cooling and heating system for commercial building retrofit: An Italian case study. *Energy Procedia* **2016**, *101*, 376–383. [CrossRef]
2. European Union. EU Energy in Figures, Statistical Pocketbook. 2012. Available online: https://op.europa.eu/en/publication-detail/-/publication/4fbba65f-6690-4c3f-878a-e4ce0bc3515c (accessed on 4 May 2020).
3. Saheb, Y.; Bodis, K.; Szabo, S.; Ossenbrink, H.; Panev, S. *Energy Renovation: The Trump Card for the New Start for Europe*; Publications Office of the European Union: Luxembourg, 2015.
4. Tettey, U.Y.A.; Dodoo, A.; Gustavsson, L. Effect of different frame materials on the primary energy use of a multi storey residential building in a life cycle perspective. *Energy Build.* **2019**, *185*, 259–271. [CrossRef]
5. Aste, N.; Del Pero, C. Impact of domestic and tertiary buildings heating by natural gas in the Italian context. *Energy Policy* **2012**, *47*, 164–171. [CrossRef]
6. Le Truong, N.; Gustavsson, L. Costs and primary energy use of heating new residential areas with district heat or electric heat pumps. *Energy Procedia* **2019**, *158*, 2031–2038. [CrossRef]
7. Directive 2010/31/EU of the European Parliament and of the Council of 19 May 2010 on the Energy Performance of Buildings. Available online: https://eur-lex.europa.eu/legal-content/EN/TXT/?uri=CELEX%3A32010L0031 (accessed on 4 May 2020).
8. Piccardo, C.; Dodoo, A.; Gustavsson, L.; Tettey, U.Y.A. Retrofitting with different building materials: Life-cycle primary energy implications. *Energy* **2020**, *192*, 116648. [CrossRef]
9. Pardo-Bosch, F.; Cervera, C.; Ysa, T. Key aspects of building retrofitting: Strategizing sustainable cities. *J. Environ. Manag.* **2019**, *248*, 109247. [CrossRef]
10. Cho, K.H.; Kim, S.S. Energy performance assessment according to data acquisition levels of existing buildings. *Energies* **2019**, *12*, 1149. [CrossRef]
11. Directive 2002/91/EC of the European Parliament and of the Council of 16 December 2002 on the Energy Performance of Buildings. Available online: https://eur-lex.europa.eu/legal-content/EN/TXT/?uri=CELEX%3A32002L0091 (accessed on 4 May 2020).
12. Act of Polish Government of 18 December 1998 on Supporting Thermomodernization Projects. Available online: http://prawo.sejm.gov.pl/isap.nsf/DocDetails.xsp?id=WDU19981621121 (accessed on 4 May 2020).
13. Regulation of the Polish Minister of Interior and Administration of 30 April 1999 Concerning Scope and Form of the Energy Audit as Well as Algorithms for Assessing the Profitability of Thermo-Modernization Project and the Patterns of the Energy Audit Cards. Available online: http://prawo.sejm.gov.pl/isap.nsf/DocDetails.xsp?id=WDU19990460459 (accessed on 4 May 2020).
14. Wojdyga, K.; Chorzelski, M. Chances for Polish district heating systems. *Energy Procedia* **2017**, *116*, 106–118. [CrossRef]
15. Woroniak, G.; Piotrowska-Woroniak, J. Effects of pollution reduction and energy consumption reduction in small churches in Drohiczyn community. *Energy Build.* **2014**, *72*, 51–61. [CrossRef]
16. Mavromatidis, L.E.; Bykalyuk, A.; Lequay, H. Development of polynomial regression models for composite dynamic envelopes thermal performance forecasting. *Appl. Energy* **2013**, *104*, 379–391. [CrossRef]
17. Życzyńska, A. The heat consumption and heating costs after the insulation of building partitions of building complex supplied by the local oil boiler room. *Eksploatacja Niezawodnosc* **2014**, *16*, 313–318.
18. El-Darwish, I.; Gomaa, M. Retrofitting strategy for building envelopes to achieve energy efficiency. *Alex. Eng. J.* **2017**, *56*, 579–589. [CrossRef]
19. Fan, Y.; Xia, X. A multi-objective optimization model for energy-efficiency building envelope retrofitting plan with rooftop PV system installation and maintenance. *Appl. Energy* **2017**, *189*, 327–335. [CrossRef]
20. Song, X.; Ye, C.; Li, H.; Wang, X.; Ma, W. Field study on energy economic assessment of office buildings envelope retrofitting in southern China. *Sustain. Cities Soc.* **2017**, *28*, 154–161. [CrossRef]

21. Fabrizio, E.; Ferrara, M.; Monetti, V. Smart heating systems for cost-effective retrofitting. In *Cost-Effective Energy Efficient Building Retrofitting, Materials, Technologies, Optimization and Case Studies*, 1st ed.; Pacheco-Torgal, F., Granqvist, C., Jelle, B., Vanoli, G., Bianco, N., Kurnitski, J., Eds.; Elsevier: Amsterdam, The Netherlands, 2017. [CrossRef]

22. Wang, Q.; Holmberg, S. Combined retrofitting with low temperature heating and ventilation energy savings. *Energy Procedia* **2015**, *78*, 1081–1086. [CrossRef]

23. Życzyńska, A. The influence of heating and hot water system on the energy performance of a building. *Rynek Energii* **2009**, *6*, 46–54.

24. Życzyńska, A. The primary energy factor for the urban heating system with the heat source working in association. *Eksploatacja Niezawodnosc* **2013**, *15*, 458–462.

25. Regulation of the Polish Minister of Infrastructure of 6 November 2008 Concerning the Methodology for Calculating the Energy Performance of the Building and a Residential Unit or Part of a Building Which Is the Whole Technical-Independent Utility and the Preparation and Presentation of Certificates of Energy Performance. Available online: http://isap.sejm.gov.pl/DetailsServlet?id=WDU20082011240 (accessed on 4 May 2020).

26. Regulation of the Polish Minister of Infrastructure of 17 March 2009 Concerning Scope and Form of the Energy Audit and the Repair Audit, Design Audits Cards, as Well as Algorithms for Assessing the Profitability of Thermo-Modernization Project. Available online: http://isap.sejm.gov.pl/DetailsServlet?id=WDU20090430347 (accessed on 4 May 2020).

27. Regulation of the Minister of Infrastructure of 6 November 2008. Amending the Regulation on the Technical Conditions to Be Met by Buildings and Their Location. Available online: http://prawo.sejm.gov.pl/isap.nsf/DocDetails.xsp?id=WDU20082011238 (accessed on 4 May 2020).

28. Regulation of the Polish Minister of Infrastructure of 27 February 2015 Concerning the Methodology for Calculating the Energy Performance of the Building or Part of a Building and the Preparation of Certificates of Energy Performance. Available online: http://prawo.sejm.gov.pl/isap.nsf/DocDetails.xsp?id=WDU20150000376 (accessed on 4 May 2020).

29. Pereira, L.D.; Raimondo, D.; Corgnati, S.P.; Gameiro Da Silva, M. Energy consumption in schools—A review paper. *Renew. Sustain. Energy Rev.* **2014**, *40*, 911–922. [CrossRef]

30. Marrone, P.; Gori, P.; Asdrubali, F.; Evangelisti, L.; Calcagnini, L.; Grazieschi, G. Energy benchmarking in educational buildings through cluster analysis of energy retrofitting. *Energies* **2018**, *11*, 649. [CrossRef]

31. Mazzola, E.; Dalla Mora, T.; Peron, F.; Romagnoni, P. An integrated energy and environmental audit process for historic buildings. *Energies* **2019**, *12*, 3940. [CrossRef]

32. Biserni, C.; Valdiserri, P.; D'Orazio, D.; Garai, M. Energy retrofitting strategies and economic assessments: The case study of a residential complex using utility bills. *Energies* **2018**, *11*, 2055. [CrossRef]

33. Mancini, F.; Nastasi, B. Energy retrofitting effects on the energy flexibility of dwellings. *Energies* **2019**, *12*, 2788. [CrossRef]

34. Zender-Swiercz, E.; Telejko, M. Impact of insulation building on the work of ventilation. *Procedia Eng.* **2016**, *161*, 1731–1737. [CrossRef]

35. Gładyszewska-Fiedoruk, K. Correlations of air humidity and carbon dioxide concentration in the kindergarten. *Energy Build.* **2013**, *62*, 45–50. [CrossRef]

Article

Benchmarks for Embodied and Operational Energy Assessment of Hellenic Single-Family Houses

Elena G. Dascalaki *, Poulia A. Argiropoulou, Constantinos A. Balaras, Kalliopi G. Droutsa and Simon Kontoyiannidis

Group Energy Conservation, Institute for Environmental Research and Sustainable Development,
National Observatory of Athens, GR-15236 Athens, Greece; litsa_arg@hotmail.com (P.A.A.);
costas@noa.gr (C.A.B.); pdroutsa@noa.gr (K.G.D.); skonto@noa.gr (S.K.)
* Correspondence: edask@noa.gr

Received: 24 July 2020; Accepted: 18 August 2020; Published: 25 August 2020

Abstract: Building energy performance benchmarking increases awareness and enables stakeholders to make better informed decisions for designing, operating, and renovating sustainable buildings. In the era of nearly zero energy buildings, the embodied energy along with operational energy use are essential for evaluating the environmental impacts and building performance throughout their lifecycle. Key metrics and baselines for the embodied energy intensity in representative Hellenic houses are presented in this paper. The method is set up to progressively cover all types of buildings. The lifecycle analysis was performed using the well-established SimaPro software package and the EcoInvent lifecycle inventory database, complemented with national data from short energy audits carried out in Greece. The operational energy intensity was estimated using the national calculation engine for assessing the building's energy performance and the predictions were adapted to obtain more realistic estimates. The sensitivity analysis for different type of buildings considered 16 case studies, accounting for representative construction practices, locations (climate conditions), system efficiencies, renovation practices, and lifetime of buildings. The results were used to quantify the relative significance of operational and embodied energy, and to estimate the energy recovery time for popular energy conservation and energy efficiency measures. The derived indicators reaffirm the importance of embodied energy in construction materials and systems for new high performing buildings and for renovating existing buildings to nearly zero energy.

Keywords: single-family houses; embodied energy; operational energy; benchmarks; renovations; energy use intensity (EUI); embodied energy intensity (EEI); energy recovery time

---

1. Introduction

Residential buildings are responsible for 12.1 million terajoule (TJ) that corresponds to 27.2% of the total final energy used in the European Union Member States (EU-28), and 677 million tonnes of carbon dioxide (MtCO$_2$) or 18.4% of total emissions, according to the latest available official statistics for 2017 [1]. According to the 2011 Census, there were about 241 million dwellings in the EU-28, averaging 96.4 m^2 per dwelling [2] and are estimated to have reached 301 million dwellings [3]. Although only some of the EU Member States collect and report the number of buildings in their Census data, it is estimated that there are about 118 million residential buildings in the EU-28, of which 68% are single-family houses (SFH), i.e., low-rise buildings with one or two floors with a total floor area of 9.6 billion m^2 (or 51% of the total), and the rest are multi-family houses (MFH) in apartment buildings [3].

More than half (54%) were built before 1980 and the wide adoption of energy efficiency regulations and codes. The long lifespan of buildings emphasizes the need to renovate aging and inefficient buildings by minimizing thermal losses through the envelope (e.g., add wall thermal insulation, replace windows in order to minimize thermal losses), decommission old and install new efficient equipment to improve performance (e.g., replace boilers), and/or add new systems that exploit renewables (e.g., solar thermal collectors).

In view of the significant role of buildings, European policies on climate and energy focus on energy efficiency and the exploitation of renewables for meeting the 2030 targets and towards a climate-neutral Europe by 2050 [4]. The main European legislative and policy tool has been the EU Directive on the energy performance of buildings (EPBD) since 2002 continuously evolves and motivates European countries to enhance their building regulations, mandating stricter requirements for new buildings and major renovations [5]. Among other requirements, the latest EPBD encourages energy efficiency and strives to accelerate the cost-effective renovation of existing buildings towards a decarbonized building stock by 2050 [6].

According to EPBD, as of January 2021, all new buildings must be nearly-zero-energy buildings (nZEB), while long-term national renovation strategies should support the renovation of existing buildings into nZEBs in the coming decades. Although there are significant differences in the way that EU Members States have defined nZEBs in their national context [6,7], the arching principle is that they have a very high energy performance, and the nearly zero or very low amount of energy required should be covered to a very significant extent from on-site or nearby renewable sources. Along these lines, nZEBs are expected to have a very low energy use intensity (EUI) that is defined as the annual energy use per unit floor area of the building (e.g., MJ/m^2), while securing the appropriate indoor environmental quality. The primary EUI in nZEBs ranges between 70 and 270 MJ/m^2, depending on the end-uses and services accounted for in the national definitions, different building sector, etc [8].

Given the long lifetime of buildings and low annual rates for new construction, the renovation of the millions of existing buildings constitutes the main driving mechanism for decarbonizing the national building stocks by 2050. Common energy conservation measures (ECM) include the addition of more thermal insulation, the replacement of low performance building elements (e.g., windows) or energy efficiency measures (EEM) that reduce energy use by replacing existing equipment with new and more energy efficient ones (e.g., boilers), and the installation of systems that exploit renewables (e.g., solar collectors), among others. However, while these measures will reduce the operational energy of the renovated buildings, from an environmental point of view, some of the resulting savings will actually be compensating for the energy used to produce the corresponding building products, e.g., thermal insulation, double-glazing windows, boiler, solar collector or other building components or systems that may be added to a building during its renovation. Accordingly, an energy recovery time will correspond to different time periods depending on the achieved building energy savings and the amount of energy used for the manufacturing of the building products used in a project.

1.1. Embodied Energy

Over the past 40 years, the concept of the embodied energy has gained a lot of attention for measuring the total energy required for the production of economic or environmental goods and services [9]. For buildings, it refers to the energy used for all direct and indirect processes associated with the production of materials, products, systems or other elements that go into the construction of a building, maintenance, and replacement or demolition at the end-of-life [10]. The indicator that is commonly used for the quantification of embodied impacts is the non-renewable primary energy use, accounted in megajoules (MJ) [11]. Using a similar indicator like the operational energy use intensity, one can calculate the embodied energy intensity (EEI) that accounts for all the energy used to manufacture building materials and produce building elements or systems per unit floor area (MJ/m^2). These key indicators can be used in the early stages of the decision-making process to make the best renovation decisions and select among alternatives by accounting for the impacts of new materials and systems that will be added or the ones that will be removed during the lifetime of buildings. Ultimately, they can facilitate major building renovation activities and be used to best navigate future activities towards a decarbonized European building stock by 2050.

A Life Cycle Assessment (LCA) evaluates the environmental impact of buildings, with respect to the energy used (or the generated emissions) for the production or manufacturing of materials, products, and systems that go into the construction, maintenance, or renovation of a building, and the energy used (or emissions) for its operation to the end of their lifetime. Although LCA is slowly becoming part of the European building industry as it is encouraged by EPBD [5] and Level(s) [12], it is still not common practice in most European countries and available data is very limited.

A Life Cycle Inventory (LCI) is a process that quantifies the data from raw material extraction, transportation, manufacturing, and distribution. There are several methods used in the LCI phase of an LCA to assess lifecycle energy that are extensively reviewed in several publications [13,14]. The most popular ones include the process-based analysis that takes into account the various inputs and outputs for all the processes during the lifecycle of a building material or product; the input-output (I-O) analysis that estimates the materials, energy use, and emissions for a given economic sector based on national statistics; or the hybrid analysis that is a combination of both. Since practically all methods have some type of limitations in terms of completeness, reliability, and specificity there is still no consensus on a globally accepted method [10]. Depending on the selected method, the calculated EE from a hybrid analysis may be from 3.8 up to 4.9 times higher than the values derived using a process-based analysis [15].

In this work, the focus is on a process-based LCA based on [16,17]. The boundaries of the analysis capture the initial and recurrent embodied energy. The initial embodied energy accounts for the total energy used by the production stage and construction processes. The most significant amounts of energy are used for the raw material extraction (i.e., quarrying or mining) and manufacturing, assembling, or other fabrication processes for delivering building construction materials, equipment, or system components to the factory gate. This is usually referred as "cradle-to-gate", which may represent up to 92% of the total lifecycle embodied energy [18]. The other basic boundary condition may include the additional energy for transportation from the factory gate to the building site and the energy used for the construction and installation. The transportation of materials and products may vary significantly depending on whether they are locally available (preferred) or imported and the mode of transport, while the construction and installation generally do not exceed 2% of the lifecycle impacts [18]. Accordingly, the initial EE that is also known as "cradle-to-site" embodied energy may account up to 99% of the total lifecycle embodied energy [18] depending on the materials and their function and use as different building elements.

The recurrent EE represents the energy consumed to maintain, repair, restore, refurbish, or replace materials and components (or systems) during the life of the building. Depending on the system boundary and analysis approach, the relative importance of the recurrent energy for residential buildings may range from 45% up to 60% over a 50-year building lifetime [10]. Sometimes, the recurrent EE may even exceed the initial EE depending on the lifetime of the building, service life of its elements and systems, and renovation practices [10,19]. The frequency of the various works and the amount of the recurrent EE depends on the lifetime of the building materials or systems, and on the need to replace some building elements or add new systems as part of energy conservation and efficiency measures during building renovations.

1.1.1. Databases and Tools

Databases are continually being developed and updated throughout the world, some of which are proprietary with a subscription fee, while others are open and have free access. Well-known and publicly available tools and data sources are annotated in [11,14]. Most of the data are available as LCI data format rather than embodied energy, which make them cumbersome to use. The most well-known commercial database is the EcoInvent, while the most commonly used public database is the inventory of carbon and energy (ICE) created by the University of Bath, while a popular LCA tool includes the SimaPro, among others [14,19].

Another potential source of information for construction product data is an Environmental Product Declaration—EPD that is elaborated in [16]. The number of companies that produce EPDs is growing with time, while some building sustainability assessment schemes also award credits for selecting materials or products that have EPDs available [20]. However, EPDs remain voluntary and their availability remains very limited. As a result, instead of full disclosure, some manufacturers still consider that EPDs include confidential and proprietary information and are reluctant to make them public.

1.1.2. Embodied and Operational Energy Intensity

As the operational energy used in high performing buildings continues to drop, accounting for the embodied energy of building construction materials and systems is gaining more importance [20]. However, published values for the embodied energy of residential buildings vary significantly as a result of different calculation approaches, system boundaries, and other uncertainties [13,19].

The embodied energy can range from 6% to 20% of the lifetime operational energy use in conventional buildings, from 11% to 33% in passive buildings, 26% to 57% in low energy buildings, and 74% to 100% in net zero energy buildings [15]. In the Netherlands, depending on the renovation rates of existing buildings, it is expected that the operational energy use of the building stock will decrease by 19% to 46% by 2050, while at over the same period, the share of embodied energy will increase by 26% with a renovation rate of 1.4% per year or by 35% with a renovation rate of 1.9% per year [14].

EU Member States are also encouraged to consider integrating the embodied energy of the materials and other sustainability benefits as part of the EPBD national cost-optimal calculations that must consider the buildings' complete lifecycle [21]. Currently, embodied energy is clearly considered in cost-optimal studies only in Lithuania, while six other member states account only some of its aspects [22]. Apparently, this topic should gain more attention to assess the cost-optimal minimum energy performance requirements for major renovations of existing building elements and technical installations.

Embodied energy is also one of the key indicators in the reporting framework for the sustainable building design, construction, and operation currently under development by the European Commission [12]. Recognizing the emerging importance of the embodied energy, the EU building stock observatory includes it as a monitored indicator for new construction, and deep and major renovations, under the technical building systems thematic area [23].

The embodied energy intensities of residential buildings for different building lifecycles can range from 0.9 to 23.1 GJ/m^2 for the initial EE, and from 1.22 to 20.4 GJ/m^2 for the recurrent EE [14]. These large variations may be attributed to inconsistent system boundaries, the use of different methods and databases, etc. Other earlier studies have reported that the embodied energy intensities for residential buildings range from 3.6 to 8.8 GJ/m^2 (averaging 5.5 GJ/m^2), while more recent works have disclosed values from 1.7 to 7.3 GJ/m^2 (averaging 4.0 GJ/m^2) for conventional constructions and 4.3 to 7.7 GJ/m^2 (averaging 6.2 GJ/m^2) for high performance buildings [24], and drop to values from 105 to 243 MJ/m^2 for low energy European residential buildings [25]. For European residential buildings, the initial EEI that have been reported in the literature for different constructions average 2.6 GJ/m^2 for timber, 4.0 GJ/m^2 for brick, 7.0 GJ/m^2 for concrete, and 13.8 GJ/m^2 for steel [19]. Compared to high-rise buildings, SFH have a higher EE intensity that ranges from 1.0 to 1.5 GJ/m^2 [26].

In Greece, there is no official database related to the environmental impacts of common building construction materials, equipment, systems, etc. As a result, relevant studies have commonly used data on the embodied energy coefficients from international databases, like the very popular, open, and free-to-use ICE database [27]; and information readily available in bibliography.

Studies have mostly focused on residential buildings, using example buildings with publicly available data sources or other international LCI databases to estimate the embodied energy [15,28]. Considering four real residential buildings (i.e., three single- and one multi-family house) as examples with representative constructions and different vintages, the initial embodied energy intensities ranged from 2.18 GJ/m^2 to 10.2 GJ/m^2 as a result of using different public databases for the EE coefficients of major construction materials [29]. For an apartment building, the calculated embodied energy intensity for renovating the thermal envelope of old buildings to the new building thermal regulation standards in different climate zones ranged from 2.42 to 2.82 GJ/m^2 using conventional materials and from 1.52 to 1.56 GJ/m^2 using more ecological materials [28].

Similar efforts have also focused on quantifying the materials and benchmarking the embodied energy in major electromechanical (E/M) installations and equipment, by auditing two real residential buildings—one house and one apartment building [30]. The main installations included a single-pipe hydronic central space heating with oil-fired boiler connected to room space radiators, a hydraulic and DHW installation with solar collectors, hot water storage tank, local split-unit heat pumps, and electrical installations (e.g., main control panels, cables, wall plugs). Using the ICE data [27], the work estimated an intensity of 421.0 MJ/m^2 for the house and 170.8 MJ/m^2 for the apartment building [30].

Progressively over the years, some have also been collecting local and national data to enhance the knowledge base on the energy aspects of LCI databases for key construction materials and major mechanical equipment. These efforts are of particular interest for commonly used materials or equipment that are usually not imported but rather mostly produced in the country and used in the construction and operation of Hellenic buildings. For example, materials like thermal insulation [31], steel rebar [32], aluminium [33], bricks [34], and E/M system components like boilers [35], solar collectors [36], etc.

Initial work that exploited an adapted Hellenic dataset for commonly used construction materials in Greece considered only four typical SFH to estimate their EE intensities that ranged from 3.2 GJ/m^2 to 7.1 GJ/m^2 [37]. This is significant when compared with the annual primary EUIs that range for the most recent building construction under the stringent national energy code from 0.3 to 0.5 GJ/m^2 in the different climate zones of Greece, or relatively less important if one considers the older pre-1980 dwellings that have high EUIs that range from 1.9 to 3.9 GJ/m^2 [38]. The present work extends previous efforts by considering 16 case studies providing a comprehensive representation of the Hellenic residential typologies for single-family houses, for characteristic building construction periods and locations in the four climate zones in Greece.

In comparison with the operational energy use, the final EUIs of residential buildings for normal climate conditions average 624 MJ/m^2 in EU-28 and 446 MJ/m^2 in Greece [23]. New buildings in the nZEB era have a very low primary energy use that range from 72 MJ/m^2 to 720 MJ/m^2 according to the different national definitions [39]. In Greece, the upper limit of the calculated primary energy use for space heating, cooling, ventilation, and DHW in nZEBs is 288 MJ/m^2 for new buildings and 342 MJ/m^2 for major renovations of existing buildings.

1.1.3. Building Materials

European residential buildings on an annual basis use about 12.1 million TJ for their operation [1], compared with an estimated embodied energy in new building materials of 2.0–2.8 million TJ [18]. Over half of the total embodied energy in building products is accounted for by steel (27.0%) and aluminium (23.9%), followed by concrete (16.5%), timber (11.6%), bricks (10.0%), glass (4.1%), copper (3.7%), aggregates (1.0%), insulation (0.9%), stone (0.6%), and clay (0.5%) [18].

The most common building materials are cementitious products, cement, wood, steel, asphalt, and brick that have different environmental footprints [40]. About 20–25% of cement is used in reinforced concrete (stone aggregates and cement as binder) and a comparable amount in mortar (sand particles and cement) that is prepared on site and used to fill the gaps between building blocks (e.g., bricks, concrete masonry units, stones) or plastering.

Among metals, the most energy intensive production processes per unit mass are for the production of aluminium [33] and steel [32]. Recycled metals have a significantly lower embodied energy, but even in this case, they have a higher EE compared to other materials [18]. On a positive note, about half of the European steel production is made from recycled scrap.

The production of thermal insulation materials is also energy-intensive, resulting in a high EE per unit [31,41]. On the other hand, thermal insulation materials play a key role in reducing heat losses through the building envelope (e.g., walls, roofs, load bearing structure) and reaching high energy performance in new and renovated buildings. As a result, considering the significant energy savings resulting over the buildings' lifetime, they outweigh the materials' cradle-to-grave impacts by 3.8 to 270 times [18]. In Greece, there are several manufacturing facilities for the production of various types of insulation materials, including polystyrene, stone wool, etc. Previous national studies have used values of 80.8 MJ/kg for expanded polystyrene, 87.1 MJ/kg for extruded polystyrene, 24.6 MJ/kg for mineral wool, and 92.1 MJ/kg for polyurethane foam [42].

The cement industry ranks in third place as an industrial energy user and in the second place as an industrial carbon dioxide emitter in the world [43]. Cement is the most energy intensive ingredient for the production of concrete and the component that has the highest contribution to its high embodied energy. The embodied energy coefficients for cement range up to 4.5 MJ/kg clinker, which is twice as much as the 2.171 MJ/kg for cement mortar (cement and sand) and about four times that of concrete (cement, gravel, and water) with 1.105 MJ/kg [44].

Clay bricks and tiles are very common building construction materials manufactured from clay or adobe soil. Bricks are commonly used for external envelope walls and internal separation walls, while tiles are used for roof or floor coverings. The brick production industry requires large amounts of raw materials and energy use for the clay extraction, crushing, screening, water mixing, shaping with machine moulding or extrusion, drying, and finally a baking process through a long furnace that is the most energy intensive stage [34]. For glazed bricks and tiles, an additional firing process is used to melt and then to adhere the glazed finish that usually contains glass on the finished surface. The embodied energy coefficient averages 3.6 MJ/kg for ordinary bricks, 4.6 MJ/kg for ceramic roof tile, and 15.6 MJ/kg for ceramic tiles [44]. An audit of a brick production plant in Greece and a follow-up LCA provided an embodied energy coefficient of 2.1 MJ/kg for cradle-to-grave, with pet-coke being the main energy source (86%), followed by diesel-oil (11%) and electricity (3%) [34].

1.2. Hellenic Residential Buildings

In Greece, residential buildings use on an annual basis 0.185 million TJ or 27.5% of the total final energy and are responsible for 17.4 MtCO$_2$ or 18.4% of total emissions, according to the latest available official statistics for 2017 [1]. According to the 2011 Census, there are about 6.4 million dwellings, averaging 88.6 m^2 per dwelling, of which more than half (52%) were built before 1980, the year that the first national thermal insulation regulation was introduced [2].

There are about 2.99 million exclusive use residential buildings [45], of which 73% are SFH and the rest MFH (Figure 1). The most common construction materials used in Hellenic buildings are reinforced concrete (58%) and brick masonry (21%), followed by natural stone (18%) that prevails in buildings constructed before 1960 and especially in historic buildings before 1920 [45]. More information on the Hellenic residential building stock characteristics are also available in [46].

EPBD transposition in Greece was initiated with a new national regulation (known as KENAK) that came into effect in 2010 [47] and was updated in 2017, following the national cost-optimal study [48], enforcing stricter requirements for the building's thermal protection and minimum energy efficiency requirements. Analysis of the energy performance certificates (EPCs) that have been issued in Greece documents that the existing building stock is energy inefficient, ranking below the current energy performance standards. About 34% of the residential buildings are ranked at the lowest energy class, while only 3% are ranked at or above the KENAK minimum energy code requirements [49].

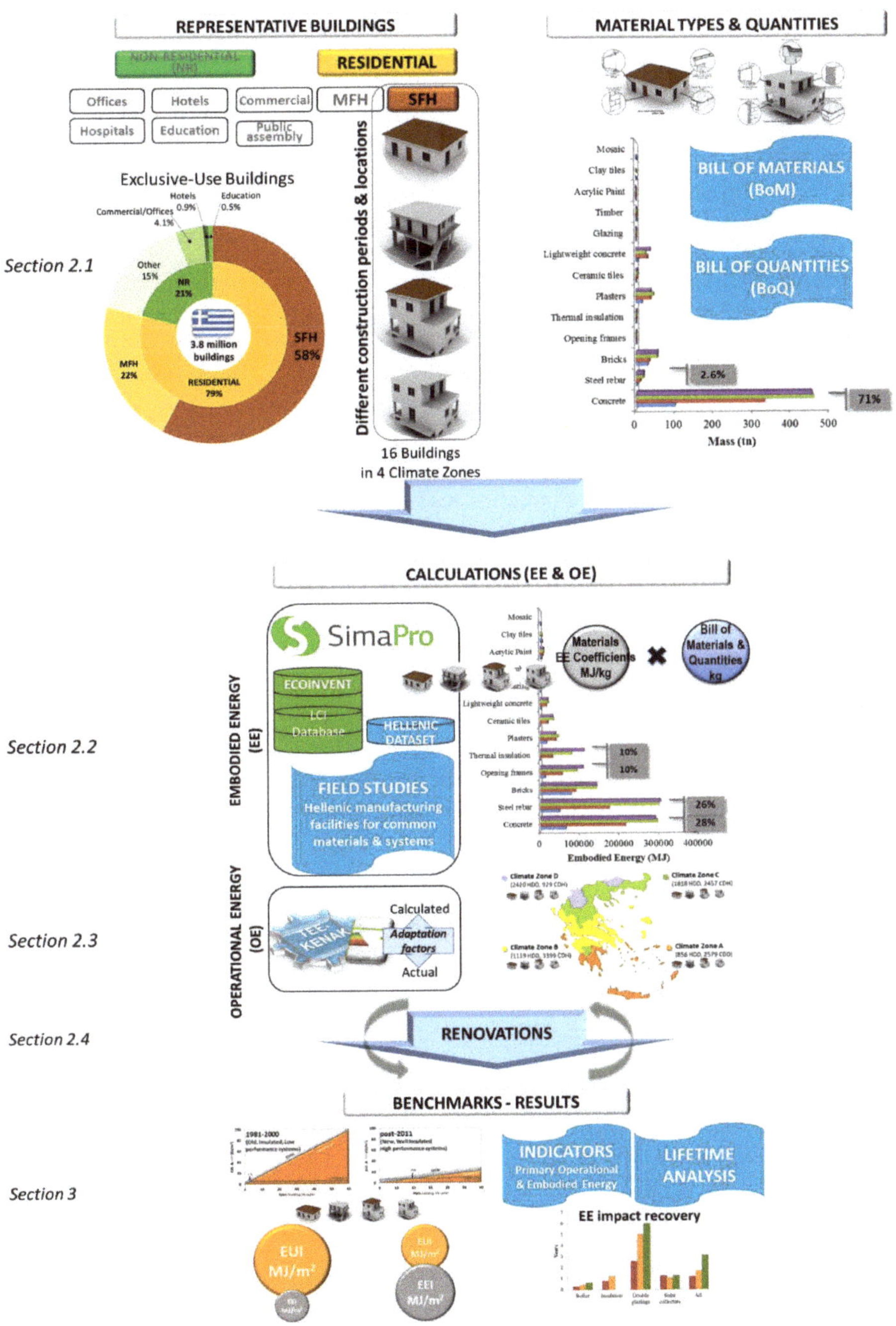

Figure 1. Methodological schematic of the work.

1.3. Work Overview

In general, performing an LCA is a time-consuming process that requires user expertise. Furthermore, differences in national production processes, system boundary considerations etc, may result in significant deviations for some materials or systems [19]. However, the challenge that many countries face is that there is no readily available locally or nationally derived data, creating a burden on building professionals to use (or misuse) complex tools that practically lack a sense of local or national context. Therefore, it is beneficial to simplify the process for the early decision-making process in building renovation or new construction. In addition, there is a need to prioritize what may be important in terms of typical materials and quantities commonly used in representative building constructions, combined with properly adapted national or local field data. This would also allow others to focus limited resources and target production processes, materials, and systems that have the most significant impacts. This knowledge can then be used to derive easy-to-use benchmarks to help stakeholders perform a first assessment and compare results against good practices or policy makers to develop regulations with realistic minimum requirements.

The vision of the present work is to:

- Derive benchmarks for assessing the EE of Hellenic buildings in order to inform and facilitate professionals in the early stages of design, construction, and renovation projects; assess the energy return from common renovation measures; and link relevant information to a building stock model.
- Enhance the knowledge base on EE of building materials in Greece and enrich the national data on EE for the production of common building materials and systems.
- Compare against international data bases and tools; gain confidence on the derived EE indicators; or identify materials and systems with high discrepancies in order to prioritize future work.

In this direction, the present work considers single-family houses that represent the largest segment (58%) of the existing building stock in Greece. Section 2 outlines the overall structure of the methodological approach. The first step is to define the typical SFH in order to identify the common construction materials, along with the assumptions for performing the operational and embodied energy calculations. The bill of materials and quantities set the priorities for the field work and the short energy audits for collecting relevant data from local manufacturing facilities. The tools used for the calculations include the national calculation engine for estimating the operational energy of the buildings, while the EE is estimated with a commonly used international tool and database that is also adapted with national data, if available from the field studies. Section 3 summarizes the main findings from the audits of local manufacturing facilities, along with other available national data for the embodied energy of materials and systems. The embodied and operational energy indicators are quantified for the different building typologies in the four national climate zones. The analysis concludes with an assessment of the energy recovery time to account for the embodied energy by the operational energy savings resulting from the implementation of energy conservation and efficiency measures. Section 4 elaborates the main findings of the work reported herein and its limitations. Finally, Section 5 outlines future work and priorities.

2. Methodology

The work focuses on single-family houses (SFH) that constitute the majority of the Hellenic building stock, considering four representative residential buildings that are organized for different tiers in terms of construction periods and location. The first step is to quantify the types of construction materials for each case study and the specific material quantities. The calculations for the embodied energy are performed using a well-known LCA software, with input data from an international lifecycle inventory database. The assessment is complemented using data from field studies collected from short energy audits of manufacturing facilities in Greece for some common building construction materials and other publicly available national data, in an effort to populate a Hellenic dataset. The calculations for the annual operational energy use for each case study are performed using the national calculation engine for assessing the building's energy performance in the four climate zones of Greece. The results are then expressed as embodied energy intensities (EEI) and energy use intensities (EUIs), i.e., energy per unit floor area (MJ/m^2), calculated for different lifetimes of the buildings. Finally, the analysis considers some popular energy conservation and efficiency measures and quantifies the time for recovering the embodied energy from the resulting annual energy savings associated with the addition of new materials (e.g., thermal insulation) and the replacement of building elements (e.g., windows) or equipment (e.g., heating systems). The overall approach and the main calculation steps are illustrated in Figure 1, and elaborated in the following sections.

2.1. Representative SFH Buildings

The main building and technical installation characteristics and other assumptions of the standalone buildings considered in the present work are summarized in Table 1. The case studies have characteristics that are common for Hellenic SFH in the different construction periods and locations, within the concept of the Hellenic typology and the building stock model for Greece [46] that is based on the European TABULA and EPISCOPE typology [50]. The buildings are organized under four construction periods (SFH-1 to SFH-4) that relate to the implementation of the different national thermal insulation regulations and the four climate zones (A to D).

The representative SFHs used in the present work have different architectural features (e.g., single- and two-storey buildings, with slab-on grade, pilotis or a non-heated basement, and with flat or tilted roof), and combine characteristic geometries (e.g., floor areas, window to wall ratios), construction materials and practices (e.g., thermal insulation of load bearing structure, different envelope thermal protection and variable degree of compliance with energy regulations), and different technical systems for heating, cooling, and domestic hot water (DHW). Accordingly, under each construction period, there are four different case-studies that refer to buildings with different number of stories, type of roof or exposure of ground floor (Table 1). As a result, the building geometries and system performances for the case studies considered here resemble the ones used in the national EPBD cost-optimal study [48]. For example, the first construction period (pre-1980) identified with SFH-1 in Table 1 includes buildings that are considered without thermal insulation and have openings with single glazing. The second construction period (1981–2000) is an intermediate period for which the SFH-2 buildings are partly insulated according to the first national thermal insulation regulation, while the third one (2000–2010) refers to the SFH-3 buildings that have full compliance. Finally, the last construction period (post-2011) refers to the SFH-4 buildings constructed under the new EPBD national regulation, which started the KENAK-era. For each building, if necessary, the description and necessary numerical values of the parameters that refer to the different climate zones (A/B/C/D) are presented in sequence if they are not the same. For example, the external walls for SFH-1 are constructed with bricks and plaster and the corresponding areas for the four case studies are (68/77/77/68 m^2) and have no thermal insulation. For SFH-4, the walls have the same construction and wall areas (114 m^2), but the thermal insulation material is different for the four climate zones (A/B/C/D) to comply with the minimum code requirements with a thickness of (4/5/6/7 cm), respectively. The BoMs includes interior walls and doors, while all building surfaces are painted. At this stage of the work, the embodied energy calculations do not include internal kitchen or bathroom accessories. In relation to the E/M, installations include only the boiler in all case studies and the solar thermal collector for DHW production in SFH-4.

Table 1. Building and technical installation characteristics for 16 single-family houses (SFH) and other assumptions used in the calculations.

	SFH-1				SFH-2				SFH-3				SFH-4			
	Climate Zones															
	A	B	C	D	A	B	C	D	A	B	C	D	A	B	C	D
Construction period	Pre-1980				1981–2000				2001–2010				post-2011			
Description	Old buildings, not thermally insulated envelope				Partly insulated according to first regulation				Fully insulated according to first regulation				Fully insulated according to KENAK-EPBD regulation			
Storeys	1	1	1	1	1	1	1	1	1	2	2	1	2	2	2	2
Slab-on grade	✓	No	no	✓	no	no	no	no	no	no	no	no	no	no	no	no
Pilotis	no	No	no	no	✓	no	✓	✓	no	no	no	no	no	no	no	no
Over unheated space	no	✓	✓	no	no	✓	no	no	✓	✓	✓	✓	✓	✓	✓	✓
Heated floor area (m^2)	80	80	80	80	130	130	130	130	150	150	150	150	155	155	155	155
Building Elements																
Load Bearing																
Reinforced concrete	✓	✓	✓	✓	✓	✓	✓	✓	✓	✓	✓	✓	✓	✓	✓	✓
Thermal insulation (cm)	no	No	no	no	no	no	no	no	4	4	4	4	5	6	7	8
Walls																
Area (m^2)	68	77	77	68	75	83	75	75	83	114	114	83	114	114	114	114
Bricks and plaster	✓	✓	✓	✓	✓	✓	✓	✓	✓	✓	✓	✓	✓	✓	✓	✓
Thermal insulation (cm)	no	no	no	no	4	4	4	4	4	4	4	4	4	5	6	7
Floors																
Reinforced concrete	✓	✓	✓	✓	✓	✓	✓	✓	✓	✓	✓	✓	✓	✓	✓	✓
Thermal insulation (cm)	no	no	no	no	no	no	no	no	no	4	4	4	2	3	4	4
Finish material - Area																
Ceramic tiles (m^2)	no	43	43	no	60	60	60	60	69	69	69	69	75	75	75	75
Mosaic-Terrazzo (m^2)	43	no	no	43	no	no	no	no	no	no	no	no	no	no	no	no
Wood (m^2)	29	43	43	29	60	60	60	60	69	69	69	69	75	75	75	75
Basement (non-heated)																
Reinforced concrete	✓	✓	✓	✓	✓	✓	✓	✓	✓	✓	✓	✓	✓	✓	✓	✓
Thermal insulation (cm)	no	no	no	no	no	no	no	no	no	no	no	no	2	3	4	4
Area (m^2)	0	97	97	0	0	147	0	0	139	100	100	139	100	100	100	100
Finish material	mosaic				mosaic				ceramic tiles				ceramic tiles			
Roofs																
Reinforced concrete	✓	✓	✓	✓	✓	✓	✓	✓	✓	✓	✓	✓	✓	✓	✓	✓
Thermal insulation (cm)	no	no	no	no	6	6	6	6	6	6	6	6	6	7	7	9
Cover material	Mosaic		Roman clay tiles		Concrete		Roman clay tiles		Concrete tiles		Roman clay tiles		Concrete tiles		Roman clay tiles	
Flat, Area (m^2)	80	80	no	no	130	no	no	no	150	88	19	no	90	90	19	19
Tilted, Area (m^2)	no	no	80	80	no	130	130	130	no	no	69	150	no	9	72	92
Windows																
Materials	Single glazing, wooden frames				Double glazing, aluminium frames				Double glazing, aluminium frames with thermal breaks				Double glazing low-e, aluminium frames with thermal breaks			
Area (m^2)	19	19	19	19	26	28	27	26	28	30	30	28	28	28	28	28
Basement																
Materials	Single glazing, wooden frames				Double glazing, aluminium frames				Double glazing, aluminium frames				Double glazing, aluminium frames			
Area (m^2)	0	0	6	0	0	10	0	0	11	11	11	11	10	10	10	10
Shading	Structural elements (e.g., balconies)				Structural elements (e.g., balconies)				Structural elements (e.g., balconies)				Structural elements (e.g., balconies)			

Table 1. *Cont.*

	SFH-1				SFH-2				SFH-3				SFH-4			
	Climate Zones															
	A	B	C	D	A	B	C	D	A	B	C	D	A	B	C	D
	Technical Installations															
Heating system																
Fuel, System		Oil-fired boiler				Oil-fired boiler				Oil-fired boiler				Oil-fired condensing boiler		
Thermal efficiency (%)	72	72	72	72	82	82	82	82	86	86	86	86	98	98	98	98
Cooling system																
System		Local split units				Local split units				Local split units				Central heat pump		
Coefficient of Performance	1.7	1.7	1.7	1.7	1.7	1.7	1.7	1.7	1.7	1.7	1.7	1.7	2.5	2.5	2.5	2.5
Domestic hot water		Electric storage tank				Electric storage tank				Electric storage tank				Winter: coupled with boiler central heating; Summer: Electric heater + Solar collectors		

	Calculation assumptions according to national regulation
Set-points	20 °C in winter, 26 °C in summer; 18-h daily operation
Internal gains	Lights: 0.1 W/m^2, People: 4 W/m^2, Equipment: 2 W/m^2
Infiltration	0.75 m^3/h/m^2 floor area
DHW consumption	27.38 m^3/bedroom at 45 °C
Lighting; Appliances	Calculations for residential buildings do not account for lighting, appliances, and similar plug loads that are not linked to occupant comfort conditions

The assumed lifetime is an important parameter since it influences the operational energy of a building and the recurrent embodied energy as a result of the maintenance needs for some building elements and systems that depends on their respective lifetimes. Referenced values used in the literature vary significantly from 50 to 150 years [13].

The lifecycle analysis in the cost-optimal studies are performed with an estimated 30-year lifetime for residential buildings, in order to secure consistency in the comparative methodology. However, this may not be representative for Hellenic buildings and generally speaking for the European building stock. According to national and European statistics, a building lifetime of 50 years is more realistic, representing about 41% of the existing Hellenic buildings [45] and about half of the European buildings [23]. Furthermore, based on empirical data, the median lifetime of Hellenic dwellings is 70 years and for European dwellings is 125 years, with a renovation cycle of 30 years and 40 years, respectively [51]. Accordingly, a sensitivity analysis has been carried out to study the influence of building service lifespans over 30, 50, and 80 years on the relative significance of embodied energy and total lifecycle energy.

Bill of Materials

The construction period of the buildings facilitates the identification and quantification of the typical materials that reflect the common construction practices used in each case study and architectural features (e.g., one- or two-story buildings, slab on grade, or pilotis) of the buildings. To a certain extent, the construction period and location of the buildings in the four climate zones also relates to the more detailed requirements of the national seismic design code and its periodic revisions that define how structures are designed and constructed to limit seismic risk. This consideration mainly impacts the amount of steel bars used in the reinforcement of the load-bearing structure.

According to the national Building Census [45], the most common construction materials among the existing building stock are concrete (57.7%) and bricks (21.5%). Since the 1980s, concrete has dominated construction from 70% in the earlier decades to about 95% for new constructions. Stone (17.6%) is also popular for single-storey buildings constructed before the 1960s and is very limited in new traditional style constructions, especially in islands and rural areas.

This step serves an additional purpose. Relevant information can be used to identify the most common materials and define the priorities for progressively collecting national data, focusing on the most common construction materials with large mass quantities in representative buildings. However, the prioritization will also need to account for the embodied energy coefficients (i.e., energy per unit mass) since for the embodied energy assessment, it is necessary to consider the product of material mass with its EE coefficient. For example, the resulting embodied energy for large quantities of a given material multiplied with a low EE coefficient would result in a low embodied energy that may not be of importance. On the other hand, small quantities of another material with a high EE coefficient would result in a high embodied energy and rank higher on the need for collecting and using national data. The output of this step can also be used to establish general criteria to simplify the inventory analysis by focusing on the most significant building elements [52]. This approach is adapted by the Spanish LCA regulation for which the lifecycle inventory analysis focuses on representative building elements [53].

The main calculation steps are summarized by Equations (1) and (2). The bill of materials (BoM) and their respective quantities (kg) constitute a mass-based inventory of all the materials that compose the construction elements or other building products. The information can also be organized in the main building elements, for example, foundations, load bearing structure, facades, etc. The specific material quantities were estimated by multiplying the material density (kg/m^3) with the volume of each building element (e.g., wall, roof, floor, pilotis) with the same construction (Equation (1)). Next, the value of the embodied energy for each constituent material was calculated by multiplying its specific mass with the material's embodied energy coefficient (i.e., MJ/kg). As shown in Equation (2), the total embodied energy is the sum of all the construction materials used in the various building elements and products, divided by the floor area of the building to obtain a normalized value of energy per unit floor area (MJ/m^2):

$$M_i = \rho_i \cdot V_i \tag{1}$$

$$EEI = \frac{\sum_{i=1}^{n} EEC_i \cdot M_i}{A} \tag{2}$$

where M_i = mass (kg) of the ith material, ρ_i = density (kg/m^3) of ith material, V_j = volume (m^3) of a building elements (j), e.g., walls, etc, EEI = embodied energy intensity that accounts for the energy used to manufacture building materials and produce products per unit floor area (MJ/m^2), EEC_i = embodied energy coefficient (MJ/kg) of ith material, i = building construction material like bricks, concrete, glazing, mortar, steel etc, n = number of different building construction materials and products, A = heated floor area of the building(m^2).

Embodied energy coefficients from eleven public sources with cradle-to-gate data are summarized in Figure 2. The specific numerical values and detailed references are presented in [29]. The scales of the y-axis in Figure 2 for the different group of materials are purposely different so that the whisker plots can display in detail the data distribution through their quartiles. The median value of the coefficients is included inside the box for each material. The width of the box shows the interquartile range between the lower 25th and the upper 75th quartile, which in most cases illustrates a large data spread out. The observed differences may originate from discrepancies in the boundary definition used in the calculation method, the national origin of the data used in the analysis, the differences in production processes that may be relevant for some materials or products, etc. As a result, some claim that since the reported embodied energy coefficients vary so much across among various studies, it may not be appropriate to simply adopt these values and use them for national calculations or comparative studies [10].

Apparently, it is always preferable to use, if available, local or national data. Furthermore, for some materials, the large variations (e.g., aluminium, insulation, tiles) or the outliers (e.g., glass, bricks) point to the ones that as a priority and progressively can initiate efforts and start collecting relevant data on a regional or national basis. For materials whose published values are rather consistent, one can use then with higher confidence in national studies, in the event that better quality data is not available.

2.2. Embodied Energy

In the present work, the system boundaries include the initial and recurrent embodied energy. The cradle-to-gate production stage is assessed using an LCA tool and is also calculated using data collected from field studies for several construction materials. For cradle-to-site, the calculations account for only the average transportation to the construction (building) site. For the recurrent EE, the use stages account for different lifetimes of the equipment that are replaced to a better condition (e.g., replace single- with double-glazing openings to reduce heat losses).

2.2.1. LCA Tool and LCI Database

The EE calculations were performed using the SimaPro life cycle assessment software package, which is one of the most popular and common LCA tools [54] with the EcoInvent v3.4 LCI database [55]. The inventory data was complemented with new information collected from field studies on the production processes of some commonly used materials that were performed during this work, to compile a Hellenic database.

In this work, embodied energy calculations were performed following the cumulative energy demand that is considered an impact category indicator under LCIA [56]. This approach investigates the primary energy use through the lifecycle of a material or a product. The energy content is quantified for non-renewable energy resources used directly or indirectly throughout the production stage, e.g., from the extraction and transfer of the raw materials to the production of the final product (cradle-to-gate). This is a standard approach implemented in EcoInvent [57] using the higher heating values for the various fuels.

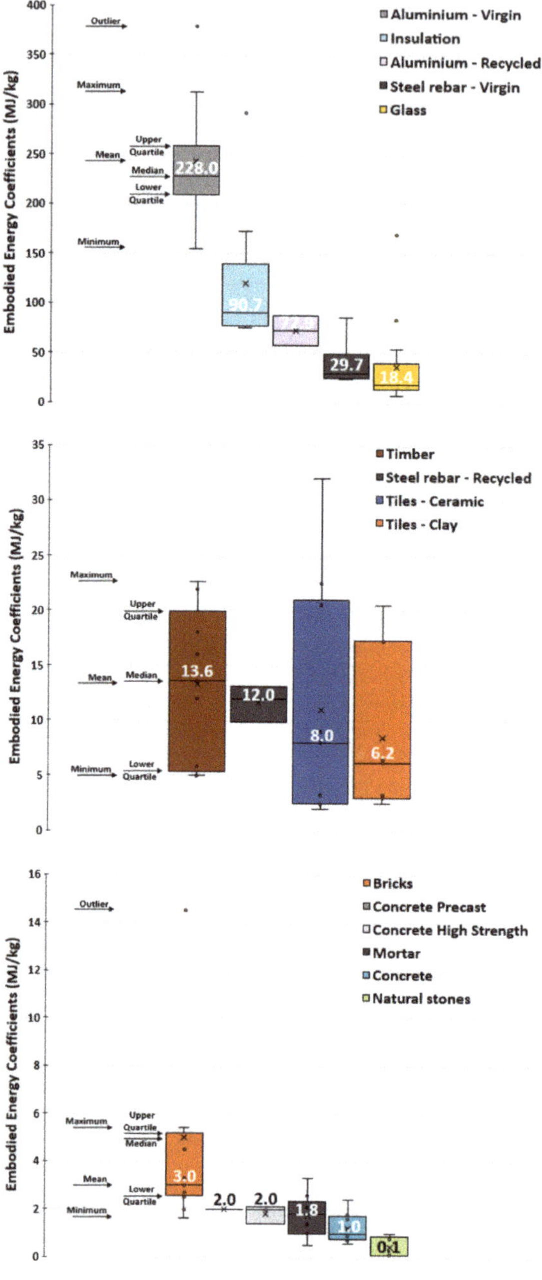

Figure 2. Range and distribution of published values for the embodied energy coefficients (MJ/kg) of commonly used construction materials.

2.2.2. Field Studies and Data Collection

In Greece, relevant studies that have collected and analysed field data on the embodied energy of locally and nationally manufactured construction materials or building products are limited, while environmental product declarations are practically not existing or not disclosed. Previous sporadic efforts have attempted to fill in this gap of knowledge by providing field data for thermal insulation materials [31], boilers for heat production [35], air-conditioning split units [30], etc.

In this work, short energy audits were performed at local facilities that produce some commonly used construction materials, for example, concrete, cement, bricks, plasters, etc. The data collection process followed a process-based analysis. The raw material quantities and energy carriers were recorded at each stage of the material's or product's manufacturing processes. The data were then processed and used to calculate the embodied energy coefficients for the respective material or product. The analysis is performed from the extraction of raw materials to the manufacturing of the final product (cradle to gate). Given that this is a time-consuming process, the campaigns were targeted to manufacturing facilities of common building materials, based on their ranking and use in Hellenic buildings. However, one needs to also take into account accessibility to these facilities and willingness of the owners to provide the necessary information.

Furthermore, a detailed national literature survey on technical papers or EPDs revealed limited information for some additional construction materials. For the materials that there was no published Hellenic data (11.in total), the SimaPro tool was used to complement the available information and generate cradle-to-gate embodied energy coefficients. The cumulative energy demand was calculated using the Hellenic fuel mix for generating electricity as follows: coal 33.16%, oil 10.83%, natural gas 26.6%, renewables 29.01%, and wastes (non-renewable) 0.40% [1]. All the results were then compiled in a Hellenic dataset that was used to investigate how they differentiate against the data from the EcoInvent database that reflect average European production processes and fuel breakdown for electricity generation. This comparison allows one to gain a better insight on how well the (default) values from the LCI database compare with the Hellenic dataset and quantify the potential differences that may be anticipated for estimating the initial and recurrent embodied energy.

2.3. Operational Energy

The annual operational energy use is calculated for the 16 building typologies. This offers the opportunity to consider different building construction materials, types and system performances that resemble the different groups of the existing building stock. The calculated values are then adapted using empirically derived factors in order to get more realistic estimates of the actual operational energy use.

2.3.1. Calculations

The building's energy demand and use are estimated using the official national software (TEE-KENAK) that was developed by the National Observatory of Athens for the Technical Chamber of Greece (TEE) in Athens, Greece for EPBD compliance [47]. The calculation engine is in accordance to the European standards (e.g., EN 13790:2008) for estimating the building's energy demand using the quasi-steady state monthly method. The software incorporates the relevant national technical libraries, weather data, and other technical specifications outlined in four supporting national technical guidelines. The calculation engine is used for issuing the energy performance certificates (EPC) and it is mandatory for assessing the energy performance of buildings of new and renovated buildings in Greece for obtaining a building permit.

The calculations are performed for standard conditions using specific assumptions and default values that are defined in the national EPDB regulation (KENAK) and the four technical guidelines. The main calculation assumptions are summarized in Table 1. The national regulation defines four climate zones [47]. Zone A in the south includes most of the Hellenic islands and part of the mainland

and is characterized by mild conditions, with a median of 856 heating degree days (HDD) and 2579 cooling degree hours (CDH). Zone D in the north parts of the mainland has the coldest conditions, with a median of 2420 HDD and 929 CDH. Finally, the primary energy conversion factors are also defined in the national regulation at 2.9 for electricity, 1.05 for natural gas, and 1.1 for oil, and these values have been adopted in the calculations performed in this work.

The selection of the TEE-KENAK as a tool, instead of a simulation tool, is intentional in order to integrate the concept of embodied energy in the overall approach and the calculation commonly used in Greece. This is the tool utilized by engineers, architects, and energy auditors for the assessment of different ECMs or EEMs that support their specific recommendations during building renovations and issuing an energy performance certificate. The same tool is regularly utilized by professionals during a new building design to confirm that the building's energy performance complies with the minimum code requirements and thus obtain a building permit.

2.3.2. Adaptation for Actual Energy Use

It is well recognized and documented in the literature that there is a gap of calculated vs. actual energy use [58] that hinders the realization of the anticipated energy savings as a result of various factors and different strategies [59]. In the present work, the calculations of the actual energy use are performed using the official calculation engine assessing the energy performance of buildings. Although the tool is well tailored for issuing EPCs, it is not suitable for predicting the actual energy use [58]. The gap of calculated and actual energy use is to be expected since the standardized calculations are performed for specific operating conditions [47]. For example, the daily use of space heating is for 18 h for the entire floor area of the dwelling, with an indoor set-point temperature at 20 °C (Table 1). On the other hand, the actual operating conditions are different. Based on information from a national survey [60], only 11% of SFH use their heating system continuously, while almost three-quarters operate their systems for less than 8 h during a typical day. Furthermore, only 7% of the houses heat their entire dwelling, since most are trying to isolate rooms that are not being used in order to reduce the operating cost.

To handle these deviations and adapt the calculated values to obtain more realistic estimates, the current work considers a simple and effective method that has been derived using data collected during the energy performance certification of buildings [58]. The EPCs in Greece are centrally issued and automatically stored in an electronic registry maintained by the Hellenic Ministry having jurisdiction. The EPCs may optionally include the actual energy use of the audited building. This is valuable information since it can provide a key for unlocking in a practical manner the relationship of the calculated with the actual energy use. Since the actual energy use from the utility bills is voluntary information, a very limited number of issued EPCs of about 3% of the total include the actual thermal and/or electrical energy use [49].

The adaptation factors are defined as the ratio of the actual to the calculated energy use from the available EPCs [58]. The analysis requires an elaborate data processing of large data sets, complemented with a comprehensive data quality control and screening, along with an elaborate manipulation for organizing the derived factors for different building typologies, e.g., building use, construction periods and location. The adaptation factors are used as multiplies for correcting the calculated energy use intensities (Equation (3)) to derive more realistic values that resemble actual energy use:

$$EUI_{a,k} = EUI_{c,k} \cdot f_k \tag{3}$$

where $EUI_{a,k}$ = adapted energy use intensity (MJ/m^2) for the kth tier, $EUI_{c,k}$ = calculated energy use intensity (MJ/m^2) for the kth tier, f_k = adaptation factor (e.g., from Appendix A) of the kth tier of the buildings for the four locations (climate zones A-D) and the four construction periods (e.g., pre-1980).

The intent for this step in the present work is not to validate the calculation tool. This approach has been derived and implemented in order to easily adapt the tool calculations to more realistic estimates of the buildings' actual operational energy. Over time, more certificates are being issued and the EPC database is enhanced. This allows to periodically update the derived adaptation factors, gain more confidence on the available data, and cover the Hellenic building typologies. The recently derived adaptation factors used in this work are summarized in Appendix A for the different building tiers, in terms of their location related to the four climate zones (A-D) and the four construction periods for the different construction periods.

2.4. Renovations

Beyond the initial embodied energy that remains the same throughout the lifetime of a building one needs to also account for the additional recurrent EE as a result of the new materials that will be added and systems that will be used to replace existing ones during building maintenance or energy renovation works [10]. Building renovations are necessary in order to maintain or even improve its performance with time. One can consider and select among different ECMs for the building envelope and EEMs for the technical installations and building services. Replacing obsolete windows or an old and inefficient boiler as a result of routine maintenance, installing wall or roof thermal insulation or a solar collector for DHW during an energy renovation, will reduce a building's operational energy use. At the same time, though, the installation of these additional building materials or equipment will influence the energy use in other lifecycle phases of a renovated building [61] and increase its total embodied energy.

The number of cycles for the recurrent EE is determined as a function of the service lifetime for different building construction materials (e.g., roof tiles), elements (e.g., windows), equipment (e.g., boiler, radiators) etc, which in turn depends on maintenance needs (e.g., painting or plastering surfaces) and actual use [10,62]. Typical values are summarized in Table 2. These time periods are expressed in years after the initial installation during which their functions and properties are in accordance to the minimum acceptable values when routinely maintained.

Table 2. Service lifetime in years of common building materials, elements, equipment, and installations.

Materials, Elements, Equipment	Service Lifetime (Years)
Window frames/Glass	30
Plastering (exterior)	50
Doors (exterior)	40
Paint	10
Laminated floor	35
Ceramic tiles	30
Mosaic	30
Timber	55
Roof tiles (clay)	30
Heat production (boiler)	20
Radiators	62
Solar thermal collectors	25

Furthermore, these cycles for the recurrent EE will also depend on the building lifetime (e.g., 50 or 80 years). For example, a heat production system that has a service lifetime of 20 years, will be replaced twice if the building's lifetime is 50 years. Likewise, there will be four cycles if the reference study period for the building extends to 80 years.

Beyond maintenance, there are several opportunities to significantly improve a building's energy performance, including renovation works and upgrades of its envelope, equipment, technical installations, and other services. Common ECMs or EEMs include the replacement of single- with double-glazing windows, the replacement of the heat production system that may also be combined with a switch to other energy carriers (e.g., from heating oil to natural gas or the use of heat pumps, the installation of solar thermal collectors), the installation of roof thermal insulation and for non-thermally protected buildings to add wall thermal insulation, etc. More aggressive and combined measures can upgrade an existing building to higher energy standards towards nZEBs. For example, a typical Swedish residential building constructed in the 1970s with 716.4 MJ/m^2 primary energy use can be renovated to the passive house standard with 169.2 MJ/m^2 by adding thermal protection, external cladding, and replacing windows [61]. As a result, the share of the embodied energy to the total operational energy savings can range from 12% to 21% over a lifetime of 50 years.

The installation of new equipment increases the building's embodied energy if one accounts for the energy used to manufacture the unit. Popular EEMs include the replacement of the heat production systems for central heating and the installation of flat solar collectors for DHW production. Small- and medium-sized boilers manufactured in Greece average an embodied energy per unit thermal capacity of 65.8 MJ/kWth for steel and 119.1 MJ/kWth for cast iron boilers, or per unit mass at about 26 MJ/kg [35].

Solar thermal collectors are among the most popular systems for exploiting solar energy and they are widely used for the production of DHW throughout the world, and in some countries for space heating with solar combi systems. The embodied energy of flat-plate solar thermal collectors based on information from several European studies averages 1.7 GJ/m^2 of collector area [63].

3. Results

The elaboration of the results follows a similar structure with the presentation in Section 2. The first subsection summarizes the results for the BoM for the 16 case studies. The second section documents the findings from the field work and the short energy audits to derive the embodied energy coefficients used in the present work and the comparison of the Hellenic data with the EcoInvent database. The calculated EEIs are then summarized for the 16 representative buildings considered in this work. The specific EUIs are also calculated and the results adapted to estimate the actual operational energy use for the different case studies. A sensitivity analysis is performed for the 16 representative SFH located in the four national climate zones, considering a 30-, 50-, and 80-year building lifetime.

3.1. Bill of Materials

Figure 2 illustrated the deviations of available information for the embodied energy coefficients of various common construction materials from different data bases. This practical information can be used to visualize and prioritize the materials for which there is overall general agreement among different data bases. This way, one may have a first direction of where to focus and what materials to consider first, in order to start collecting relevant data, if possible. When the mean (marked with "x" inside the box plots in Figure 2) is very close to the median value (the 50th percentile horizontal line and numerical value inside the box plots in Figure 2), the available data are normally distributed. Otherwise, the numerical value of the median that is shown inside each box is more representative and may be used especially for materials that the available data are skewed or for the ones that they include outliers. For most materials (e.g., thermal insulation, tiles, bricks), the values above the median are more spread out or the upper whisker is longer than the lower, which means that the distribution is skewed toward higher coefficients. For others, e.g., aluminium, the data distribution is symmetric around the median, which is illustrated by the median value that is centered in the box between the upper and lower quartiles and the two whiskers have similar length. The outliers are identified when their values deviate by 50% from the interquartile range, e.g., aluminium-virgin, insulation, glass, bricks. Outliers may indicate materials with high data variability under special or local conditions that may also deserve special attention, if they are not in error.

The embodied energy intensity depends on the type and the quantities of the various construction materials. The detailed calculations were performed for all 16 case studies. The results are summarized in Figure 3 as average values for the four construction periods. The total mass quantities of all materials normalized per unit floor area of the representative buildings range from 2123 kg/m^2 to 4643 kg/m^2. In all construction periods, the prevailing materials are concrete and bricks that average 72.6% and 9.4% of the overall building mass, respectively. The share of plasters averages 13.1% of the building mass, and steel rebar about 2.4%. For some material, there are significant variations among the construction periods. For example, mosaic averages 3.1% in pre-1980 and drops below 0.5% in 1981–2000 buildings, while thermal insulation material start at a low 0.08% in 1981–2000 buildings and reach 0.19% in post-2011 constructions.

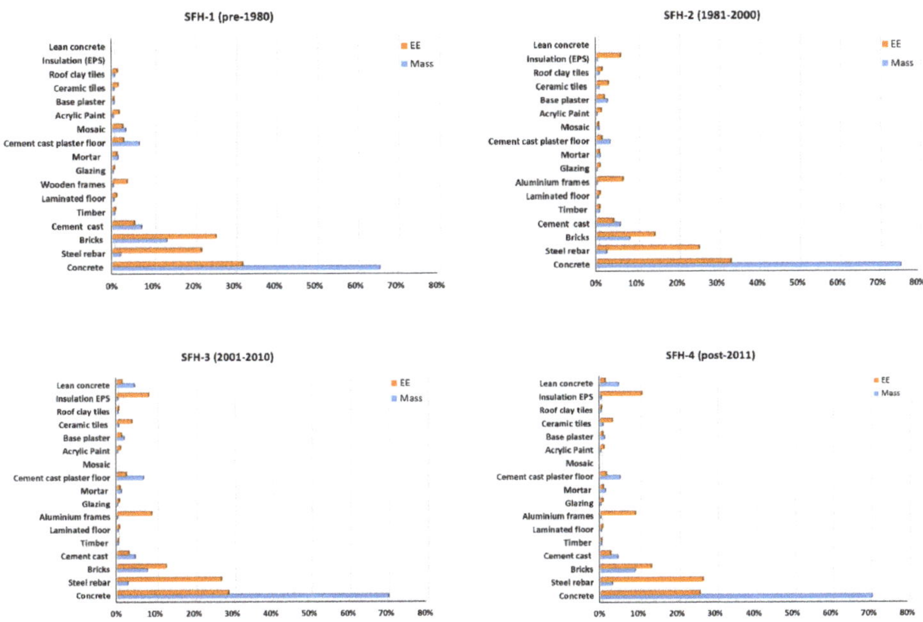

Figure 3. Percentage of the total building mass and embodied energy for the materials used in the case studies and averaged for the four construction periods.

Another way to look at the results is by aggregating the various material quantities used for the main structural elements of the buildings. Figure 4 illustrates the average material quantities for all the case studies that correspond to the four construction periods. The load bearing structures (i.e., foundations, columns and beams, roofs, floors) dominate with about 73% to 84% of the material quantities, which is expected for complying with the national seismic design code.

3.2. Hellenic Dataset

The compiled Hellenic dataset with the derived material EE coefficients includes 28 entries that cover a representative range of building construction materials that are used in Hellenic buildings. Most of the information is based on the data collected from the field surveys during this work. In some cases, supplementary national data from the literature was used to compile the results derived using the SimaPro LCA software [54] and complemented by the EcoInvent LCI [64], where necessary. The main results for the derived embodied energy per unit mass (MJ/kg) of various materials for the cradle-to-gate processes are summarized in Table 3, along with an overview of the production processes at the facilities considered during the field work and the other available public resources.

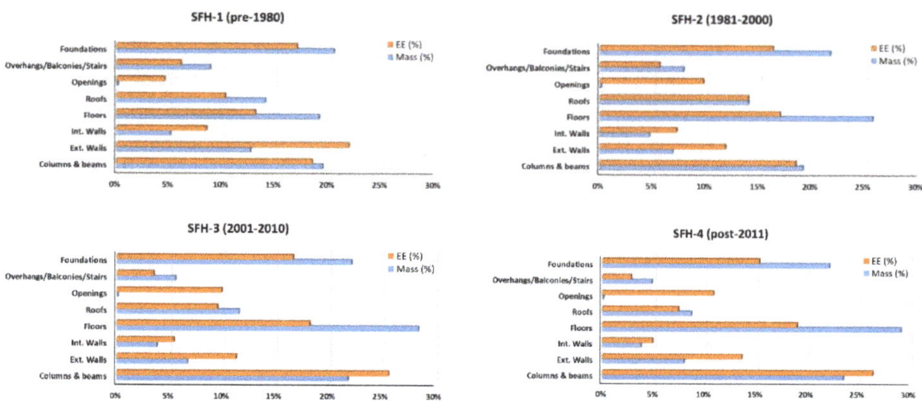

Figure 4. Percentage of the total building mass and embodied energy for the aggregated materials used in the main structural elements of the case studies and averaged for the four construction periods.

The values derived and presented in Table 3 are mostly within the range of cradle-to-gate embodied energy coefficients published in various European databases [65]. Comparing these values against the ones from EcoInvent provide some insight on anticipated impacts and differences from using the different data. The differences range from +55% for stone wool insulation to −137% for cement base plaster. Part of these differences are due to the specific primary energy conversion factors that are used for electricity (e.g., 2.9 in Greece) and the fuel mix accounted for in EcoInvent LCI. In some cases, there are different manufacturing processes. For example, in Greece, the production of cement mortar and base plaster uses limestone sand, while the production process described in EcoInvent LCI considers the use of silica sand that is a more energy-intensive process. Similarly, in Greece, the production of steel rebar uses mainly scrap material, whereas the production process in EcoInvent LCI is mainly based on 84% raw steel that demands larger amounts of energy. On the other hand, considering the energy intensive production process of aluminium and the high electricity consumption, the derived EE coefficient is higher since the calculations in EcoInvent LCI are based on a fuel mix for electricity generation mainly from hydro.

The available national data or EPDs that refer to the production of insulation materials is limited. The value listed in Table 3 refers to stone wool. However, during the follow-up calculations for the case studies with insulated building elements, the material considered refers to extruded polystyrene since this is the most commonly used material in Hellenic buildings. The value used for the calculations was derived at 100.85 MJ/kg from EcoInvent for extruded polystyrene.

3.3. Energy Use Intensities

The annual operational energy use was calculated independently for each case study in the different tiers. The calculations were performed for each of the 16 case studies presented in Table 1 for all four climate zones (A-D). The estimates were then adapted with the corresponding correction factors that are presented in Appendix A using Equation (3). The adapted EUIs represent more realistic estimates of the actual energy use for the representative SFHs.

Table 3. Production processes and derived embodied energy coefficients (MJ/kg) in the Hellenic dataset from the field surveys (cradle-to-gate) or calculated from other Hellenic data sources using EcoInvent.

Material	Overview of Production Process	Field Survey	Other Sources	Coefficient (MJ/kg)	Difference from EcoInvent
Aluminium	Primary aluminium ingot production from a Hellenic manufacturing facility; bauxite ore from national mines and imports is used. Bauxite is refined into alumina that is then treated in a high temperature kiln to produce aluminium oxide that next undergoes an electrolytic reaction to smelt the material into aluminium. The processes are very energy intensive using large amounts of electricity for the electrolysis and fuel for the combustion in the kilns. Using recycled raw materials that can be melted at a lower temperature results in energy savings. For aluminium window profiles, the calculations are based on primary aluminium ingot production (60%) and recycled aluminium (40%) for 1 m^2 window from EPDs [66] with coated profiles and 24 mm thermal breaks at 2.59 GJ/m^2.		Hellenic data [33] and EcoInvent	209.5	43.4%
Bricks (burnt clay)	Clay is extracted from a local quarry and transported by tracks (30 km roundtrip) to the facility; water is pumped from a well at the site. The raw material is crushed, grounded, and screened re-grounded, if necessary. The fine material is mixed with water and finally shaped with machine moulding. The formed bricks are dried in enclosed spaces that are heated with a biomass (olive pit pellet) furnace and then pass through an oil-fired kiln for baking. The final product is 8 × 10.5 × 19 cm and weight 1150 gr/unit. Used clay and intermediate products are recycled back to the raw material feeders. Defective bricks and other scrap are also crushed and used as ceramic powder for use in lime mortars or gravel for road construction projects.	✓		2.62	−12.5%
Cement	Facility for the production CEM II/A-M Portland-composite cement. Raw materials are extracted from local quarries (e.g., limestone at 1.2 km roundtrip, clay or sandstone at 28 km) are mixed with small quantities of alumina and iron oxide. A fine powder of the raw materials is mixed (e.g., limestone 1250 kg/tonne clinker, clay minerals 300 kg/tn, liquid ammonia 20 kg/tn and other alternative materials 20 kg/tn) and fed into a high temperature dry-process kiln. The clinker is grinded and mixed with calcium sulphate (gypsum) and minor quantities of additional constituents (e.g., pozzolana, fly ash) to produce the known grey powder material. Electricity is mainly used in the stage of crushing and pre-homogenization of raw materials, as well as for the rotating furnaces. Fuel oil, petcoke, and coal are also used during the calcification phase in rotary kilns.	✓		1.02	52.2%
Concrete (ready-mix)	The facility for the production of ready-mix concrete receives the cement from a manufacturing site located 100 km away and the transportation is made by trucks. For cement production and transportation to the site, the value becomes 3.3 MJ/kg cement. The quarry for mining the aggregates is within a radius of 9 km from the treatment plant and are transported by trucks. The aggregates are processed, crashed, cleaned, and stored at the ready-mix concrete production site and include gravel (16–31.5 mm diameter), washed gravel (12–31.5 mm), fine limestone (6–12 mm), and sand (<6 mm).	✓		0.65	0.5%
Concrete (lean)	Material with less than 10% cement content.	✓	EcoInvent	0.47	14.0%
Insulation (stone wool)	Production of stonewool products (30–200 kg/m^3) from a Hellenic manufacturing facility. Raw materials include amphibolite (70% p.w.), bauxite (8% p.w.), lime calcium oxide (8% p.w.), and dolomite (4% p.w.) that are transported to the plant and reuse of its own recycled stonewool material (10% p.w.). The facility uses an immersed electric arc furnace for melting the raw materials, followed by fiberizing, injection of binders, and curing. Other materials used for the final product include the production of various resins, auxiliary materials (silicone, mineral oil), materials used as back cover (e.g., aluminium foil) and packaging.	✓	Hellenic data [31,41] and EcoInvent	37.8	55.1%
Mosaic	Mosaic and Gabriel mosaic floors were very popular cast floors in the past, as economical, easy, and long-lasting floor cover, for both interior and exterior spaces. The raw materials include cement mortar made of white cement and small pieces of marble (waste). After the material have solidified and dried, the surface is sanded and polished.	✓	EcoInvent	1.04	3.9%
Cement mortar and base plaster	Facility for the production of various types of ready to use cement-based mortars. Various modes of transportation are used for the raw materials (e.g., sand, limestone, cement). Coefficients calculated for different mixing ratios for (a) pressed cement screed for walls and flooring; (b) material used in masonry construction to fill the gaps between the bricks and blocks.	✓		0.72 0.78	−136.5% −88.5%
Roof (clay) tiles	Similar production process with burnt clay bricks. The formed roman roof tiles are dried for longer periods (24–36 h) using heat and exhaust gases from an oil-fired, LPG, or biomass (olive pit pellet) furnace, followed by an additional firing process for surface treatment in order to melt and adhere the glazed surface finish. Final product 42 × 17.5 cm and weight 3850 gr/unit.	✓		3.69	−12.5%
Steel rebar	Steel production is a major industrial and energy intensive activity. Iron ore, limestone, and coke can be fed into a furnace for melting into cast iron and the carbon content can be controlled through further smelting. The production considered here is based on the use of ferrous scrap that is mostly imported and an electric arc furnace to produce weldable bars and rods to strengthen concrete in load-bearing building construction. Production satisfies the domestic market demand, although some facilities have faced major obstacles during the recession and slowdown in construction activities.		Hellenic data [32] and EcoInvent	16.02	−36.7%

The resulting actual energy use intensities are summarized in the matrix on the right in Figure 5. For example, the representative building from the first construction period (e.g., pre-1980) described in Table 1 for the climate zone C is identified as SFH-1C. This corresponds to an adapted annual EUI of 1.449 GJ/m^2 that is identified in bold. Similar calculations are performed as if the building was also located in climate zones A, B, and D; the corresponding results are included in the summary of the EUIs in Table 1 under the corresponding heading for climate zones from A to D. For all 16 case studies, the values average 0.570 GJ/m^2 in the south (climate zone A) and reach 0.991 GJ/m^2 in the north (climate zone D).

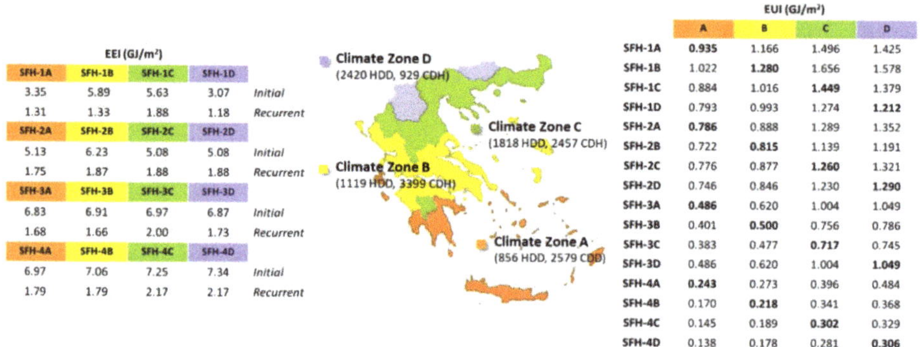

	EUI (GJ/m^2)			
	A	B	C	D
SFH-1A	**0.935**	1.166	1.496	1.425
SFH-1B	1.022	**1.280**	1.656	1.578
SFH-1C	0.884	1.016	**1.449**	1.379
SFH-1D	0.793	0.993	1.274	**1.212**
SFH-2A	**0.786**	0.888	1.289	1.352
SFH-2B	0.722	**0.815**	1.139	1.191
SFH-2C	0.776	0.877	**1.260**	1.321
SFH-2D	0.746	0.846	1.230	**1.290**
SFH-3A	**0.486**	0.620	1.004	1.049
SFH-3B	0.401	**0.500**	0.756	0.786
SFH-3C	0.383	0.477	**0.717**	0.745
SFH-3D	0.486	0.620	1.004	**1.049**
SFH-4A	**0.243**	0.273	0.396	0.484
SFH-4B	0.170	**0.218**	0.341	0.368
SFH-4C	0.145	0.189	**0.302**	0.329
SFH-4D	0.138	0.178	0.281	**0.306**

Figure 5. Primary initial EE and recurrent EE (for an 80-year building lifetime) embodied energy intensities (EEIs) and annual energy use intensities (EUIs) expressed in GJ/m^2 for all the 16 case study buildings (Table 1) in the four climate zones of Greece, including the average heating degree days (HDD) and cooling degree days (CDD) in each zone.

3.4. Embodied Energy Intensities

The EEI for each material was calculated using the corresponding quantities from the BoM and the embodied energy coefficients from the Hellenic dataset (Table 3). The prevailing materials in terms of their contribution to the total embodied energy of the buildings include concrete that averages 30.4%, followed by steel rebar at 25.0%, which are the materials with the largest quantities and high EE coefficients. On the other hand, although some materials may be encountered at low quantities, they may have high EE coefficients and as a result, a relatively high contribution in the total building's embodied energy. For example, thermal insulation materials contribute by as much as 10.7% for SFH-4 buildings (Table 1), which correspond to recent construction periods with stricter thermal envelope requirements that mandate the use of more insulation. The use of aluminium window frames, which have dominated recent construction practices, may have a low mass contribution of about 0.02%, but the embodied energy contributes up to 9.1% for SFH-4 buildings.

Considering the main building structural elements illustrated in Figure 4, the embodied energy of the load-bearing structures dominates from 58.9% up to 69.9% of the total. The openings (e.g., windows, balcony doors) have a minute mass contribution, but a notable share in the total embodied energy of the buildings. The values range from 9.7% up to 10.8% since the case studies SFH-2 to −4 use aluminium frames (Table 1).

The calculation of the initial embodied energy was extended to include transportation from the factory gate to the building site, corresponding to the cradle-to-site. Accordingly, the calculations assumed an average distance of 100 km for transportation from the factory gate to the building site for practically all building products. In the case of concrete, the assumed distance was 40 km.

The results of the calculated initial and recurrent embodied energy for the different building construction materials in the 16 case studies are summarized in Figure 5. The embodied energy intensities have been calculated for each case study and remain the same for all locations. The value of the recurrent EE was calculated for different building lifetimes, but the value included in Figure 5 refers to the case of a building lifetime that corresponds to 80 years.

At this stage, the work has focused on construction materials that constitute the most significant percentage of a building's embodied energy, and from the technical installations include only the heat generation equipment and solar collectors, where applicable. This is a reasonable first step, since previous studies have quantified that the building envelope represents most of the embodied energy compared to the other E/M installations of residential buildings, estimated at about 2460 MJ/m^2 and 99 MJ/m^2, respectively [67].

3.5. Lifetime Calculations

The recurrent EE is estimated for the building's lifetime. The renovation of the building envelope (e.g., thermal insulation, double glazing) is assumed to meet the code minimum requirements that correspond to the post-2010 KENAK standards for the maximum U-values of the building components. Similarly, the renovations of the equipment and technical installations considered in this work include EEMs, like the replacement of the heat generation equipment with a more efficient boiler and the installation of solar thermal collectors for DHW. As previously noted, these renovation measures will impact the cycles for recurrent EE depending on their respective service time (Table 2) during the building's lifetime. The annual energy savings as a result of the improved energy performance of the renovated buildings are calculated for each measure and then considered all together as a combined scenario in each one of the 16 representative buildings in the four climate zones. The calculated values for the operational energy consumption are also adapted to obtain more realistic estimates of the actual energy consumption using the derived adaptation factors (Figure 6) that correspond to the representative buildings. The results are then compared to the total EE of the various building products used in each case study in order to estimate the corresponding energy recovery time.

In all cases, the calculations of the recurrent energy were simplified to consider that the building materials or products will be replaced with others that comply with the current standards and technical characteristics. In other words, with the current approach, there is no consideration of stricter building standards and technological advances of new materials and equipment that will become available in the market during the following decades. In this case, one may reasonably expect that the building's operational energy use may be lower in the coming decades, which would mean that the EE share would be higher.

3.5.1. Embodied and Operational Energy Use

The relative importance of the embodied energy for the 16 case studies can be significantly different. The time it will take for the adapted operational energy use to match the initial EE of the buildings, if there are no changes in construction and manufacturing practices, is quantified in Table 4 for each building. The minimum and maximum values are identified with the grey highlighted cells. The results confirm that the embodied energy is becoming increasingly important for the high-performance buildings in the south climate zones. For example, it will take over 53 years for a building from the post-2010 tier (SFH-4D) with a low annual energy use to match its initial EE, in the event that such a building would operate in climate zone A. On the other hand, the high annual energy use of an old building from the pre-1980 tier (SFH-1A) will take just over two years if it operates in climate zone C to match the building's initial EE.

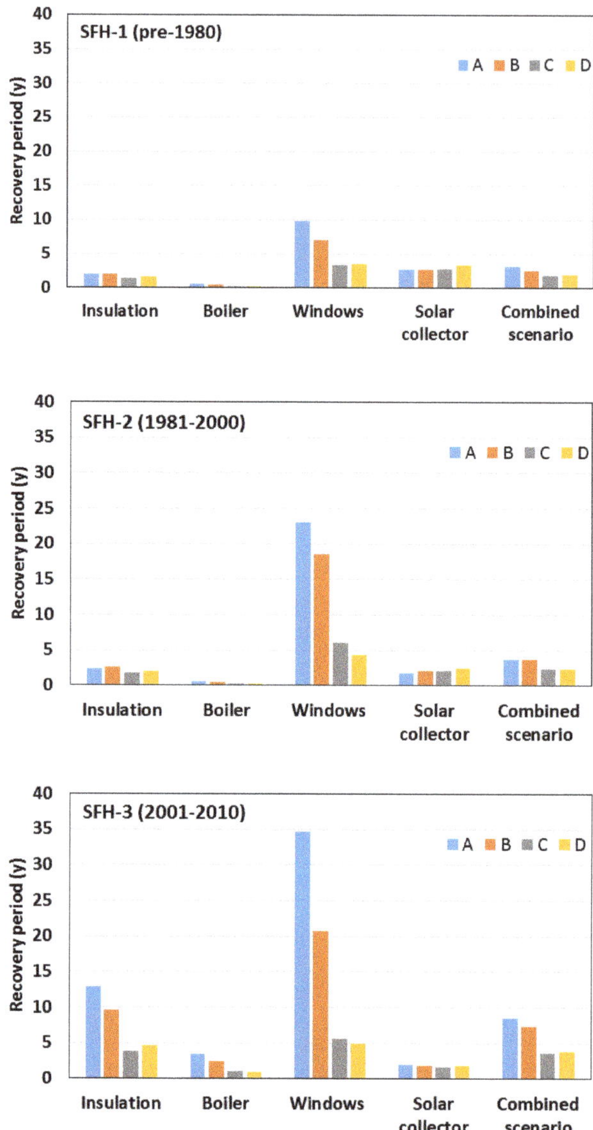

Figure 6. Energy recovery time (years) for individual energy conservation and efficiency measures and a combined scenario for the different SFH case studies in the four climate zones.

Table 4. Time (years) for the adapted annual operational energy use to match the initial EE for the 16 case studies in the four climate zones.

Case Study	Climate Zone			
Buildings	A	B	C	D
SFH-1A	3.6	2.9	2.2	2.4
SFH-2A	6.5	5.8	4.0	3.8
SFH-3A	14.1	11.0	6.8	6.5
SFH-4A	28.7	25.6	17.6	14.4
SFH-1B	5.7	4.6	3.5	3.7
SFH-2B	8.6	7.6	5.5	5.2
SFH-3B	17.2	13.8	9.1	8.8
SFH-4B	41.6	32.4	20.7	19.2
SFH-1C	6.3	5.5	3.9	4.1
SFH-2C	6.5	5.8	4.0	3.8
SFH-3C	18.2	14.6	9.7	9.4
SFH-4C	49.8	38.4	24.0	22.0
SFH-1D	3.9	3.1	2.4	2.5
SFH-2D	6.8	6.0	4.1	3.9
SFH-3D	14.1	11.1	6.8	6.5
SFH-4D	53.2	41.2	26.1	24.0

3.5.2. Energy Recovery Time

The annual energy savings resulting from the implementation of various ECMs or EEMs for each one of the 16 case studies were calculated for the different tiers. The most significant energy savings are anticipated from the addition of thermal insulation on the building's opaque elements (e.g., walls, roof) to reduce space heating, which actually represents the highest percentage of total energy use. The addition of thermal insulation is assumed to meet the minimum code requirements of the new 2017 KENAK for the opaque elements of the building envelope. Among the most popular ECMs in existing residential buildings is the renovation of windows, although it may not be the most effective measure for optimizing the building's energy performance [49]. The popularity of this measure is actually driven by a combination of several other factors that go beyond the improved energy performance and thermal comfort, e.g., improved aesthetics, sense of security, better acoustical comfort.

Two additional EEMs are also considered. One refers to the installation of a new boiler for space heating with a more efficient unit, and one for the installation of solar thermal collector to cover 60% of the domestic hot water demand, according to the minimum code requirements of KENAK. Each measure was first assessed individually and then all together as a combined scenario for the 12 case studies. This includes the tiers for the pre-1980 buildings and the following two construction periods, but excludes the renovation of the most recent constructions under the post-2010 buildings that would not be currently considered as cost-effective measures.

The calculated annual operational energy for each case study considered in all four climate zones were adapted for more realistic estimates of the actual energy savings and then compared with the resulting increase of the embodied energy. This is an estimate of the time it would take for the anticipated annual savings to counterbalance the invested embodied energy of the building products used for the ECMs or EEMs considered in the analysis.

For the addition of thermal insulation to the opaque building elements of the various case study buildings in all climate zones, the shortest energy recovery time of 1.1 years occurred for the SFH-1A building operating under climate zone C conditions (Table 5). However, this will stretch to over 21 years for buildings with good thermal performance from the third tier, operating in the warm, south climate zone, i.e., SFH-3A under climate zone A conditions. For windows, the corresponding periods are relatively much longer. The shortest energy recovery time of 2.3 years occurred for SFH-2D under the coldest, north comate zone D. Once again, the longest period for over 37 years would occur from a building with a good thermal performance from the third tier (i.e., SFH-3A), operating in the warm, south climate zone A that even exceeds their theoretical service lifetime of this building product. The results for the EEMs are reasonable ranging from 0.2 years up to 4.6 years for replacing the heat production unit with a new, energy efficient condensing boiler, and from 1.6 years up to 3.5 years for the use of solar collectors for DHW. Finally, the combined scenario with all the individual measures considered together for major renovations of the case study buildings would range from a minimum of 1.6 years for the oldest buildings in tier SFH-1A and SFH-1B, up to a maximum of 9.5 years, for the relatively recent constructions in tier SFH-3C buildings.

Table 5. Energy recovery time (years) for individual energy conservation and efficiency measures and a combined scenario for the different case studies.

Buildings	Climate Zone				Climate Zone				Climate Zone				Climate Zone				Climate Zone			
	A	B	C	D	A	B	C	D	A	B	C	D	A	B	C	D	A	B	C	D
	Insulation				Windows				Boiler				Solar Collector				Combined			
SFH-1A	1.5	1.3	1.1	1.3	9.7	6.9	3.6	3.7	0.6	0.4	0.3	0.3	2.6	2.5	2.7	3.3	2.5	2.1	1.6	1.8
SFH-2A	2.1	2.3	1.5	1.7	21.0	16.1	5.6	4.9	0.5	0.4	0.2	0.2	1.7	2.0	2.1	2.4	3.2	3.2	2.0	2.1
SFH-3A	21.4	14.8	5.6	5.6	37.1	22.1	5.5	4.8	2.5	1.8	0.8	0.8	2.0	1.9	1.6	1.9	9.0	7.9	3.9	3.9
SFH-1B	1.6	1.4	1.2	1.4	9.9	6.9	3.7	3.7	0.5	0.4	0.2	0.2	2.8	2.7	2.9	3.5	2.5	2.1	1.6	1.8
SFH-2B	2.1	2.7	2.0	2.2	22.6	17.4	6.1	4.5	0.7	0.6	0.3	0.3	1.7	2.0	2.1	2.4	4.1	4.1	2.6	2.6
SFH-3B	15.1	10.5	2.5	5.4	32.8	19.8	5.7	5.0	4.2	2.9	1.3	1.1	2.0	1.9	1.6	1.9	9.5	8.0	3.4	4.2
SFH-1C	2.8	3.0	1.7	1.9	9.9	7.1	2.7	2.9	0.6	0.4	0.3	0.3	2.8	2.7	2.9	3.5	4.1	3.3	2.0	2.1
SFH-2C	2.6	2.7	1.8	1.9	24.9	19.0	6.4	5.7	0.5	0.5	0.3	0.2	1.7	2.0	2.1	2.4	3.8	3.7	2.3	2.4
SFH-3C	12.3	9.9	5.4	5.4	33.6	19.6	5.6	5.0	4.6	3.2	1.3	1.2	2.0	1.9	1.6	1.9	9.5	8.1	4.3	4.3
SFH-1D	2.3	2.0	1.7	1.9	9.9	7.0	3.6	3.7	0.7	0.5	0.3	0.3	2.7	2.7	2.8	3.4	3.5	2.8	2.1	2.3
SFH-2D	2.7	2.8	1.8	2.0	23.7	21.6	6.1	2.3	0.5	0.5	0.3	0.2	2.0	2.3	2.3	2.5	3.9	3.8	2.3	2.4
SFH-3D	2.4	3.0	2.1	2.4	35.3	21.5	5.4	4.8	2.5	1.8	0.8	0.8	2.0	1.9	1.6	1.9	6.1	5.3	2.9	2.9

According to published results from the assessment of various ECMs under average European climate conditions, the energy recovery time ranges from 0.1 to 2.0 years for the addition of more thermal insulation and 1.4 to 2.1 years for the installation of thermal solar collectors [20]. The results for US buildings have been reported to range from around 3–5 years for old residential buildings constructed before the introduction of energy codes and 1.6–3.2 years for existing buildings in compliance with some energy codes [68].

4. Discussion

One of the main objectives of this work was to quantify and prioritize building construction materials in representative Hellenic single-family houses. The vision is to prepare for the emerging trend of lifecycle analysis of building performance and to facilitate relevant work at the early stages of the decision-making process to account for the embodied energy of building materials and products.

4.1. Hellenic Dataset

As it was anticipated, the data collection from local manufacturing facilities proved a time-demanding process. The major obstacle was that most of the time, the process stumbled on the owner's reluctance to allow access or release the necessary data since it is considered "proprietary" information. Another obstacle is that the voluntary EPDs remain very scarce, either because they are not public information or manufacturers have not yet considered the benefits for developing EPDs and disclosing relevant data.

Overall, the availability of national and local information continues to be very scarce in Greece and other European member states. This should be taken into consideration during the development of European mandates and regulations on LCA calculations like LEVEL(s) [12].

4.2. Embodied and Operational Energy Use

The present work elaborated 16 representative Hellenic single-family houses. The estimated initial EE averages only 3–9%% of the total operational energy use over 80-year lifetime, for old buildings that are not thermally insulated, i.e., SFH-1 for the pre-1980 tier. On the other hand, the work also confirmed the relative high importance of embodied energy for high energy performance buildings reaching 18–67% of the operational energy use for the most recent tier building constructions, i.e., SFH-4 for post-2010 buildings.

The results agree with the findings from other European studies. For example, in Belgium, the embodied energy of non-thermally insulated and inefficient residential buildings represents only 4% compared to the high total primary energy use over a 30-year lifetime, while for nZEBs, the embodied energy is 50% higher than the lifetime operational energy use [67]. Similar work in the Netherlands has reported that for an average residential building construction, the embodied energy represents 10% to 12% of the total operational energy use for typical homes and reaches 36% to 46% in energy-efficient homes [69].

4.3. Renovations and Energy Recovery Time

The present work assessed some common energy conservation and efficiency measures for the renovation of existing buildings to the minimum code requirements of KENAK. The measures included the addition of thermal envelope insulation, the replacement of windows, and for the E/M installations the replacement of heat production system with a high efficiency boiler and the addition of solar thermal collector for DHW. The energy recovery time for the individual measures and a combined scenario that incorporates all of them are illustrated in Figure 6. In all cases, the replacement of windows has the longest energy-recovery time, ranging from 3.4 years up to 35 years. The adapted operational energy savings range from 0.015 GJ/m^2 to 0.259 GJ/m^2, while the EEI range from 0.539 GJ/m^2 to 0.652 GJ/m^2. On the other end, the measure with the shortest recovery period is the replacement of the boiler, averaging from about 1 year up to 3.5 years for the more recent building tier. The operational energy savings range from 0.008 GJ/m^2 to 0.305 GJ/m^2, while the embodied energy ranges from 0.038 GJ/m^2 to 0.071 GJ/m^2. The addition of thermal insulation on the building's envelope results to annual energy savings from 0.007 GJ/m^2 to 0.749 GJ/m^2 and an embodied energy intensity from 0.118 GJ/m^2 to 1.040 GJ/m^2. The annual energy savings from the installation of solar thermal collectors to cover 60% of the DHW range from 0.053 GJ/m^2 to 0.084 GJ/m^2, while the EEI is estimated from 0.114 GJ/m^2 to 0.214 GJ/m^2. Finally, for the combined scenario of all the individual measures applied together, the adapted annual energy savings range from 0.093 GJ/m^2 to 1.123 GJ/m^2, while the combined EEI ranges from 0.808 GJ/m^2 to 1.928 GJ/m^2.

4.4. Future Work

Parallel efforts and similar analysis as the one elaborated in this work can focus on building electromechanical installations, equipment, and systems. Priority should be given to ones that are usually part of building renovations, including upgrades to higher performance equipment, the installation of new equipment that exploit renewables, etc. At this stage of the work, the service lifetime of the equipment is simply used to estimate the relevant cycles for replacing them and estimating the recurrent EE, without accounting for any improved future performance. However, beyond this quantitative assessment, it would be more realistic to also have a qualitative assessment and account for an improved performance that may also reduce the operational energy. For example, replacing a heat production system once it reaches its service lifetime would be more realistic to account for a higher thermal efficiency of a new boiler after a 20- or 40-year cycle. Technological advances that improve equipment's performance will also have a positive impact on the building's performance and lower the operational energy use.

The case studies considered in this work are detached single-family houses that constitute 73% of the residential building stock. A similar approach can be followed in the future for multi-family houses. Other building types, e.g., semidetached and row houses, may also be considered in the future, although detailed statistical data are not currently available for a building stock analysis. For practical purposes, the anticipated differences will be limited as a result of the slightly lower material quantities used for the intermediate load-bearing walls, and the lower operational energy use. The non-residential building sector is more difficult to handle since it is very heterogeneous, including various building typologies, functions, sizes, etc., making it more challenging to organize the available data and define the different tiers.

Future work should also consider an improved efficiency and/or primary energy factor (PEF) for electricity generation in the following decades. For example, reflecting on technological progress and the growing share of renewable energy sources in the electricity generation sector, the default European average PEF for electricity used in the European Energy Efficiency Directive (EED 2018/2002) has been lowered from 2.5 to 2.1 and will be revised every four years.

A PEF of 2.9 for electricity was considered in this work, according to the national technical guidelines that were published in 2010. Apparently, the value is outdated and should be revised with a different fuel mix for electricity generation to make more realistic future projections. As illustrated in Figure 7, the time evolution of the PEF in Greece, calculated as the ratio of (primary energy consumption + imports)/final electricity consumption, is following a dropping trend. Beyond the average historic trend line, the projections may be based on the data over the last decade, 2008–2017 (ambitious trend), to reflect more aggressive efforts towards the renewables target (e.g., >32% share of renewables in final energy use), fuel switching from lignite to natural gas, and the overall decarbonization efforts of energy transformation. Greece had recently announced a phase-out of coal-fired power generation by 2028. Even if there may be some delay as a result of the expected economic recession following the COVID-19 pandemic, these developments are of particular importance for the operational and embodied energy use, primary energy, and emission reduction, as buildings move into the era of electrification.

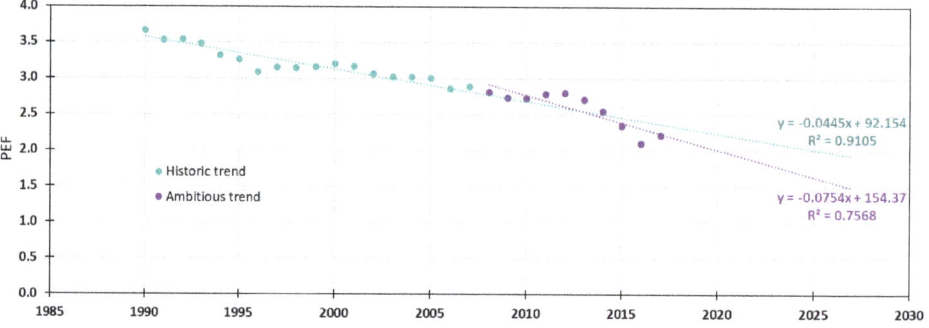

Figure 7. Time evolution of total primary energy factors for electricity generation in Greece (Data source: Energy balances from [1]).

Future work should also account for technological advances in industry and improved production methods and more energy-efficient processes for the production of materials, systems, and other building products. More emphasis may also be placed on recycling and the use of more environmentally-friendly materials. Finally, the system boundaries should be extended to include the end-of-life stage, i.e., the demolition embodied energy.

5. Conclusions

Assessing the embodied energy of buildings during deep renovations is an emerging issue of particular importance in the era nZEBs and decarbonized building stock by 2050. However, readily available national information and data that can be used for calculations and initial benchmarking is limited in some countries. In this direction, this work has set up and implemented a methodology for assessing the operational and embodied energy of national building stocks and for providing some guidance in the design and renovation of buildings. The results can be used by professionals as benchmarks for a first assessment of similar buildings' energy performance and how their different design solutions or renovations compare against the baselines. This preliminary insight information can also be used to customize energy conservation measures or reconsider specific energy efficiency measures that merit further consideration and more detailed analysis. The results are also of interest to policy makers that need to establish relevant baselines, set priorities and targets, and monitor progress for sustainable buildings.

As a first step, the work considered single-family houses that represent the most common building use. Short energy audits were performed in local manufacturing facilities in an effort to enhance the knowledge base on embodied energy coefficients of common construction materials in Greece. The analysis was performed using 16 case studies, representing various buildings at different tiers in terms of characteristic building construction periods and locations related to the four national climate zones. The total (initial and recurrent) embodied energy and the adapted annual energy use indicators for Hellenic SFH range from 4.26 to 9.50 GJ/m^2 and 0.14 to 1.66 GJ/m^2, respectively. Depending on the lifecycle of buildings, the embodied energy recovery times for major renovations as a combined scenario of various energy conservation and efficiency measures range from 1.6 to 9.5 years.

The method and accumulated data from this work is also useful for assessing other types of buildings. On-going work considers multi-family houses and representative non-residential buildings that reflect the largest percentage in terms of floor areas (e.g., offices, schools). In addition, there is a need to enhance the Hellenic database of EE coefficients with more local and national data of building construction material and system production processes. Information from the EPDs will be of great value, provided that they become mandatory in order to increase their availability. From the experience gained during this work, it is evident that the related industry in Greece has not yet adapted to the voluntary mandate for developing and releasing EPDs, while some even consider this type of data as proprietary information.

A similar approach focusing on embodied carbon or GHG emissions is also an interesting evolution. Again, efforts should optimize impacts from both operational and embodied emissions. Along these lines, this work needs to account for the European and national efforts to decarbonize power plants that influence operational energy and carbon emissions, along with improved production processes that reduce primary energy use and improve the carbon footprint of the relevant industrial and manufacturing processes.

Author Contributions: Conceptualization, C.A.B. and E.G.D.; methodology, E.G.D. and C.A.B.; formal analysis, P.A.A., E.G.D., C.A.B., K.G.D., and S.K.; investigation, P.A.A. and E.G.D.; data curation, P.A.A. and E.G.D.; writing—original draft, E.G.D.; writing—review and editing, C.A.B., P.A.A., K.G.D., and S.K.; visualization, E.G.D. and P.A.A.; supervision, E.G.D.; project technical coordinator, E.G.D. All authors have read and agreed to the published version of the manuscript.

Funding: This research was funded by the Operational Programme "Competitiveness, Entrepreneurship and Innovation" (NSRF 2014–2020) of the Hellenic Ministry of Development & Investments, and co-financed by Greece and the European Union (European Regional Development Fund). The APC was also funded by the same sources.

Acknowledgments: This work was performed in the frame of the project "THESPIA II—Foundations of synergistic and integrated management methodologies and tools for monitoring and forecasting of environmental issues and pressures" (MIS 5002517), implemented under the action, "Reinforcement of the Research and Innovation Infrastructure". The paper reflects the views only of the authors who have made every effort to prepare this material for the benefit of the public in light of current and available information. It does not represent the opinion of the European Union or the Hellenic Ministry of Development & Investments. Neither the European Union and Hellenic institutions and bodies nor the authors may be held responsible for the use which may be made of the information contained therein. The authors gratefully acknowledge the collaboration of the owners and technical staff at the production sites that participated in the current field work. The national EPC repository (buildingcert) has been developed and maintained by the Hellenic Ministry of Environment & Energy (YPEN) in collaboration with the Centre for Renewable Energy Sources. The authors wish to acknowledge YPEN for allowing access to the EPC database. The analysis presented herein does not necessarily reflect the opinion of the Ministry.

Conflicts of Interest: The authors declare no conflict of interest. The funders had no role in the design of the study; in the collection, analyses, or interpretation of data; in the writing of the manuscript, or in the decision to publish the results.

Abbreviations

A	Heated floor area of a building (m^2)
BoM	Bill of Materials
DHW	Domestic Hot Water
ECM	Energy Conservation Measures
EE	Embodied Energy
EEI	Embodied Energy Intensity, i.e., total embodied energy per unit floor area (e.g., GJ/m^2)
EEM	Energy Efficiency Measures
EPBD	Energy Performance of Building Directive
EPC	Energy Performance Certificate
EPD	Environmental Product Declaration
EUI	Energy Use Intensity, i.e., operational energy per unit floor area (e.g., GJ/m^2)
HDD	Heating Degree Days (K.Days)
CDH	Cooling Degree Hours (K.Hours)
ICE	Inventory of Carbon and Energy
KENAK	Hellenic national regulation on the energy performance of buildings
LCA	Life Cycle Assessment
LCI	Life Cycle Inventory
MFH	Multi-Family Houses
nZEB	nearly Zero Energy Buildings
SFH	Single-Family Houses

Appendix A. Adaptation Factors for EUIs

The derived adaptation f-factors from 1734 EPCs are given in Figure A1. The clustered data of the calculated and actual energy use are organized are analysed for the different tiers of the four construction periods and the four climate zones. The 45°dashed lines (i.e., x = y) identify the (ideal) cases when the calculated and actual energy use are in perfect agreement. The large data scatter reflects the anticipated differences in the calculated EUIs resulting from variations in the building construction and systems performance, the location, weather data, etc. The variability of the actual EUIs reflect the different occupants' interactions with the space heating system (e.g., actual operating periods, indoor temperature set-points) for buildings in the same tier. In principle, a low actual energy use should not be automatically interpreted as a high performing building, unless one considers that the prevailing indoor environmental quality is acceptable.

The average and median adaptation factors, i.e., defined as the ratio of the actual to the calculated energy use for space heating, are inserted as legends in Figure A1 for each tier. Each bullet corresponds to an SFH and the number of cases in each tier are also identified in each graph. The 45° dashed lines (i.e., x = y) identify the (ideal) cases when the calculated and actual energy use are in perfect agreement. Cases below the diagonal show a prebound effect (i.e., the actual EUI is lower than the calculated) and are more evident for low performance buildings (high calculated EUIs). Cases over the diagonal show a rebound effect (i.e., the actual EUI is higher than the calculated) and are more common for high performance buildings (low calculated EUIs).

The average adaptation factor is 0.536 (median 0.501), which means that the actual energy use for space heating and DHW is 46.4% (49.9%) lower than the calculated value. The results are in-line with similar values published in the literature that also document large deviations of actual and calculated energy use from 30% to 47% [70,71].

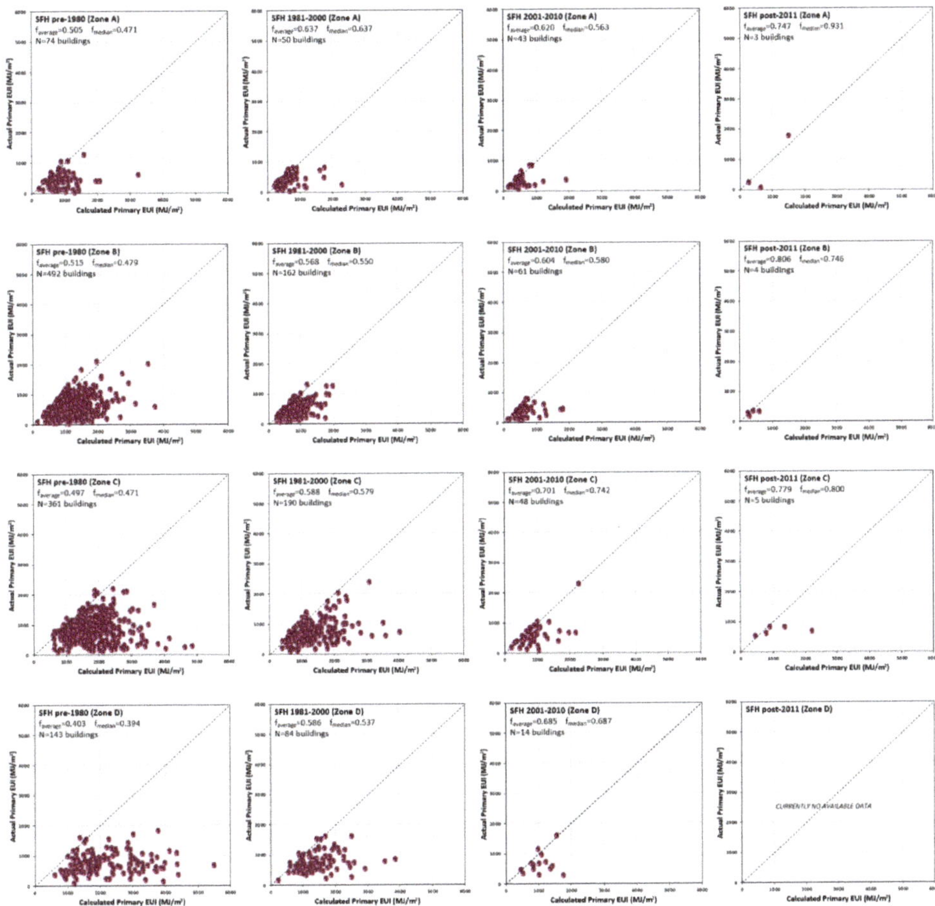

Figure A1. Calculated and actual space heating and DHW primary energy use intensity (MJ/m^2) from 1734 SFHs in Greece, grouped for the different tiers (four climate zones A–D and four construction periods).

Currently, the available data for new buildings are missing (SFH post-2011 in Zone D) or have not yet reached a sufficient population (e.g., SFH post-2011 in Zone A), so they should be used with caution. For the case of the SFH-4D that corresponds to new constructions (post-2011) in the north climate zone D, there is currently no available data from the EPCs. In this case, the adaptation factor used in the calculations was taken as the median value of 0.853 that corresponds to the data for the post-2010 time period.

References

1. Eurostat. *Energy Statistical Country Datasheets*; European Commission, DG Energy, Unit A4: Brussels, Belgium, 2019; Available online: https://ec.europa.eu/eurostat/web/energy/data/energy-balances (accessed on 21 May 2020).
2. European Commission. *Census Hub—European Statistical System*; European Commission: Brussels, Belgium, 2011; Available online: https://ec.europa.eu/CensusHub2 (accessed on 21 May 2020).

3. Pezzutto, S. Open Data Set for the EU28. Hotmaps Project, D2.3 WP2 Report. 2019. Available online: https://www.hotmaps-project.eu/wp-content/uploads/2018/03/D2.3-Hotmaps_for-upload_revised-final_.pdf (accessed on 21 May 2020).

4. EU Climate Strategies & Targets. Available online: https://ec.europa.eu/clima/policies/strategies/2050_en (accessed on 21 May 2020).

5. *Directive (EU) 2018/844 of the European Parliament and of the Council of 30 May 2018 Amending Directive 2010/31/EU on the Energy Performance of Buildings and Directive 2012/27/EU on Energy Efficiency*; European Commission: Brussels, Belgium, 2018; Available online: http://data.europa.eu/eli/dir/2018/844/oj (accessed on 21 May 2020).

6. Filippidou, F.; Jimenez Navarro, J.P. *Achieving the Cost-Effective Energy Transformation of Europe's Buildings*; EUR 29906 EN; Publications Office of the European Union: Luxembourg, 2019; p. 50. [CrossRef]

7. Ferrara, M.; Monetti, V.; Fabrizio, E. Cost-Optimal Analysis for Nearly Zero Energy Buildings Design and Optimization: A Critical Review. *Energies* **2018**, *11*, 1478. [CrossRef]

8. D'Agostino, D.; Mazzarella, L. What is a Nearly zero energy building? Overview, implementation and comparison of definitions. *J. Build. Eng.* **2019**, *21*, 200–212. [CrossRef]

9. Chen, J.; Zhou, W.; Yang, H. Is Embodied Energy a Better Starting Point for Solving Energy Security Issues?—Based on an Overview of Embodied Energy-Related Research. *Sustainability* **2019**, *11*, 4260. [CrossRef]

10. Dixit, M.K. Life cycle recurrent embodied energy calculation of buildings: A review. *J. Clean. Prod.* **2019**, *209*, 731–754. [CrossRef]

11. Frischknecht, R. *Basics for the Assessment of Embodied Energy and Embodied GHG Emissions for Building Construction, Guideline for Designers and Consultants—Part 1*; IEA EBC Annex 57; Lützkendorf, T., Balouktsi, M., Eds.; Institute for Building Environment and Energy Conservation: Tokyo, Japan, 2016; Available online: http://www.iea-ebc.org/Data/publications/EBC_Annex_57_Guideline_for_Designers_Part_1.pdf (accessed on 21 May 2020).

12. Building Sustainability Performance—Level(s). Available online: https://ec.europa.eu/environment/eussd/buildings.htm (accessed on 21 May 2020).

13. Vilches, A.; Garcia-Martinez, A.; Sanchez-Montanes, B. Life cycle assessment (LCA) of building refurbishment: A literature review. *Energy Build.* **2017**, *135*, 286–301. [CrossRef]

14. Azari, R.; Abbasabadi, N. Embodied energy of buildings: A review of data, methods, challenges, and research trends. *Energy Build.* **2018**, *168*, 225–235. [CrossRef]

15. Chastas, P.; Theodosiou, T.; Bikas, D. Embodied energy in residential buildings-towards the nearly zero energy building: A literature review. *Build. Environ.* **2016**, *105*, 267–282. [CrossRef]

16. European Committee for Standardization. *Sustainability of Construction Works. Environmental Product Declarations. Core Rules for the Product Category of Construction Products*; EN 15804:2012 + A2:2019; European Committee for Standardization: Brussels, Belgium, 2019.

17. European Committee for Standardization. *Sustainability of Construction Works—Assessment of Environmental Performance of Buildings—Calculation Method*; EN 15978:2011; European Committee for Standardization: Brussels, Belgium, 2011.

18. Herczeg, M.; McKinnon, D.; Milios, L.; Bakas, I.; Klaassens, E.; Svatikova, K.; Widerberg, O. *Resource Efficiency in the Building Sector*; ECORYS Nederland BV: Rotterdam, The Netherlands, 2014; Available online: http://ec.europa.eu/environment/eussd/reports.htm (accessed on 21 May 2020).

19. Dixit, M.K. Life cycle embodied energy analysis of residential buildings: A review of literature to investigate embodied energy parameters. *Renew. Sustain. Energy Rev.* **2017**, *79*, 390–413. [CrossRef]

20. Balouktsi, M.; Lützkendorf, T. Energy Efficiency of Buildings: The Aspect of Embodied Energy. *Energy Technol.* **2016**, *4*, 31–43. [CrossRef]

21. Chastas, P.; Theodosiou, T.; Kontoleon, K.J.; Bikas, D. The Effect of Embodied Impact on the Cost-Optimal Levels of Nearly Zero Energy Buildings: A Case Study of a Residential Building in Thessaloniki, Greece. *Energies* **2017**, *10*, 740. [CrossRef]

22. Berardi, U. Definitions, Design Methodologies, Good Practices, and Case Studies. In *Handbook of Energy Efficiency in Buildings*, 1st ed.; Asdrubali, F., Desideri, U., Eds.; Butterworth-Heinemann Elsevier Ltd.: Oxford, UK, 2018; Chapter 3.2 ZEB and NZEB; p. 836. [CrossRef]

23. BSO, EU Building Stock Observatory; European Commission, Energy. Available online: https://ec.europa.eu/energy/en/topics/energy-efficiency/buildings/eubuildings (accessed on 21 May 2020).

24. Azari, R. Life Cycle Energy Consumption of Buildings; Embodied + Operational. In *Sustainable Construction Technologies, Life-Cycle Assessment*; Tam, V.W.Y., Le, K.N., Eds.; Elsevier: Amsterdam, The Netherlands, 2019; Chapter 5; pp. 123–144. [CrossRef]

25. Hu, M. A Building Life-Cycle Embodied Performance Index—The Relationship between Embodied Energy, Embodied Carbon and Environmental Impact. *Energies* **2020**, *13*, 1905. [CrossRef]

26. Becalli, M.; Cellura, M.; Fontana, M.; Longo, S.; Mistretta, M. Energy retrofit of a single-family house: Life cycle net energy saving and environmental benefits. *Renew. Sustain. Energy Rev.* **2013**, *27*, 283–293. [CrossRef]

27. Hammond, G.P.; Jones, C.I. *Inventory of Carbon and Energy (ICE)*; Department of Mechanical Engineering, University of Bath: Bath, UK, 2011.

28. Alexandri, E.; Androutsopoulos, A. Energy Upgrade of Existing Dwellings in Greece; Embodied Energy Issues. *Proc. Environ. Sci.* **2017**, *38*, 196–203. [CrossRef]

29. Balaras, C.A.; Argiropoulou, P.; Koubogiannis, D.; Syngros, G. Operational Energy Savings & Embodied Energy in Hellenic Residential Buildings. In Proceedings of the EinB2016–5th International Conference "Energy in Buildings 2016", Athens, Greece, 12 November 2016; ASHRAE Hellenic Chapter and Technical Chamber of Greece: Athens, Greece, 2016; p. 237. Available online: http://www.ashrae.gr/Proceedings/EinB2016_PROCEEDINGS.pdf (accessed on 21 May 2020).

30. Koubogiannis, D.G.; Balaras, C.A. Embodied energy in electro-mechanical installations of Hellenic dwellings. In Proceedings of the 3rd International Conference Energy in Buildings, Athens, Greece, 2 September 2014; ASHRAE Hellenic Chapter and Technical Chamber of Greece: Athens, Greece, 2014. Available online: http://www.ashrae.gr/EinB2014/EinB2014_Proceedings.pdf (accessed on 21 May 2020).

31. Papadopoulos, A.M.; Giama, E. Environmental performance evaluation of thermal insulation materials and its impact on the buildings. *Build. Environ.* **2007**, *42*, 2178–2187. [CrossRef]

32. Halyvourgiki Steel Production, Greece. Available online: http://www.halyvourgiki.gr (accessed on 21 May 2020).

33. Mytilineos Holdings. *Sustainability Report*; Mytilineos Holdings: Athens, Greece, 2016; Available online: https://www.mytilineos.gr/Uploads/ETHSIA_DELTIA/csr_reports/Sustainability_Report_2016_GR.pdf (accessed on 21 May 2020).

34. Koroneos, C.; Dompros, A. Environmental assessment of brick production in Greece. *Build. Environ.* **2007**, *42*, 2114–2123. [CrossRef]

35. Koubogiannis, D.; Nouhou, C. How much Energy is Embodied in your Central Heating Boiler? *IOP Conf. Ser. Mater. Sci. Eng.* **2016**, *161*, 012094. [CrossRef]

36. Koroneos, C.J.; Nanaki, E.A. Life cycle environmental impact assessment of a solar water heater. *J. Clean. Prod.* **2012**, *37*, 154–161. [CrossRef]

37. Dascalaki, E.; Argiropoulou, P.; Balaras, C.A.; Droutsa, K.G.; Kontoyiannidis, S.; Koubogiannis, D. On the share of embodied energy in the lifetime energy use of typical Hellenic residential buildings. *IOP Conf. Ser. Earth Environ. Sci.* **2020**, *410*, 012070. [CrossRef]

38. Dascalaki, E.; Argiropoulou, P.; Balaras, C.A.; Droutsa, K.G.; Kontoyiannidis, S. Analysis of the embodied energy in building lifecycle assessment of Hellenic residential buildings. In Proceedings of the 50th International HVAC&R Congress and Exhibition, SMEITS—KGH Srbije, Belgrade, Serbia, 4–6 December 2019.

39. European Commission, Energy. Nearly Zero-Energy Buildings. Available online: https://ec.europa.eu/energy/topics/energy-efficiency/energy-efficient-buildings/nearly-zero-energy-buildings (accessed on 21 May 2020).

40. Scrivener, K.L.; Vanderley, M.J.; Gartner, E.M. Eco-Efficient cements: Potential economically viable solutions for a low-CO2 cement-based materials industry. *Cem. Concr. Res.* **2018**, *114*, 2–26. [CrossRef]

41. Chadiarakou, S. The compliance of building materials to the European & National standards. In Proceedings of the 2nd International Conference Energy in Buildings—Northern Hellas 2015, Thessaloniki, Greece, 9 May 2015; ASHRAE Hellenic Chapter and Technical Chamber of Greece, Central Macedonia Section: Thessaloniki, Greece, 2015. Available online: http://www.ashrae.gr/EstaK2015/EstaK2015_Chadiarakou.pdf (accessed on 21 May 2020).

42. Anastaselos, D.; Giama, E.; Papadopoulos, A.M. An assessment tool for the energy, economic and environmental evaluation of thermal insulation solutions. *Energy Build.* **2009**, *41*, 1165–1171. [CrossRef]

43. IEA. *Technology Roadmap—Low-Carbon Transition in the Cement Industry*; International Energy Agency: Paris, France, 2018; Available online: https://www.iea.org/reports/technology-roadmap-low-carbon-transition-in-the-cement-industry (accessed on 21 May 2020).

44. Bribián, I.Z.; Capilla, A.V.; Usón, A.A. Life cycle assessment of building materials: Comparative analysis of energy and environmental impacts and evaluation of the eco-efficiency improvement potential. *Build. Environ.* **2011**, *46*, 1133–1140. [CrossRef]

45. ELSTAT. *2011 Buildings Census*; Hellenic Statistical Authority: Athens, Greece, 2015. Available online: http://www.statistics.gr/census-buildings-2011 (accessed on 21 May 2020).

46. Dascalaki, E.G.; Balaras, C.A.; Kontoyiannidis, S.; Droutsa, K.G. Modeling Energy Refurbishment Scenarios for the Hellenic Residential Building Stock Towards the 2020 & 2030 Targets. *Energy Build.* **2016**, *132*, 74–90. [CrossRef]

47. Dascalaki, E.G.; Balaras, C.A.; Gaglia, A.G.; Droutsa, K.G.; Kontoyiannidis, S. Energy performance of buildings—EPBD in Greece. *Energy Policy* **2012**, *45*, 469–477. [CrossRef]

48. Energy Performance of Buildings Directive. EU Countries' 2018 Cost-Optimal Reports. Available online: https://ec.europa.eu/energy/en/topics/energy-efficiency/energy-performance-of-buildings/energy-performance-buildings-directive/eu-countries-2018-cost-optimal-reports (accessed on 17 January 2020).

49. Droutsa, K.G.; Kontoyiannidis, S.; Dascalaki, E.G.; Balaras, C.A. Mapping the Energy Performance of Hellenic Residential Buildings from EPC (energy performance certificate) Data. *Energy* **2016**, *98*, 284–295. [CrossRef]

50. Visscher, H.; Dascalaki, E.; Sartori, I. Towards an energy efficient European housing stock: Monitoring, mapping and modelling retrofitting processes: Special issue of energy and buildings. *Energy Build.* **2016**, *132*, 1–3. [CrossRef]

51. Sandberg, N.H.; Sartori, I.; Heidrich, O.; Dawson, R.; Dascalaki, E.; Dimitrou, S.; Vimm, T.; Filipiddou, F.; Stegnar, G.; Šijanec Zavrl, M.; et al. Dynamic building stock modelling: Application to 11 European countries to support the energy efficiency and retrofit ambitions of the EU. *Energy Build.* **2016**, *132*, 26–38. [CrossRef]

52. Malmqvist, T.; Glaumann, M.; Scarpellini, S.; Zabalza, I.; Aranda, A.; Llera, E.; Díaz, S.; Bribián, I.Z. Life cycle assessment in buildings: The ENSLIC simplified method and guidelines. *Energy* **2011**, *36*, 1900–1907. [CrossRef]

53. Soust-Verdaguer, B.; Llatas, C.; Garcia-Martinez, A. Simplification in life cycle assessment of single-family houses: A review of recent developments. *Build. Environ.* **2016**, *103*, 215–227. [CrossRef]

54. PRé Consultants, B.V. SimaPro Analyst Release 8.5.2.0. Available online: https://www.pre-sustainability.com/simapro (accessed on 21 May 2020).

55. Wernet, G.; Bauer, C.; Steubing, B.; Reinhard, J.; Moreno-Ruiz, E.; Weidema, B. The ecoinvent database version 3 (part I): Overview and methodology. *Int. J. Life Cycle Assess.* **2016**, *21*, 1218–1230. [CrossRef]

56. Frischknecht, R.; Wyss, F.; Knöpfel, S.B.; Lützkendorf, T.; Balouktsi, M. Cumulative energy demand in LCA: The energy harvested approach. *Int. J. Life Cycle Assess.* **2015**, *20*, 957–969. [CrossRef]

57. Frischknecht, R.; Jungbluth, N. *Overview and Methodology*; Ecoinvent Report No. 1; Swiss Centre for Life Cycle Inventories: Dübendorf, Switzerland, 2007; Available online: https://www.ecoinvent.org/files/200712_frischknecht_jungbluth_overview_methodology_ecoinvent2.pdf (accessed on 21 May 2020).

58. Balaras, C.A.; Dascalaki, E.G.; Droutsa, K.G.; Kontoyiannidis, S. Empirical Assessment of Calculated and Actual Heating Energy Use in Hellenic Residential Buildings. *Appl. Energy* **2016**, *164*, 115–132. [CrossRef]

59. Zou, P.X.W.; Wagle, D.; Alam, M. Strategies for minimizing building energy performance gaps between the design intend and the reality. *Energy Build.* **2019**, *191*, 31–41. [CrossRef]

60. ELSTAT. *Survey on Energy Consumption in Households 2011–2012*; Hellenic Statistical Authority: Athens, Greece, 2013. Available online: http://www.statistics.gr/statistics/-/publication/SFA40/ (accessed on 21 May 2020).

61. Piccardo, C.; Dodoo, A.; Gustavsson, L.; Tettey, U.Y.A. Retrofitting with different building materials: Life-Cycle primary energy implications. *Energy* **2020**, *192*, 116648. [CrossRef]

62. Balaras, C.A.; Droutsa, K.; Dascalaki, E.; Kontoyiannidis, S. Deterioration of European Apartment Buildings. *Energy Build.* **2005**, *37*, 515–527. [CrossRef]

63. Hugo, A.; Zmeureanu, R. Residential Solar-Based Seasonal Thermal Storage Systems in Cold Climates: Building Envelope and Thermal Storage. *Energies* **2012**, *5*, 3972–3985. [CrossRef]

64. EcoInvent v.3.4. Available online: https://www.ecoinvent.org (accessed on 21 May 2020).

65. Dixit, M.K.; Fernandez-Solis, J.L.; Lavy, S.; Culp, C.H. Identification of parameters for embodied energy measurement: A literature review. *Energy Build.* **2010**, *42*, 1238–1247. [CrossRef]

66. ALUMIL. *Environmental Product Declaration, European Aluminium*; Alumil 7; Alumil: Thessaloniki, Greece, 2017.
67. Verbeeck, G.; Hens, H. Life cycle inventory of buildings: A contribution analysis. *Build. Environ.* **2010**, *45*, 964–967. [CrossRef]
68. Shirazi, A.; Ashuri, B. Embodied Life Cycle Assessment (LCA) Comparison of Residential Building Retrofit Measures in Atlanta. *Build. Environ.* **2020**, *171*, 106644. [CrossRef]
69. Koezjakov, A.; Urge-Vorsatz, D.; Crijns-Graus, W.; van den Broek, M. The relationship between operational energy demand and embodied energy in Dutch residential buildings. *Energy Build.* **2018**, *165*, 233–245. [CrossRef]
70. Sunikka-Blank, M.; Galvin, R. Introducing the prebound effect: The gap between performance and actual energy consumption. *Build. Res. Inf.* **2012**, *40*, 260–273. [CrossRef]
71. Dineen, D.; Gallachoir, B.P.O. Exploring the range of energy savings likely from energy efficiency retrofit measures in Ireland's residential sector. *Energy* **2017**, *121*, 126–134. [CrossRef]

Article

Human Comfort-Based-Home Energy Management for Demand Response Participation

Herie Park

Division of Electrical and Biomedical Engineering, Hanyang University, Seoul 04763, Korea;
bakery@hanyang.ac.kr or park.herie@gmail.com

Received: 18 April 2020; Accepted: 11 May 2020; Published: 14 May 2020

Abstract: The residential building sector is encouraged to participate in demand response (DR) programs owing to its flexible and effective energy resources during peak hours with the help of a home energy management system (HEMS). Although the HEMS contributes to reducing energy consumption of the building and the participation of occupants in energy saving programs, unwanted interruptions and strict guidance from the system cause inconvenience to the occupants further leading to their limited participation in the DR programs. This paper presents a human comfort-based control approach for home energy management to promote the DR participation of households. Heating and lighting systems were chosen to be controlled by human comfort factors such as thermal comfort and visual comfort. Case studies were conducted to validate the proposed approach. The results showed that the proposed approach could effectively reduce the energy consumption during the DR period and improve the occupants' comfort.

Keywords: home energy management system; human comfort factor; thermal comfort; visual comfort; demand response

---

1. Introduction

A building energy management system (BEMS) measures, analyzes, reports, and optimizes the demand and supply of energy in a building for saving its energy and cost, and for realizing the occupant's comfort [1]. As the BEMS can globally control the demand and supply of energy inside a building, it should be installed in green energy buildings to balance their energy generation and consumption. The BEMS can be integrated into both residential and non-residential buildings by focusing on the energy-saving potentials of these buildings [2].

Moreover, the BEMS can generate the profile of demand resources of the buildings for realizing their participation in demand response (DR) programs. The demand resources in DR programs are proactive resources that are assured by modifying the energy consumption pattern of occupants in the buildings. The energy saving accomplished by the occupants is considered as a form of energy resource and is traded in power markets by load aggregators. The load aggregators need to collect a quantity of demand resources from individual DR participants and provide the individuals with benefits of energy savings and economic profits from power markets [3]. The power markets and utilities take advantage of demand resources by reducing energy production and operational costs of the grid, and by improving the reliability of power systems through demand resources [4].

In industrial and commercial buildings, several electricity customers have participated in the DR programs with the help of BEMS, and they followed fixed building control strategies. Most of the occupants of these buildings were the general public, and therefore, services were provided to them via BEMS operation. In other words, the occupants did not have any authority to control the building, and they generally accepted the hierarchical control schedule of the buildings via energy management systems. Therefore, the energy management of these buildings was relatively simple and effective for participating in DR programs. However, the fast cut-off in these buildings can cause malfunction, failure, or life-shortening of mechanical and electrical equipment and their installations [5,6].

In contrast, household-level occupants found it difficult to participate in DR programs due to several barriers such as lack of individual smart infrastructures, home energy management systems (HEMS) and smart meters to monitor and control their appliances, technical mismatches among systems, and their small energy resource quantity [3]. However, aggregated demand resources of households are still attractive to utility operators and load aggregators, as these resources are flexible and effective during peak hours [7,8]. Several researchers have investigated HEMS algorithm for DR participation [9–13]. They formulated optimization problems to minimize the electricity cost of the households [9–11], physical and operational constraints of appliances [12], and the cost of energy and thermal discomfort [13]. These investigations focused on the generation of appropriate input scenarios for the households and their load scheduling.

Even though these algorithms allow the household occupants to participate in DR programs owing to the advanced logic and reasonable scheduling of households, it is not easy for them to allow unidirectional and formulated energy management. For example, a unidirectional HEMS oriented to the reduction of energy consumption causes inconvenience to occupants via unwanted interruptions or strict guidelines on the usage of building sub-systems such as washing machines, refrigerators, computers, rice cookers, electrical ovens, televisions, lighting fixtures, and other home appliances. Further, the occupants at home tend to actively interact with their energy management system and with its connected systems. Therefore, it is important to develop a bi-directional HEMS which follows the occupants' behavioural patterns and their preferences for the usage of appliances and equipment to reduce their fatigue [13,14]. As one of the methods to preserve the user's comfort and to reduce the peak load and electricity bill at smart homes, control algorithms of home appliances, renewables, electric vehicles, and energy storage system based on the quality of experience (QoE) were presented for the HEMS [15,16]. Even though these investigations did not consider the DR programs, the proposed algorithms may further cover the DR programs.

More specifically, building environmental conditions such as indoor/outdoor temperature, meteorological and thermal response characteristics of the building, as well as the occupants' preferences should be considered to control a heating, ventilation, and air-conditioning (HVAC) system. While the operational response characteristic of the HVAC system is faster, the thermal response of the building is slower due to its slower thermal dynamics. It is considered to be one of the advantages of using the HVAC system as an important demand resource. This means that switching on/off the HVAC system does not directly influence the thermal satisfaction of occupants and thus, they have a relatively smaller amount of risks to participate in DR programs and to reply to DR events by using HVAC loads. In addition, the acceptable operating range of HVAC systems contributes to large energy savings owing to their relatively larger power capacity than that of small appliances in buildings. However, despite its potentials as a demand resource, there are few studies on the HVAC system for DR programs in residential buildings, whereas considerable effort has been devoted to this system in industrial and commercial buildings [14,17,18].

Apart from the occupant's thermal comfort, visual comfort was also considered as an important environmental issue in buildings. For achieving low energy consumption and high thermal and visual comforts, natural lighting by using large glazing windows was investigated in different buildings [19–22]. The illuminance level, luminance, and glare indices affecting the visual comfort in these buildings were used as a control input of these buildings [22,23]. However, although much research on natural lighting in the analysis of building energy performance and visual comfort were conducted, there is still less study to apply the natural lighting to HEMS algorithms. As a related study by Li et al. [24] suggested a smart lighting control methodology integrated in a smart home control system to save energy consumption at home with respect to the design of lighting systems for buildings. They confirmed in detail the feasibility and the performance of the lighting control algorithm in the smart home control system. Owing to the effectiveness of the lighting system and the HVAC system for both energy management and user's comfort, these systems should be prior-candidates integrated in the HEMS.

To maintain the satisfaction of occupants, HEMS implementing technologies are required to sense, monitor the environmental conditions and occupants' individual data, store and reproduce the data as knowledge and information of the overall building system. As such, home appliances and equipment should finely be controlled by HEMS by considering the influence of the occupants' preference and fatigue. They should be also considered as demand resources for the purpose of energy savings of the household during DR events. The contributions of this paper are highlighted as follows.

- The human comfort factors including both thermal and visual comforts were integrated into the HEMS algorithm in this study. As mentioned above, the effectiveness of the lighting system and the HVAC system in terms of energy savings and user's comfort have been confirmed in several investigations. However, there has rarely been the investigations dealing with both systems in different perspectives of energy savings and human thermal and visual comfort integrated into the HEMS.
- To reduce the response fatigue of occupants and to slow down the negative effect of curtailment of appliances during DR events, an HVAC system was proposed as a prior demand resource in the household due to its slower thermal dynamics and higher power capacity than those of other appliances. The HVAC system was controlled to follow the DR events as well as the occupant's thermal comfort in the household.
- In addition to the occupants' thermal comfort, a visual comfort-based-control algorithm for a dimming artificial lighting system using natural lighting was integrated into HEMS to improve the visual comfort and energy savings of the household.
- Simulation case studies were performed using Matrix Laboratory (MATLAB)/Simulink® to verify the effectiveness of the proposed approach.

The rest of this paper is organized as follows. Section 2 suggests the comfort-based HEMS algorithm and the human comfort factors considered in the proposed algorithm. Section 3 describes a building integrated HEMS to present the thermal and visual dynamics of the building and its HVAC and lighting systems controlled by HEMS. Section 4 elaborates on case studies to test and validate the proposed approach. Finally, Section 5 draws the conclusions of this study.

2. Comfort-Based Home Energy Management System

2.1. Main Features of the Comfort-Based HEMS

The HEMS accomplishes the sensing, processing, monitoring, and reproducing of information to control the electric and thermal energy fluxes of the building by appropriately combining the operations of electrical and thermal appliances. In this study, human comfort factors such as the thermal and visual comforts were integrated into the HEMS algorithm. The thermal condition of the indoor environment was controlled by the predicted mean vote (PMV) and the predicted percentage of dissatisfied (PPD) values. The PMV-based control strategy of a heating system as an HVAC system was integrated into the proposed HEMS. Even though several metrics of user comfort were defined and proposed in the literature [15,16,25], this study only considered the PMV and PPD as the evaluation index of user's thermal comfort in order to present that the suggested HEMS algorithm could be useful to control an HVAC system to participate in DR programs preserving user's thermal comfort. Then, a dimmable light-emitting diode (LED) lighting system was applied to the proposed HEMS to efficiently use the daylight as a natural lighting and to satisfy the visual comfort of occupants. Illuminance inside a building was the factor that assessed the visual comfort of the occupants in this study.

The comfort-based control algorithm comprises the following sequences. At first, the presence of occupants was detected to assess the necessity of operation of heating/cooling and lighting systems. If any occupant was detected inside the target space, the PMV values were calculated. The thermal condition of the household was then controlled by the predefined ranges of the PMV values. Because the main idea of this study was the availability of the proposed HEMS algorithm during the occurrence of DR events, the acceptable ranges of the PMV values was adjusted and the control command signals of the heating/cooling system depended on these values. At the same time, the lighting system was controlled for the purpose of visual comfort and energy savings.

2.2. Thermal Comfort

Thermal comfort is defined as 'the expression of mind that expresses satisfaction with the thermal environment', according to the American society of heating, refrigerating, and air-conditioning (ASHRAE) [26]. It is a personally determined sensation that has a large discrepancy among different people. It is also difficult to quantify the value and analyze it. Several researchers have investigated the parameters influencing thermal comfort and have developed models to express a common range of thermal comfort among people. Among the models, the PMV model introduced by Fanger is the most used model to assess thermal sensation of the human body with respect to the environment [25]. The thermal comfort was assessed by considering six parameters including the indoor air temperature, mean radiant temperature, relative humidity, air velocity, clothing, and metabolic rate of the occupant. As a result of the evaluation, PMV was obtained and PPD was expressed as a function of the PMV value as follows [27]:

$$PMV = [0.303 \cdot \exp(-0.036 \cdot M) + 0.028] \cdot L \tag{1}$$

$$PPD = 100 - 95.0 \cdot \exp\left(-0.03353 \cdot PMV^4 - 0.2179 \cdot PMV^2\right) \tag{2}$$

where M is the metabolic rate (W/m^2) and L is the thermal load of the human body (W/m^2). The PPD value is symmetric with respect to thermal neutrality, whereas the PMV value is zero. However, as shown in Equation (2), a minimum rate of dissatisfaction of 5% always exists. The index of the PMV is listed in Table 1. The PPD as a function of the PMV is depicted in Figure 1.

Table 1. Index of predicted mean vote (PMV).

Index	−3	−2	−1	0	1	2	3
Sensation	Cold	Cool	Slightly Cool	Neutral	Slightly Warm	Warm	Hot

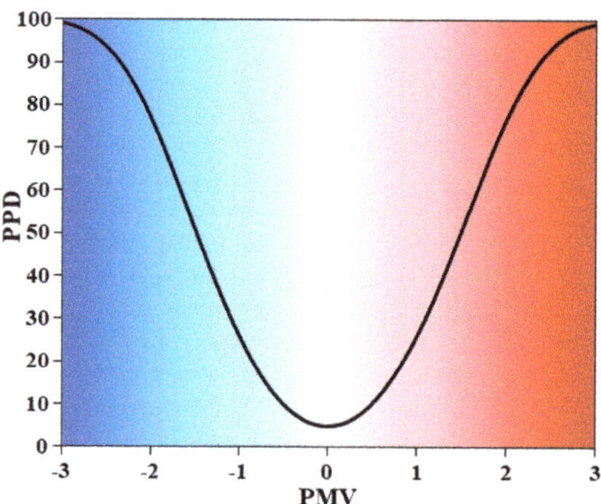

Figure 1. Percentage of dissatisfied (PPD) as a function of PMV.

Acceptable ranges of PMV and PPD variations were defined by the ASHRAE, the European Committee for Standardization (CEN), and the International Organization for Standardization

(ISO) [26,28–31]. ASHRAE Standard 55-2013 [26] indicates an acceptable thermal environment for general comfort. It recommends that the PPD is inferior by 10% and the PMV is between −0.5 and +0.5. This indicated that 10% of the people were dissatisfied [26,28]. The CEN Standard EN 15251 and ISO EN 7730 recommend four categories by prescribing a maximum PPD value for the body and the corresponding PMV range [29–31]. Category I commonly describes a high level of expectation for spaces occupied by very sensitive and fragile persons. Categories II and III explain a normal and moderate expectation, respectively. Category IV is applicable only for a limited period of the year. The first three recommended PPD and PMV ranges are listed in Table 2.

Based on the above recommended thermal comfort values in the standards, appropriate values of PMV and PPD could be selected to control the heating/cooling system of the building.

Table 2. Three categories of thermal comfort.

Category	PPD	PMV Range
I	<6	−0.2 < PMV < +0.2
II	<10	−0.5 < PMV < +0.5
III	<15	−0.7 < PMV < +0.7

2.3. Visual Comfort

The European standard EN 12665 [32] defines the visual comfort as 'the subjective condition of visual well-being induced by the visual environment'. The visual environment inside a building is formed by both natural and artificial lighting. To immediately respond to an occupant's visual demand, the visual comfort index should be easily measured for on-demand services. The quantity, quality, uniformity, and glare of light are the four main factors that evaluate the visual environment.

Among them, the value of the illuminance, which was found to be independent of the light source, was easily and immediately measured. The illuminance is a physical quantity, and is measured in lux (lx) at a point on a given surface by an illuminometer, an illuminance sensor, or by a smartphone that has an illuminance measurement application and its calibration algorithm. It is defined as the ratio between the luminous flux incident on an infinitesimal surface and the area of that surface. It has threshold values for each task or space inside a building. The illuminating engineering society (IES), the commission Internationale de l'Éclairage (CIE), the U.S. general services administration (GSA), the chartered institution of building services engineers (CIBSE), and most countries have their own recommended lighting levels [32,33]. An example of the recommended illuminance levels indicated in the illuminance recommendations [34] is listed in Table 3.

Table 3. Illuminance threshold values of tasks.

Visual Tasks	Illuminance (lx)
Resting, causal visual tasks	10–50
Simple activities, ordinary visual tasks	100
Visual tasks of high contrast and large size	300
Visual tasks of high contrast and small size/ Visual tasks of low contrast and large size	500
Difficult visual tasks of low contrast and small size	1000
Severe visual tasks	Above 3000

3. Building Integrated HEMS

3.1. Thermal Dynamics of the Building

Based on the first principle of thermodynamics, a heat balance of a building with a HVAC system could be developed based on the first principle of thermodynamics as follows [35]:

$$\frac{dU_{stored}}{dt} = \dot{Q}_{HVAC} - \dot{Q}_{envelope} \tag{3}$$

$$\frac{dU_{stored}}{dt} = C_{building}\frac{dT_{building}}{dt} \tag{4}$$

$$\dot{Q}_{HVAC} = \dot{Q}_{heating} - \dot{Q}_{cooling} - \dot{Q}_{ventilation} \tag{5}$$

$$\dot{Q}_{envelope} = -\frac{1}{R_{building}}\left(T_{building} - T_{outside}\right) \tag{6}$$

where U_{stored} is the stored heat energy inside the building (J), $\dot{Q}_{HVAC}$ is the heat flow (W) through the HVAC system including heating flow $\dot{Q}_{heating}$ (W), ventilation flow $\dot{Q}_{ventilation}$ (W), and cooling flow $\dot{Q}_{cooling}$ (W). However, in heating mode of the HVAC system, the cooling flow $\dot{Q}_{cooling}$ (W) is zero, while the heating flow $\dot{Q}_{heating}$ is zero in cooling mode of the system. $\dot{Q}_{envelope}$ is the heat loss (W) through the building envelope, $C_{building}$ is the thermal capacitance (J/kg·°C) of the building, $R_{building}$ is the thermal resistance (°C/W) of the building $T_{building}$ and $T_{outside}$ are the indoor temperature and outdoor temperature (°C) of the building, respectively. Focusing on the control command of the HVAC system, the thermal dynamics of the building could be established as follows:

$$C_{building}\frac{dT_{building}}{dt} = \delta(t)\dot{Q}_{HVAC} - \frac{1}{R_{building}}\left(T_{building} - T_{outside}\right) \tag{7}$$

where $\delta(t)$ is the binary control command of the HVAC system. The conventional control strategy of the HVAC system is to follow the acceptable temperature range of the building. The control command of the system follows the temperature of the building, as depicted in Figure 2.

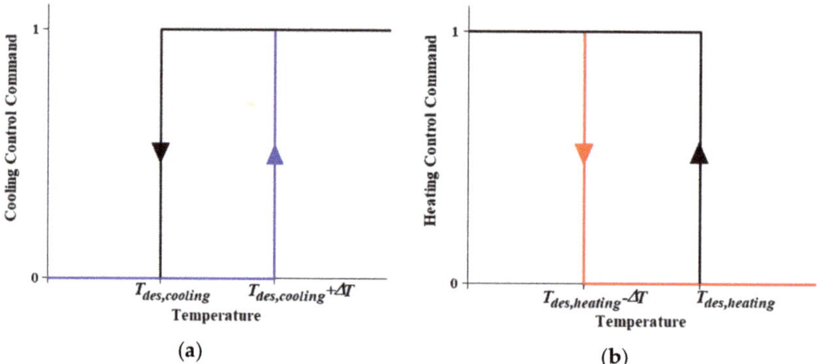

Figure 2. Control command of the heating, ventilation, and air-conditioning (HVAC) system based on acceptable temperature range: (**a**) cooling process and (**b**) heating process.

When the temperature was at the desired temperature, the system turned off. When the difference between the present temperature and the desired temperature was beyond a specified threshold, the system turned on. However, the proposed control strategy of this study focused on human comfort. The control command based on the thermal comfort index PMV is depicted in Figure 3. The control command was determined by the acceptable range of the PMV. When the PMV value reached the desired PMV, the system turned off. When the difference between the values of the present PMV and the desired PMV was beyond a specified threshold, the system turned on.

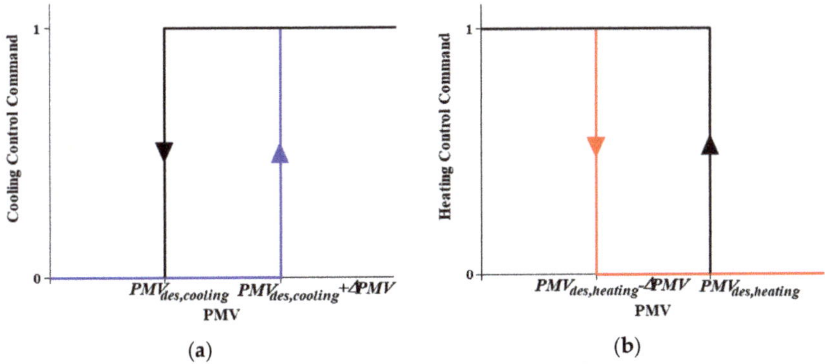

Figure 3. Control command of the HVAC system based on acceptable PMV range: (**a**) cooling process and (**b**) heating process.

3.2. Visual Comfort-Based Control of Artificial Lighting Systems

Artificial lighting systems are designed based on their power consumption, illumination, working area, and functionality. In this study, a dimmable LED lighting fixture was selected for both visual comfort of the occupants and low energy consumption of the building. The required illumination of the artificial lighting, $E_{art,\ total}$ is expressed as follows:

$$E_{art,total} = \eta(t)\kappa(t)E_{art} + E_{aux} \tag{8}$$

$$\eta(t) = 0(E_{nat} \geq E_{ref}) \tag{9}$$

$$\eta(t) = \frac{E_{ref} - E_{nat}}{E_{art}}(E_{nat} < E_{ref}) \tag{10}$$

where $\eta(t)$ is the lighting control command, E_{nat} and E_{art} are the illumination (lx) by natural lighting and artificial lighting, respectively. E_{aux} is the auxiliary illumination (lx) for night time or for emergency. E_{ref} is the illumination reference (lx) of the building. $\kappa(t)$ is the binary switch following the time schedule of the lighting system. $\kappa(t)$ was 1 when the occupants were present, whereas, it was 0 during both night time as well as when the occupants were absent. However, during night time or in emergency, occupants may use the auxiliary illumination as indicated in Equation (8). The required artificial illumination of the building and its operating time schedule is depicted in Figure 4.

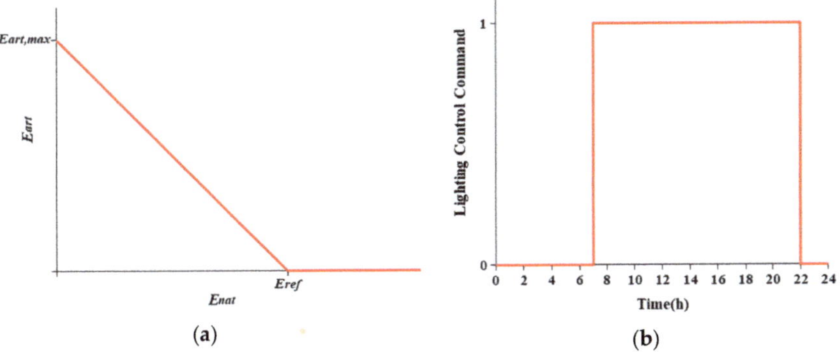

Figure 4. The required artificial illumination of the building. (**a**) Artificial lighting versus natural lighting and (**b**) control command based on a time schedule.

4. Case Study

4.1. Simulation Conditions

To investigate the effect of human comfort-based algorithm-integrated HEMS, a numerical study on building energy performance was conducted using MATLAB/Simulink®. A simple *RC*-lumped building model for a house was developed based on the first principle of thermodynamics. Three cases were selected to investigate the HEMS function for DR participation and human comfort. Each scenario of the cases is listed in Table 4. The simulation of each scenario was conducted assuming that the house was occupied during the simulation and that the DR command was provided from 9 to 12 h and 18 to 22 h for a week during winter to avoid peak loads and peak electricity pricing. The meteorological data provided by the French Technical Research Center for Building (CSTB) were used [36,37]. The time step of the simulation was in a minute. A heating system was used for heating the space and was considered as a demand resource for replying to the DR commands. If the simulation is conducted during summer, a cooling system may be preferred. In addition, a lighting system is operated for the occupants' visual comfort during the day and night time. To ensure the occupants' visual comfort, the lighting system inside the house was controlled to keep the illuminance above 300 lx from 7 to 22 h [34]. The illuminance data was calculated by using a simple window model and a lighting system model integrated in a building simulation tool developed by the CSTB [36].

Table 4. Scenario description of case studies.

Case	Scenario	Regulation Method	Regulation Range	Metabolic Heat
1	Scenario 1	Temperature	Without DR: 19–21 °C With DR: 19–21 °C	115 W
2	Scenario 2 A	Temperature	Without DR: 19–21 °C With DR: 18–20 °C	100 W
	Scenario 2 B	Temperature	Without DR: 19–21 °C With DR: 18–20 °C	115 W
	Scenario 2 C	Temperature	Without DR: 19–21 °C With DR: 18–20 °C	130 W
3	Scenario 3 A	PMV	Without DR: −0.2–0 With DR: −0.5−−0.2	100 W
	Scenario 3 B	PMV	Without DR: −0.2–0 With DR: −0.5−−0.2	115 W
	Scenario 3 C	PMV	Without DR: −0.2–0 With DR: −0.5−−0.2	130 W

Case 1 represents a basic case followed a predefined temperature range without any additional adjustment of a heating system for responding to DR. The temperature-based control strategy was applied to case 1, as depicted in Figure 2. The temperature ranges of the heating system in case 1 was fixed from 19 to 21 °C. There was only one scenario with a metabolic heat of 115 W in case 1.

Meanwhile, cases 2 and 3 followed the DR commands by controlling the heating systems. In case 2, a heating system was controlled by a temperature range similar to that of case 1 without the DR command. However, the heating system was controlled in the temperature range of 18–20 °C with the DR command. To reduce the power consumption of the heating system during the DR period, the indoor temperature remained relatively lower than that in the normal conditions without DR events. The heating system in case 3 was regulated by PMV ranges as depicted in Figure 3. The PMV ranged from 0 to −0.2 without DR events, and from −0.2 to −0.5 with DR events. The range of PMV was first set from 0 to −0.2 to ensure the occupants' thermal comfort when there was no DR command. When a DR command was provided during a certain period, the PMV value within the range of −0.2 to −0.5 was applied as acceptable ranges at normal expectation. There were three types of scenarios in each of cases 2 and 3. However, they differed from the aspect of metabolic heat quantity. Scenario A represented a low metabolic heat of 100 W. Scenario B showed a moderate metabolic heat of 115 W. Scenario C indicated a high metabolic heat of 130 W.

4.2. Simulation Results

The indoor and outdoor temperatures of the house in scenario 1 of case 1 is depicted in Figure 5. The outdoor temperature varied from −2 °C to 6 °C during the simulation, and this was applied to all the scenarios of this study. In scenario 1 of case 1, the heating system of the house was always set from 19 to 21 °C, although there were DR commands from 9 to 12 h and from 18 to 22 h. In scenario 1, the amount of heat energy consumed during a week was 1166 kWh.

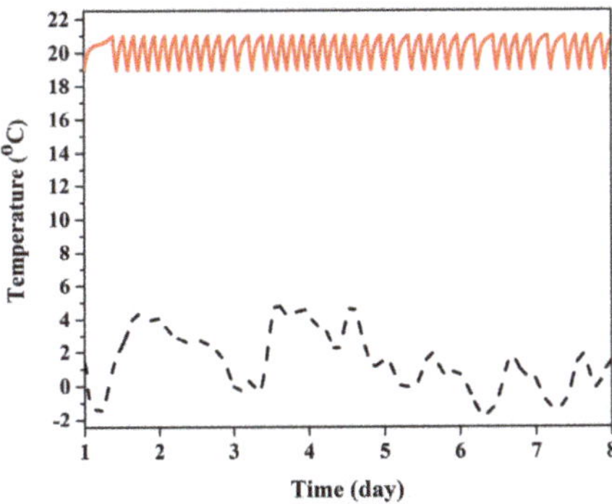

Figure 5. Indoor (—) and outdoor (- -) temperatures in scenario 1.

The zoomed indoor temperature of scenario 1 for reference and that of scenario 2 B for comparing the DR participation with and without a heating system as demand resources is depicted in Figure 6a. In these scenarios, a metabolic heat of 115 W was applied. In scenario 2 B, the temperature ranges of the house were modified when DR commands arrived during a DR period. As stated above, the DR events were sent to the house for 9–12 h and 18–22 h. During this period, the temperature was set from 18 to 20 °C.

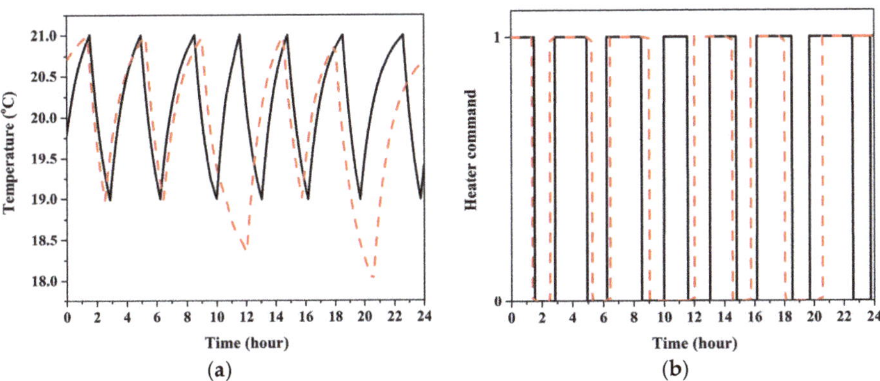

Figure 6. Comparison of indoor temperature and heater signal: (**a**) temperature and (**b**) heater command (—: scenario 1 and - -: scenario 2 B).

The heater command of scenarios 1 and 2 B based on the temperature-based control strategy is depicted in Figure 6b. The indoor temperature varied (increased and decreased) with respect to the thermal dynamics of the building once the power was on and off, respectively. The indoor temperature was set below 19 °C during the DR period in scenario 2 B and therefore, the heater command was not rapidly reversed compared to the heater command of scenario 1. It permitted the curtailing of the heater's power consumption during DR periods and reduced the energy consumption of the house while there were DR commands. The amount of heat energy consumed during a week in scenario 2 B was 1136.6 kWh. In this scenario, less amount of heat energy was required compared to that of the heat energy in scenario 1.

This result showed that lower temperature ranges for a heating system reduced the energy consumption of the house and realized its participation in DR programs with a heating system as the demand resource. However, this did not mean that the temperature ranges ensured the thermal comfort of the occupants. The occupants dissipated different amounts of metabolic heat according to their physical properties, physical activities, and conditions. The indoor temperature of the house in scenarios 2 A, B, and C distinguished by different quantity of the metabolic heat of occupants is depicted in Figure 7. As the heating system of the house operated in fixed temperature ranges in these scenarios, all the evolutions of indoor temperatures for the three scenarios were in the range of 19–21 °C in a non-DR period and 18–20 °C in a DR period.

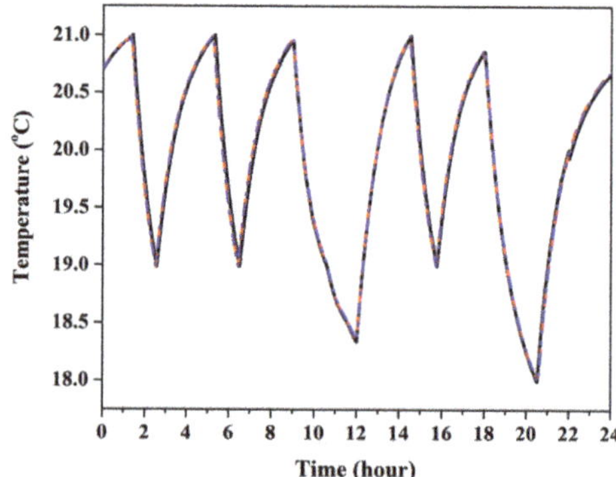

Figure 7. Comparison of indoor temperature (—: scenario 2 A, - -: scenario 2 B, and — -: scenario 2 C).

Therefore, the problem of thermal comfort increased as depicted in Figure 8. Under the same indoor temperature conditions, occupants dissipated different metabolic heat with different thermal comfort values. Although the occupants could successfully participate in the DR programs by controlling the temperature ranges of heating systems, it had a negative effect on personal satisfaction. This may induce a short continuity of DR participation in households. Therefore, during a DR period, a simple curtailment of power cannot be recommended as a single solution for all the occupants. To satisfy the thermal comfort of occupants as well as to participate in DR programs with an HVAC system, a PMV-based-control strategy has to be proposed as stated in the previous section. From the perspective of energy saving, the quantities of energy consumption in scenarios 2 A, B, and C were 1138.6, 1136.6, and 1133.1 kWh, respectively, during the given period. As higher metabolic heat contributed in heating the space, the energy consumption of the house was more in scenario 2 C than those in 2 A and B.

The evolution of the indoor temperature of the house in scenarios 3 A, B, and C is depicted in Figure 9. Although the difference in temperature was more than 3 °C among the three scenarios of the same case, the thermal comfort of occupants in each scenario was ensured at the expected PMV ranges between −0.2 and 0 during the non-DR period and between −0.5 and −0.2 during the DR period as depicted in Figure 10. In scenario 3 A, the indoor temperature was higher than those in other scenarios. It explains that the lowest metabolic heat requires the more quantity of heat in the house for improving occupant' thermal comfort. In other words, the occupant dissipated lower heat feels colder than the occupants dissipating higher metabolic heat. Therefore, in a house where an occupant dissipating lower metabolic heat, more heat should be supplied to this house for thermal satisfaction of the occupant. In contrast, the indoor temperature was lower in scenario 3 C as the occupant dissipated more heat and thus feels warmer or hotter than others. This led to a higher PMV value of the occupant even at a lower temperature. This indicated that the occupants dissipating different quantities of heat have a significant gap with the satisfaction on the indoor thermal condition. This result supported the previously obtained results depicted in Figures 7 and 8. Moreover, the highest metabolic heat required less heating energy in the house. The quantities of energy consumption during the given period in scenarios 3 A, B, and C were 1263, 1094, and 920 kWh, respectively. This showed that thermal comfort could be obtained by controlling a heating system based on the expected PMV values of the occupants. As the curtailment of heating power depended on the thermal comfort of the occupants, their satisfaction in DR programs could be ensured by maintaining their thermal comfort.

To achieve the visual comfort of occupants and energy savings of the building, a dimming artificial lighting system was integrated into the building model. The illuminance level of the house and natural illuminance are depicted in Figure 11. When the occupant was awake from 7 to 22 h, the minimum artificial illuminance was maintained at 300 lx. Therefore, the artificial illuminance was controlled by dimming the artificial lighting with respect to the recommended illumination level of 300 lx, a lighting system operated for the occupants' visual comfort during the day and night time. The dimming level of the lighting system is depicted in Figure 12. Compared to the system without dimming control, this system reduced 27% of the lighting energy. However, in this study, glare occurrence was not considered. In further studies, a blind system for avoiding glares would be investigated.

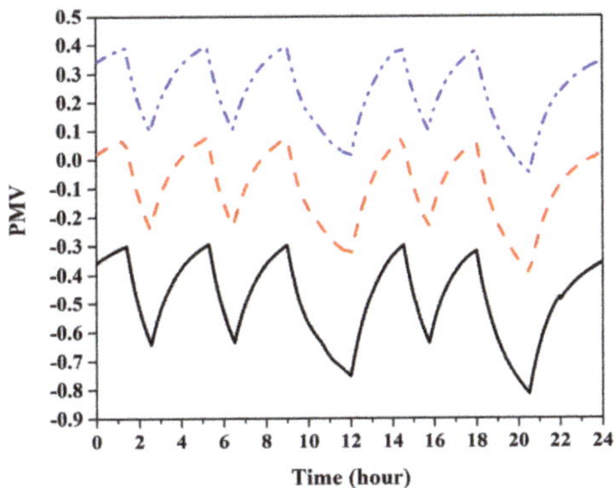

Figure 8. Comparison of PMV (—: scenario 2 A, - -: scenario 2 B, and — -: scenario 2 C).

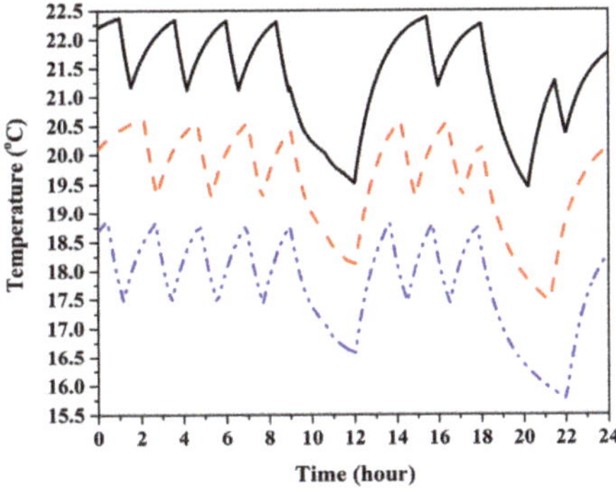

Figure 9. Comparison of indoor temperature (—: scenario 2 A, - -: scenario 2 B, and — -: scenario 2 C).

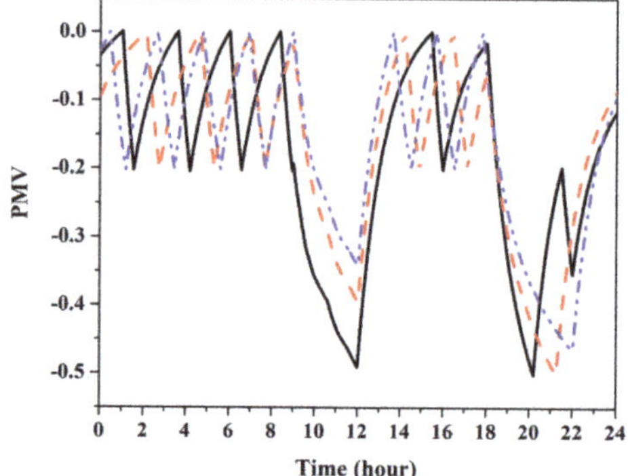

Figure 10. Comparison of PMV (—: scenario 2 A, - -: scenario 2 B, and — -: scenario 2 C).

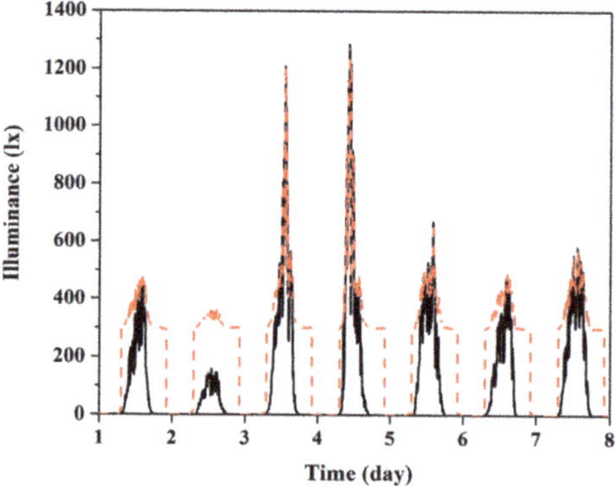

Figure 11. Illuminance (—: natural illuminance and - -: indoor illuminance).

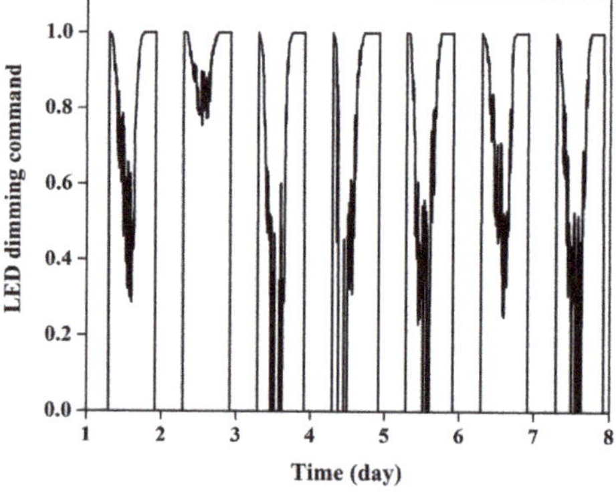

Figure 12. Dimming level of lighting system.

5. Conclusions

Home appliances and equipment should be finely controlled by HEMS by considering the influence of occupant's preferences and behaviour patterns. To promote the occupant's participation in DR programs, this paper has presented an algorithm based on human comfort factors that can be integrated into the HEMS. The proposed algorithm included PMV and PPD values for the thermal comfort of the occupants, and illuminance level for their visual comfort. To reduce response fatigue of occupants and the curtailment of small appliances, the HVAC system with a relatively larger power capacity and faster dynamics was preliminarily selected to be used as the demand resource of a household. The HVAC system could be controlled to follow the DR events and to reduce the energy consumption of the household. To perform the proposed approach, the thermal comfort-based-control algorithm for a heating system in HEMS was proposed to improve the occupants' thermal comfort. At the same time, a dimming artificial lighting system (LED) was integrated into the visual comfort-based-control algorithm of the lighting system in HEMS to improve the visual comfort and energy savings of the household. Simulation studies were carried out using MATLAB/Simulink® to verify the effectiveness of the proposed approach. Because the simulation was conducted during winter, the heating process of the HVAC system was controlled. The results showed that the heating system with PMV-based control significantly improved the thermal comfort of the occupants during both the DR and non-DR periods compared to the temperature-based control in scenarios with different metabolic heat dissipation of occupants. In addition, visual comfort based on illuminance and energy savings of a lighting system controlled by the proposed approach in HEMS was also validated. Further study includes advanced control strategies of the HVAC system by using thermal inertia of the households before and after DR events and a blind system control for avoiding glare occurrences.

Funding: This research was funded by Basic Science Research Program through the National Research Foundation of Korea (NRF) funded by the Ministry of Education, grant number 2017R1D1A3B03035693.

Conflicts of Interest: The author declares no conflict of interest. The funders had no role in the design of the study; in the collection, analyses, or interpretation of data; in the writing of the manuscript, or in the decision to publish the results.

References

1. Aste, N.; Manfren, M.; Marenzi, G. Building automation and control systems and performance optimization: A framework for analysis. *Renew. Sustain. Energy Rev.* **2017**, *75*, 313–330. [CrossRef]
2. Martirano, L.; Parise, G.; Greco, G.; Manganelli, M.; Massarella, F.; Cianfrini, M.; Parise, L.; Frattura, P.L.; Habib, E. Aggregation of Users in a residential/commercial building managed by a building energy management system (BEMS). *IEEE Trans. Ind. Appl.* **2019**, *55*, 26–34. [CrossRef]
3. Park, H.; Bae, S.; Chang, M.; Park, S.H.; Yoon, G.G. A community-scale energy management system for demand response participation of households with DERs and EVs. In Proceedings of the 2019 IEEE 4th International Future Energy Electronics Conference (IFEEC), Singapore, 23–25 November 2019; pp. 1–5.
4. Palensky, P.; Dietrich, D. Demand side management: Demand response, intelligent systems, and smart loads. *IEEE Trans. Ind. Inf.* **2011**, *7*, 381–388. [CrossRef]
5. Shoreh, M.H.; Siano, P.; Shafie-khah, M.; Loia, V.; Catalão, J.P.S. A survey of industrial applications of Demand Response. *Electr. Power Syst. Res.* **2016**, *141*, 31–49. [CrossRef]
6. Eissa, M.M. Developing incentive demand response with commercial energy management system (CEMS) based on diffusion model, smart meters and new communication protocol. *Appl. Energy* **2019**, *236*, 273–292. [CrossRef]
7. Liu, Y.; Xial, L.; Yao, G.; Bu, S. Pricing-based demand response for a smart home with various types of household appliances considering customer satisfaction. *IEEE Access* **2019**, *7*, 86463–86472. [CrossRef]
8. Yan, X.; Ozturk, Y.; Hu, Z.; Song, Y. A review on price-driven residential demand response. *Renew. Sustain. Energy Rev.* **2018**, *96*, 411–419. [CrossRef]
9. Samuel, O.; Javaid, S.; Javaid, N.; Ahmed, S.H.; Afzal, M.K.; Ishmanov, F. An efficient power scheduling in smart homes using jaya based optimization with time-of-use and critical peak pricing schemes. *Energies* **2018**, *11*, 3155. [CrossRef]
10. Du, Y.F.; Jiang, L.; Li, Y.; Wu, Q. A robust optimization approach for demand side scheduling considering uncertainty of manually operated appliances. *IEEE Trans. Smart Grid* **2018**, *9*, 743–755. [CrossRef]
11. Javaid, N.; Ahmed, A.; Iqbal, S.; Ashraf, M. Day ahead real time pricing and critical peak pricing based power scheduling for smart homes with different duty cycles. *Energies* **2018**, *11*, 1464. [CrossRef]

12. Huang, Y.; Wang, L.; Guo, W.; Kang, Q.; Wu, Q. Chance constrained optimization in a home energy management system. *IEEE Trans. Smart Grid* **2018**, *9*, 252–260. [CrossRef]
13. Shafie-Khah, M.; Siano, P. a stochastic home energy management system considering satisfaction cost and response fatigue. *IEEE Trans. Ind. Inf.* **2018**, *14*, 629–638. [CrossRef]
14. Park, H.; Rhee, S.B. IoT-based smart building environment service for occupants' thermal comfort. *J. Sens.* **2018**, *2018*, 1757409. [CrossRef]
15. Lee, S.; Choi, D.-H. Energy management of smart home with appliances, energy storage system and electric vehicle: A hierarchical deep reinforcement learning approach. *Sensors* **2020**, *20*, 2157. [CrossRef] [PubMed]
16. Li, M.; Li, G.-Y.; Chan, H.-R.; Jiang, C.-W. QoE-aware smart home energy management considering renewable and electric vehicles. *Energies* **2018**, *11*, 2304. [CrossRef]
17. Vand, B.; Martin, K.; Jokisalo, J.; Kosonen, R.; Hast, A. Demand response potential of district heating and ventilation in an educational office building. *Sci. Technol. Built Environ.* **2019**, *26*, 304–319. [CrossRef]
18. Yao, L.; Huang, J.-H. Multi-objective optimization of energy saving control for air conditioning system in data center. *Energies* **2019**, *12*, 1474. [CrossRef]
19. Hawila, A.A.W.; Merabtine, A.; Troussier, N.; Bennacer, R. Combined use of dynamic building simulation and metamodeling to optimize glass facades for thermal comfort. *Build. Environ.* **2019**, *157*, 47–63. [CrossRef]
20. Hee, W.J.; Alghoul, M.A.; Bakhtyar, B.; Elayeb, O.; Shameri, M.A.; Alrubaih, B.; Sopian, K. The role of window glazing on daylighting and energy saving in buildings. *Renew. Sustain. Energy Rev.* **2015**, *42*, 323–343. [CrossRef]
21. Vanhoutteghem, L.; Skarning, G.C.J.; Hviid, C.A.; Svendsen, S. Impact of façade window design on energy, daylighting and thermal comfort in nearly zero-energy houses. *Energy Build.* **2015**, *102*, 149–156. [CrossRef]
22. Huang, X.; Wei, S.; Zhu, S. Study on daylighting optimization in the exhibition halls of museums for chinese calligraphy and painting works. *Energies* **2020**, *13*, 240. [CrossRef]
23. Kolokotsa, D.; Tsiavos, D.; Stavrakakis, G.S.; Kalaitzakis, K.; Antonidakis, E. Advanced fuzzy logic controllers design and evaluation for buildings' occupants thermal–visual comfort and indoor air quality satisfaction. *Energy Build.* **2001**, *33*, 531–543. [CrossRef]
24. Li, M.; Lin, H.-J. Design and implementation of smart home control systems based on wireless sensor networks and power line communications. *IEEE Trans. Ind. Electron.* **2015**, *62*, 4430–4442. [CrossRef]
25. Enescu, D. Models and indicators to assess thermal sensation under steady-state and transient conditions. *Energies* **2019**, *12*, 841. [CrossRef]
26. American Society of Heating, Refrigeration and Air Conditioning Engineers. *Standard 55: Thermal Environmental Conditions for Human Occupancy*; ASHRAE: New York, NY, USA, 2013.
27. Fanger, P.O. *Thermal Comfort: Analysis and Applications in Environmental Engineering*; Danish Technical Press: Copenhagen, Denmark, 1970.
28. ISO7730. *Moderate Thermal Environments—Determination of the PMV and PPD Indices and Specification of the Conditions for Thermal Comfort*; ISO: Geneva, Switzerland, 1984.
29. Kao, J.C.M.; Sung, W.-P.; Chen, R. *Green Building, Materials and Civil Engineering*; CRC Press: Boca Raton, FL, USA, 2014.
30. Nicol, F.; Wilson, M. An overview of the European Standard EN 15251. In Proceedings of the Adapting to Change: New Thinking Comfort, London, UK, 9–11 April 2010; pp. 1–13.
31. CEN/TC 156. *Indoor Environmental Input Parameters for Design and Assessment of Energy Performance of Buildings—Addressing Indoor Air Quality, Thermal Environment, Lighting and Acoustics*; CEN/TC: Brussels, Belgium, 2006.
32. EN 12665. *Light and Lighting—Basic Terms and Criteria for Specifying Lighting Requirements*; European Committee for Standardization: Brussels, Belgium, 2011.
33. Mahoney, E. Using Luminance for Design and Evaluation of Energy Efficient and Sustainable Luminous Environments to Supplement Current Illuminance-Based Design Codes. Master's Thesis, University of Kansas, Lawrence, KS, USA, 2019.
34. Grondzik, W.T.; Kwok, A.G.; Stein, B.; Reynolds, J.S. *Mechanical and Electrical Equipment for Buildings*; Wiley: Hoboken, NJ, USA, 2011.
35. Park, H. Dynamic Thermal Modeling of Electrical Appliances for Energy Management of low Energy Buildings. In Electric Power. Ph.D. Thesis, Université de Cergy-Pontoise, Cergy-Pontoise, France, 2013.

36. SIMBAD. *Building and HVAC Toolbox*; entre Scientifique du Bâtiment: Marne-la-Vallée, France, 2005.
37. Ruellan, M.; Park, H.; Bennacer, R. Residential building energy demand and thermal comfort: Thermal dynamics of electrical appliances and their impact. *Energy Build.* **2016**, *130*, 46–54. [CrossRef]

Article

Mainstreaming Energy Communities in the Transition to a Low-Carbon Future: A Methodological Approach

Sara Torabi Moghadam [1,*], Maria Valentina Di Nicoli [1,*], Santiago Manzo [2] and Patrizia Lombardi [1]

[1] Interuniversity Department of Regional and Urban Studies and Planning, Politecnico di Torino, 10125 Turin, Italy; patrizia.lombardi@polito.it

[2] Department of Management and Production Engineering (DIGEP), Politecnico di Torino, 10129 Turin, Italy; santiago.manzo@studenti.polito.it

* Correspondence: sara.torabi@polito.it (S.T.M.); mariavalentina.dinicoli@polito.it (M.V.D.N.)

Received: 13 January 2020; Accepted: 19 March 2020; Published: 1 April 2020

Abstract: Innovations in technical, financial, and social areas are crucial prerequisites for an effective and sustainable energy transition. In this context, the construction of a new energy structure and the motivation of the consumer towards a change in their consumption behaviours to balance demand with a volatile energy supply are important issues. At the same time, Consumer Stock Ownership Plans (CSOPs) in renewable energies sources (RESs) have proven to be an essential cornerstone in the overall success of energy transition. Indeed, when consumers acquire ownership in RES, they become prosumers, participating in the phase of production and distribution of energy. Prosumers provide benefits by (1) generating a part of the energy they consume, (2) reducing their overall expenditure for energy, and (3) receiving a second source of income from the sale of excess production. Supporting Consumer Co-Ownership in Renewable Energies (SCORE) is an ongoing Horizon 2020 project with the aim of overcoming the usage of energy from fossil sources in favour of RES, promoting the creation of energy communities (EC) and facilitating co-ownership of renewable energies (RE) for consumers. SCORE hereby particularly emphasises the inclusion of women, low-income households, and vulnerable groups affected by fuel poverty that are as a rule excluded from RE investments. In this framework, the main goal of the present study is to illustrate the general procedure and process of EC creation. In particular, this paper focuses on the description of the methodological approach in implementing the CSOP model which consists of three main phases: the identification and description of selected buildings (preparation phase), the preliminary and feasibility analysis phase, and finally the phase of target group involvement. SCORE first started in three pilot regions in Italy, Czech Republic, and Poland, and later, with the aim of extending the methodology, in various other cities across Europe. In this study, Italian pilot study sites were chosen as a case study to develop and test the methodology.

Keywords: energy community (EC); renewable energy sources (RESs); citizen involvement; co-ownership in renewable energies

---

1. Introduction

Nowadays, due to global environmental problems (e.g., climate change and the increase of greenhouse gas emissions) it is necessary to follow a "decarbonization process" for an energy transition. Energy transition means not only a move away from energy from fossil sources in favour of renewable ones, but also an improvement of the energy efficiency related to the energy production and an awareness of energy consumption by building users and citizens [1,2].

In this regard, energy community (EC) initiatives seem to be the way through which it is possible to provide a concrete response to the aforementioned environmental issues. Moreover, the EC represents a new model which considers energy as well as economic and social perspectives. This emerging concept leads to positive implications in different areas such as global CO_2 emission reductions with the reduction of local pollutants for an improvement in external air quality, and economic developments (e.g., creation of job self-sufficiency, reduction of energy poverty, and community cohesion) [3]. In addition, apart from shifting towards a new market no longer founded on large centralized plants fuelled by fossil fuels but towards small-centralized plants powered by renewable energy sources (RESs), in this emerging system the consumer plays an active role. The consumers' willingness to actively participate in decisions together with the production, distribution, and consumption of energy from RESs represents a key element in the EC definition. The EC emerges from bottom-up willpower in which municipalities, small and medium-sized enterprises, and citizens, located in a specific area, share the willingness to self-produce, self-consume, and exchange energy from renewable energy sources among different users in different end-use buildings [3,4].

It is clear that participation is the core topic of community projects, but the main and innovative issue addressed in this work is the inclusion of several target groups. Indeed, usually these projects are undertaken by men who are middle aged with a higher income, whereas the involvement of women, low-income households, and vulnerable groups affected by fuel poverty is uncommon and as a rule these groups are excluded from renewable energies (RE) investments [5].

This new paradigm has to be supported by a legislative framework in order to allow the birth and proliferation of these communities. Currently, the allowed energy model in Italy is based on a "one-to-one configuration" from a single energy system to a single end-consumer. The case of a single-family house with a photovoltaic system installation for personal consumption or the case of a condominium with a photovoltaic system installation for the satisfaction of only common loads (e.g., elevator, lighting of common areas, etc.) fall in this typology. The "one-to-many configuration", from a single system to multiple end-consumers (between different buildings with different end-uses) is allowed with the support of new legislative framework (Figure 1).

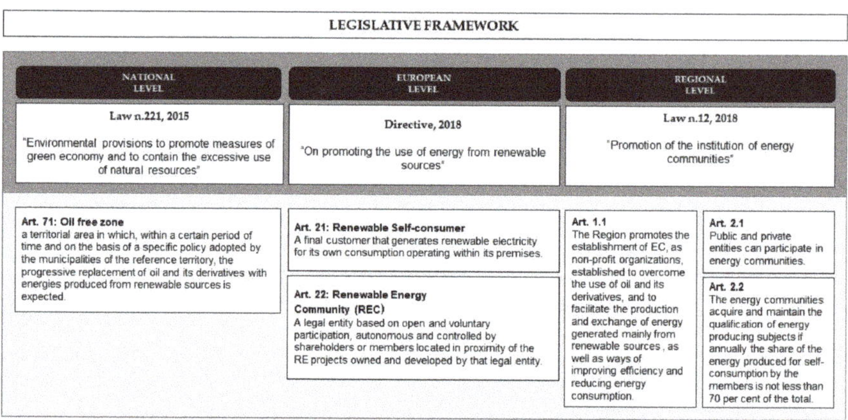

Figure 1. The legislative framework of energy communities.

In this context on 30 November 2016, the European Commission presented the "Clean Energy for All Europeans" package, also known as the "Clean energy package", which includes several measures in the fields of energy efficiency, renewable energy, and internal energy market power [6]. Among all, two directives are important since they significantly address EC issues: (1) Renewable Energy Directive II (RED II, 2018) and (2) the new directive on the new rules of electricity market (2019). The RED II overall target is for 32% of energy consumption to be supplied by RESs. In addition, the new Directive describes the 2020 national targets for each country, taking into account the renewable energy production potential for the next years and the actual production level. In this way, each EU country defines how to reach the targets through National Energy Action Plans. Moreover, under RED II, Member States, when transposing the new rules into national law, will have to ensure that private consumers of RE in the same building are authorized to organise among themselves the exchange of the RE produced on their sites. In this regard, an innovative energy model is founded, overcoming mono-directional consumption by passive consumers from energy produced by large-scale industrial producers. Furthermore, RED II enables different stakeholders to join RECs that produce energy (e.g., electricity) for self-consumption and to share the energy produced within the REC. In this way, the "prosumer" has an active role in the production and consumption of energy from REC, sanctioning the right of citizens and communities to produce, store, and consume energy from RES.

In Italy, at the national level, recognition of ECs can be found in the 2017 National Energy Strategy (SEN) containing the ten-year plan of the Italian Government to manage the change in the energy system. The SEN, in fact, places the figure of the consumer at the centre, considering it the "engine of the energy transition, to be used in a greater involvement of the demand to the markets through the activation of the demand response, the opening of the markets to the consumers and self-producers the regulated development of energy communities". Furthermore, Law 221 of 2015, "Environmental provisions to promote measures of green economy and to contain the excessive use of natural resources", within article 71 establishes the possibility of creating areas free from dependence on fossil fuels, in a so-called "oil-free zone". These territorial areas have the possibility of encouraging experiments, including those on new forms of association.

Following the new regulatory framework, the Piedmont Region has displayed the willpower to "promote the birth of energy communities as non-profit organizations". Indeed, the Piedmont Region is the first Italian region, through the Regional Law of 3 August 2018, n. 12 ("Promotion of the institution of energy communities"), that encourages the new paradigm related to energy communities. This law launches these communities as non-profit organisations in which public and private subjects can take part. They are established to promote the energy transition facilitating the production and exchange of energy generated mainly from RESs as well as to pave the way for an improvement of energy efficiency (EE) and a reduction of energy consumption. According to this law, the municipalities that intend to set up an EC must adopt a specific protocol of understanding, drawn up on the basis of criteria that must be indicated by a subsequent regional implementing provision.

The Region, through future ad hoc incentives, has committed to financially support the establishment of energy communities. This may also involve agreements with Italian Regulatory Authority for Energy and Networks (ARERA – Autorità di Regolazione per Energia Reti e Ambiente), in order to optimize the management and use of energy networks. The regional law also provides for the establishment of a permanent technical panel between the ECs and the region in order to acquire data on the reduction of energy consumption, on the amount of self-consumption, and on the share of use of renewable energy, and to identify the methods for more efficient management of energy networks. This action represents an important step in the direction of energy self-sufficiency and the construction of a new model of virtuous territorial cooperation.

Within this framework, the establishment of a cooperative is particularly advantageous as it permits the delegation of contracts to members of the community since the Law does not prescribe a specific legal form for this type of energy community. A cooperative can be set up by at least nine members and it is characterised as follows:

- It is a legal entity and its functioning is regulated through its statutes;
- The assembly decides on everything and appoints a board of directors;
- The rule of "one member, one vote" is applied;
- Responsibility can be (and is almost always) limited, avoiding an intermingling with the shareholders' personal assets;
- It is an organisation which, although it can make profits, has the primary aim of delivering benefits to its members, for example by providing goods and services on better terms than on the market or carrying out the activities of their corporate purpose;

- The number of members is variable, as is the capital, which simplifies membership entries and exits. In addition to the share capital, the shareholders can lend money to the company depending on the establishment of a social loan regulation. This activity is not considered to be a collection of savings from the public and is therefore not subject to capital market regulation rules;
- Citizens, as members of the cooperative, control the operations of the EC of which the cooperative is the owner or holds majority shares. Moreover, the citizens may also hold minority stakes in other companies.

The present study presents the results of an ongoing research project, which focuses mainly on the engagement of private and/or public consumers towards sustainable energy transition. The purposes could be summarized as follows:

- Facilitating consumers to become prosumers of RE, firstly in three pilot regions (Italy, Poland, and the Czech Republic), and secondly in cities across Europe after the pilot projects. This involves the application of Consumer Stock Ownership Plans (CSOPs), utilising established up-to-date best practice by inclusive financing techniques combined with energy efficiency measures.
- Encouraging local authorities and consumers, demonstrating the positive impact co-ownership has on consumer behaviour, and showing the ability of this democratic participation model to include women as well as those of low-income households, in particular the unemployed.
- Empowering consumers and municipalities in a capacity-building program through the launch of an interactive online "RE Prosumer Investment Calculator" and seminars in the five partner countries (Germany, Italy, Bulgaria, Poland, and the Czech Republic).
- Formulating policy recommendations to promote prosumership and to remove barriers for consumers to become active market players at the EU and national levels.

Considering the emerging regulatory framework and the European projects supporting the birth and creation of the new energy–economic–social system, the objective of this study is to make a contribution explaining how elements are important for the creation of EC through a real case study application, and reflecting which elements facilitate or do not facilitate the creation of these communities.

In this framework, the main goal of the present study is to illustrate the comprehensive procedure and process of the EC creation. In particular, this paper focuses on the description of the methodological approach in implementing the CSOP model, which consists of three main phases: the identification and description of selected buildings (preparation phase), the preliminary and feasibility analysis, and finally target group involvement. Supporting Consumer Co-Ownership in Renewable Energies (SCORE) first started in three pilot regions in Italy, the Czech Republic, and Poland, and later, with the aim of extending the methodology, was carried out in various other follower cities across Europe. The Italian pilot studies were chosen as a case study to develop and test the methodology.

The paper is divided as follows. Section 2 describes the details of methodological framework consisting in the succession of three main phases of EC creation. Section 3 illustrates the case study, which is used for testing the effectiveness of the proposed methodological framework. The results and discussions are presented in Section 4. Finally, conclusive remarks are discussed in Section 5 and future developments are identified.

2. Methodological Framework

The methodological framework of the creation of the CSOP model consists in a process determined by succession of three major phases, in which each phase (and sub-phase) is fundamental since its output represents the starting point for the next step. The first phase (I) is preparation, which includes building identification and data collection. The second phase (II) consists in the preliminary and feasibility analysis proposing different energy retrofit alternatives in order to shift from fossil fuels to renewable ones, to reach a reduction of energy consumption, and to increase the efficiency of the building envelope and the energy system. This second phase employs multi-criteria analysis (MCA) to select the best alternative based on the key performance indicators (KPIs) considering different stakeholders' opinions. The final phase (III) is target group involvement, in which citizens and public and private entities will be a part of financial model. It is helpful to break the study down into the main elements that frame it to understand the research process steps employed. To this end, in Figure 2 a schematic flowchart of the methodological approaches of the research is shown. Consequently, for each phase, the relative outputs and proposed methodologies are shown in a detailed way.

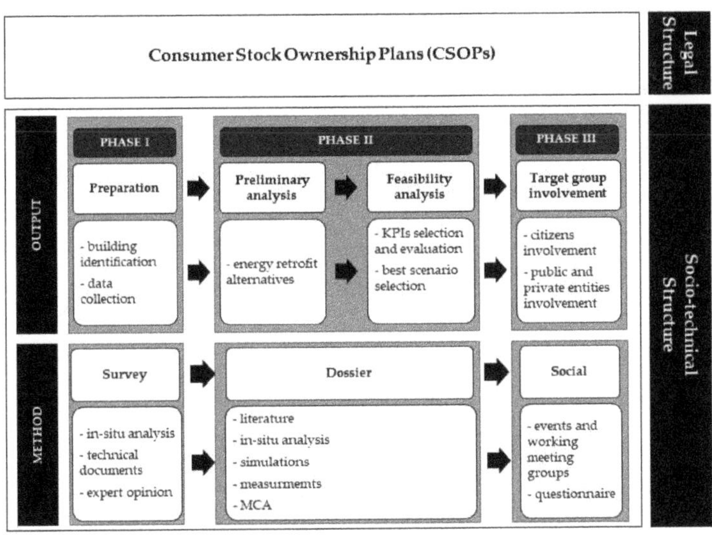

Figure 2. Flow-work: different steps of work. MCA: multi-criteria analysis; KPIs: key performance indicators.

2.1. Phase I: Preparation

Within the first phase, preparation, the different pilot case studies are identified, and their characteristics are described. Basically, the data is collected through filling in two pre-defined surveys. In order to fulfil the first phase different in-situ analyses have been done. Moreover, relative technical documents and expert opinions have been considered to compile the surveys.

(1) The first survey regards the investment identification of RESs, which is composed of five main sections [7]. This survey collects a general description of the buildings considered for each pilot study case, describing the current situation (i.e., geometry and energy plant systems) and the design one (i.e., planned project in terms of RES and financial aspects). The first section identifies the building characteristics (e.g., building ownership, building construction year, year of the last refurbishment, heat and domestic hot water (DHW) distribution system operator, average of consumption expenses, total number of dwellings or offices, total official number of inhabitants/employees, number of floors, total usable area, and total roof area). The second section investigates the existing conventional energy sources or external supplier (e.g., type of energy sources, installed power or purchased power if the district heating (DH) network is present). The third section describes the existing RESs, for example the type of energy sources and installed power. In the fourth part, the planned RESs are investigated. Finally, the fifth section is dedicated to the planned structure of financial sources for RES investment (e.g., type of financial sources and percentage of overall costs). Since the target is on a local scale, the definition of the building's database is crucial.

(2) The second survey reports the data in terms of energy costs and tariffs for the actual situation, for the use of non-renewable energy sources. The aim of this survey is to collect information about

the use of non-renewable energy sources. Specifically, average consumption fee data (€/GJ) are reported (e.g., annual consumption (GJ), historical data for oil and natural gas cost (€/GJ), and the average fixed fee (€/month)).

2.2. Phase II: Preliminary and Feasibility Analysis

The second phase, which consists of preliminary and feasibility analysis, is investigated within the specific document, called the "dossier" [8]. The dossier represents a guideline in order to illustrate the collected information and data related in order to improve and to increase the energy efficiency of the pilot building. Additional data are collected for defining different refurbishment measures, which are described in dossier using simulation and measurements approaches. Issues addressed in the detailed dossiers are as follows:

(1) Energy impact assessment of the current situation, which determines the energy needs and energy uses for space heating, DHW, and lighting and equipment through collecting the measured data and in-situ analysis. Also, energy analysis has been assessed after implementing retrofitting measures through the building energy simulation model. At least two different refurbishment alternatives (for each case study) have been proposed. The retrofit alternatives concern the envelope system, the energy system installing RES, and the control system.
(2) Environmental impact assessment illustrates the strategies to minimize the environmental impact with each alternative.
(3) Economic and financial assessment of the investment costs.

Finally, in order to select the best scenario, MCA has been implemented for which KPIs are first defined. In particular, the use of an MCA assesses the best refurbishment alternative, considering different KPIs. The choice of the KPIs to identify the most feasible and sustainable project is made as per previous work [7]. The KPIs have been defined based on three main steps. The first step is performed through a comprehensive review of the existing literature [9,10]. In the second step, the number of KPIs is reduced as a result of five internal discussion rounds among relevant experts. In the third step, the final set of KPIs is selected through a participatory workshop in which the playing card method was employed [11]. Finally, the MCA allows us to define the best alternative considering different indicators, in order to determine the most feasible one. Once the best alternative is defined, in order to proceed to the effective realization of the project, it is necessary to define a business plan. The business plan allows us to assess the economic profitability of the selected project and whether it can be increased to optimize economic feasibility.

2.3. Phase III: Target Group Involvement

The first two phases have a technical character aimed at defining the best alternative; instead, in the third phase, the social aspects are examined in depth to describe and define the new financial model based on co-ownership (CSOP). As mentioned in the introduction, one of the purposes of the project is to encourage the active role of consumers (private or public users). Indeed, the users undertake a crucial role in the EC, not only as simple consumers but also prosumers, participating actively in the phases of decision, dissemination, production, and distribution of energy. In addition, considering the future role of EC in the energy market, it is necessary to understand the institutional setting based on financial participation schemes that (1) confer ownership rights in RE projects, (2) involve "active" consumers with the specific attention on vulnerable ones, and (3) and consider local or regional areas. Since user participation is the core topic of EC creation, the purpose of the third phase is to involve several target groups. Although the previous community projects are widespread, the inclusion of all citizens is not complete [5]. Moreover, these types of projects are usually performed by men who are middle aged and with a higher income, whereas vulnerable groups (affected by fuel poverty), women, and low-income households are excluded from RE investments [5]. As such, a social analysis will be conducted through a specific action plan to collect information through, first of all, (1) events and a work group, and then, (2) surveys and questionnaires. These analyses help in understanding the citizens' subjective willingness to engage in local energy initiatives. At the same time, the aim of the social analyses is to obtain objective data about users' characteristics in order to identify the main drivers that favour/hinder their participation. As mentioned before, the citizens' involvement took place through three steps:

(1) Info events: Meetings with local institutions and organizations that work in the area in order to transmit the project objectives and dialogue on how to include citizens, without neglecting those belonging to vulnerable groups.

(2) Workshops: This second method allowed us to inform invited citizens about a specific topic and create a semi-structured debate with them. Specifically, with the support of local authorities, known and recognized in the area, a diverse group of citizens were invited with the aim of giving them some fundamental notions about the project topic, such as the meaning of energy transition, the use of energy from renewable sources, the energy community, and the share ownership plan by consumers. At the same time, the educational moments were alternated with moments of learning verification through answers to questions or specific activities in order to express their thoughts and create a constructive debate. This is a semi-structured method in which people are free to express themselves.

(3) Administering a specific questionnaire: through this method, the interviewees were asked to choose only one answer among those proposed; this method is more restrictive than the previous one. In particular, the results obtained in the "workshop meeting group" made it possible to define the questionnaire which in its final version is composed of five macro parts including detailed information:

    a.   Attitude and willingness information: level of degree interest towards the EC project;

    b.   Feeling related to community identity information: level of feeling related to trust, satisfaction, pride, hope, disgust, shame, fear, boredom;

    c.   Technical information: building type and age, type of heating system, efficiency of the energy plant or building envelope;

    d.   Socio-economic information: personal and family income, family composition, building construction year, and building ownership;

    e.   Socio-demographic information: age, gender, education level, nationality, marital status, and municipality.

The questionnaire was given citizens in a specific context and the data analysis produced a division of citizens into population segments that shared common features. The study allowed us to understand and, subsequently, to promote the users' cooperation to become co-owners of the new energy plant system. The definition of different population segments highlighted the clusters that were interested in or would like to be part of community project but, for different reasons, did not have the possibility of participating (e.g., women, low-income households, vulnerable groups affected by energy poverty, etc.).

Then, at the end of the whole process, on the basis of technical and social analysis, the CSOP Operating Company was established, including each population segment through ad hoc policies in order to facilitate their participation. In the Italian case studies, the financing model could be represented by the following scheme in Figure 3.

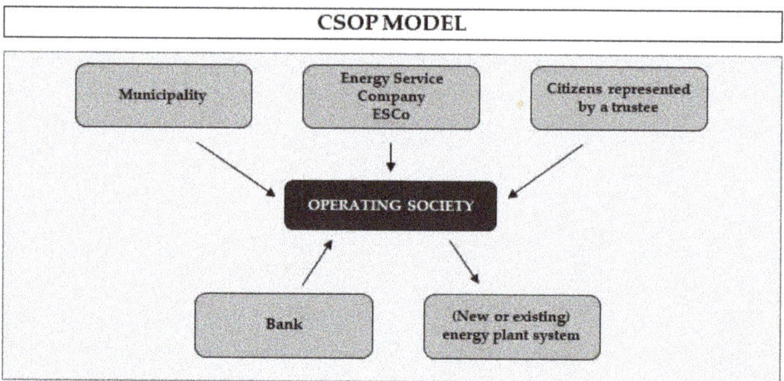

Figure 3. Financing of a renewable energy (RE) plant and energy efficiency (EE) measures through a CSOP, authors' elaboration.

Once all the three above mentioned phases (socio-technical structure) were concluded, the financial CSOP model (legal structure) could be implemented. The creation of CSOPs enables consumers, especially the vulnerable, to have a co-ownership stake in a utility they use and thus to become prosumers. Moreover, investments can be made into any kind of utility, for instance, energy, water, transportation. Moreover, CSOPs contribute to the energy transition and climate change mitigation by facilitating local, decentralized production through investment in renewable energy installations. Interestingly, in the CSOP model [12,13] different actors become as owners of the new energy plant system of RE, as shown in Figure 3, and the main elements are as follows:

- The participation in decision-making is possible through the trustee, who represents the citizens interested in CSOP, while individual consumer-shareholders may execute control rights on a supervisory board or advisory council. Therefore, the model is of consumer-centred investment for general services, providing participation both financially and in regard to management decisions.
- Municipalities, small and medium-sized enterprises (SMEs) and other local stakeholders are permitted as co-investors. CSOPs avoid personal liability for consumer-shareholders.
- The Operating Society invests in new or existing RE plants and operates on behalf of different actors as co-owners.
- It is possible to demand loans from banks;
- New RE plants supply energy to consumers at fixed price and generate revenues from excess production sold to the grid.

3. Case Study

The Susa Valley (45°8′12″ N, 7°3′29″ E, from 300 to 3.612 m asl), was selected as a pilot case study. It is one of the widest and deepest Italian alpine valleys. It extends for about 100 km in length, belonging to the Metropolitan Region of Turin via the western part of Piedmont region of northern Italy to the border of France. In Susa Valley, 39 municipalities have been settled, characterized by different locations, territorial extensions, and demographic sizes. The different morphological, altitudinal, and climatic characteristics have contributed to differentiate the development of the territory, aggregating municipalities into four geographical areas: the Oulx area, Susa area, Condove area, and Avigliana area. The population is over 90,000, and 30% of the valley's inhabitants live in the main towns Avigliana, Bardonecchia, Bussoleno, and Susa.

Ten municipality pilot projects have been chosen in Susa Valley as case studies where the implementation involves substituting the existing heating system fuelled by diesel oil and natural gas. The new planned systems will be fed by local biomass, wood chips instead of pellets or wood blocks that are the typical solution for small individual boilers. To avoid repetitions in this paper of these 10 projects, one representative project is analysed in the following EE analysis; the remaining projects have similar properties to those analysed. Table 1 shows the selected municipality pilot projects and their relative buildings indicating existing and planned heating systems.

Table 1. Pilot case studies. DH: district heating; LGP: Liquid Propane Gas.

No.	Municipality (City)	Number	Building	Existing Energy Sources for Heating	Type of Installation
1	Oulx	1.a 1.b 1.c 1.d 1.e 1.f 1.g	School and gym Nursery Gym Municipality Touristic office Social activity building Building (residential)	Oil and natural gas boiler (individual generators)	DH network (biomass)
2	Novalesa	2.a 2.b 2.c	Abbey Private building 1 Private building 2	Oil and LGP boiler (individual generators)	DH network (biomass)
3	Rueglio	3.a 3.b	Municipality Retirement house	Oil boiler (individual generators)	DH network (biomass)
4	San Giorio di Susa (building scale)	4.a	Multi-use room and bar	Natural gas boiler (individual generators)	DH network (biomass)
5	San Giorio di Susa (city scale)	5.a	Private residential buildings	Individual oil stove	DH network (biomass)
6	Villar Dora	6.a 6.b	School and gym Kindergarten	Natural gas boiler	DH network (biomass) and solar thermal collectors
7	Susa	7.a	DH network	Oil and natural gas boiler (individual generators)	DH network (biomass)
8	Bardonecchia	8.a	DH network	Oil and natural gas boiler (individual generators)	DH network (biomass)
9	Bussoleno	9.a	DH network	Natural gas boiler (individual generators)	DH network (biomass)
10	Almese	10.a 10.b 10.c	Sport (facilities) buildings Middle school Private buildings	Natural gas boiler (individual generators)	DH network (biomass)

3.1. RESs in Susa Valley

Currently in the Susa Valley, 75% of the pilot projects originate from fossil fuels, while 25% of energy is produced by RESs, mostly from biomass. Although there is a vast quantity of local biomass sources in the region, the biomass used is not produced locally but imported from other European and non-European countries. Moreover, the majority of the imported biomass is not certified and cannot be statistically quantified since it is subjected to the grey market. Notably, in Susa Valley, 11 public buildings have already been connected to new biomass heating systems. These can play a significant role of replicators for the future sub-pilots.

3.2. Energy Poverty in Susa Valley

One of the presented issues in Susa Valley is energy poverty. Energy poverty is defined as the lack of access to energy or a difficulty in paying for necessary energy, which leads to a decline in living conditions [14]. Groups vulnerable to energy poverty are not located in a particular area, but are rather spread over the municipal territory. Some areas of the Susa Valley, due to their geographical position and therefore lack of sunny exposure (specifically the north slope of the Dora), are not very attractive for housing. Hence, vulnerable households are located in these areas since the rent or the housing costs are low. Some associations work with these vulnerable groups. For example, Consorzio Intercomunale Socio-Assistenziale (Con.I.S.A.), Cooperativa Sociale Amico (COOPAMICO), and Caritas are three entities that operate on the Susa Valley territory, dealing with people in difficulty in order to help them with issues involving poverty, unemployment, and access to social services. With respect to energy behaviour and efficiency, vulnerable households tend to use older, less energy efficient stoves and consequently fossil fuels due to their low prices. The planned energy community facilitates the replacement of old utilities and the provision of locally sourced wood chips as fuel.

3.3. Implementing the EC Project in Susa Valley

The main foreseen project activity in Susa Valley is to implement the new plant system fuelled by local and certified biomass with an existing heating system, fuelled by diesel oil and natural gas. In some cases, a DH network might be developed (see Table 1). The idea is to substitute fossil fuels, imported by external countries, with local wood chips. This leads to generation of positive economic externalities for the territory since fuel will be provided by the local forest, leading to a sustainable path. Indeed, the replacement of fossil sources with local wood chips entails (1) lower costs for energy, (2) a high share (>80%) of energy cost remaining on the territory, and (3) lower CO_2 emissions (closed carbon cycle). As mentioned above, the project aim is to create a RE community employing the CSOP model in the whole Susa Valley. Moreover, project sets a specific focus on low-income households and women to become co-owners and co-investors in RE CSOPs. For this reason, the Susa Valley action plan focuses specifically on the involvement of citizens and particularly vulnerable groups, as well as other residents, SMEs, and municipalities. These main project activities will be undertaken in 10 municipalities (Table 1). It is planned to extend the energy community created within SCORE to all 39 municipalities in Valley Susa. On one hand, the majority of the buildings identified in Susa Valley are public, which provides economic security. On the other hand, the sub-pilot study in San Giorio di Susa with activities at the city scale deals with residential buildings, and is of crucial importance regarding citizen involvement. Incorporating residential buildings leads to involving citizens directly in the energy community.

4. Results and Discussions

4.1. Phase I: Preparation

As was explained there is a process of phases and methods to obtain the final results. This phase, "Building identification and data collection", illustrates how the methodology is applied on one out of the 10 case studies. Oulx was the pilot study chosen due to the prior approval and engagement of the municipality, and the variety of possible refurbishment alternatives proposed. The aforementioned aspects will enrich the procedure of selection. The first phase involved the preparation, and therefore, the collection of data and information, as shown in the workflow (Figure 2). Table 2 illustrates the main significant data collected regarding Oulx pilot project through the questionnaire prepared within phase I. As shown in Table 2 below, each pilot study building (detailed in Table 1) in Oulx has been described through the following information: building ownership (private or public) and building function (residential or non-residential), building construction year, the latest refurbishment year, average heat and domestic hot water (DHW) expenses, total number of building zones (dwellings or offices), total number of users (inhabitants or employees), total usable area, and finally, average annual energy consumption.

Table 2. Oulx data and information collection.

No.	Ownership and Function	Construction Year	Latest Refurbishment Year	Average Heat and DHW Expenses (€/year)	Total Number of Zones	Total Number of Users	Total Usable Area (m^2)	Average Annual Consumption (MWh)
1.a *	Public; non-residential (educational)	1958	2018 (seismic)	57,915	27	250	2800	300
1.b *	Public; non-residential (educational)	1988	none	5585	1	50	270	
1.c *	Public; non-residential (sportive)	NA	NA	NA	1	220	NA	
1.d	Public; non-residential (administrative)	1980	2016 (windows)	13,831	10	26	660	150
1.e	Public; non-residential (services)	1995	none	14,669	3	6	700	150
1.f	Public; non-residential (services)	First years of 1900	2016 (structural)	3,000	3	2	300	30

* The three buildings are the subject of the energy analysis in order to reach the nearly Zero Energy Building (nZEB) conditions. nZEBs are buildings that have very high energy performance, pursuant to Directive 2018/844 on the energy performance of buildings. As per O.J. L 156/75 the nearly zero or very low amount of energy required should be covered to a significant extent by energy from renewable energies sources (RESs), including RE produced on-site or nearby (cf. recital (7) and Annex I point 2 of the Directive). DHW: domestic hot water.

4.2. Phase II: Preliminary and Feasibility Analysis

4.2.1. Preliminary Analysis (Energy Retrofit Alternatives)

As mentioned above, Phase II, preliminary, and feasibility analyses were performed through dossier documents. This phase started with the general description and historical information of the buildings involved, the current situation regarding the energy sources, and a brief investigation of the planned RES. During the analysis, the physical properties of the materials used for the construction of the building (walls, roofs, slabs, windows) were acquired. Specifically, the Oulx dossier investigated the school complex that is the subject of energy retrofitting in order to access the "Conto Termico" (a package of incentives and concessions set up with an Italian ministerial decree to promote measures to improve the EE of existing buildings and to encourage the production of RE). Later, a small DH network will be installed to cover also the adjacent buildings. The school area includes three different buildings (Figure 4):

1.a. An elementary and middle school building, with a basement floor and three overlying floors in elevation.

1.b. A gym that has only a ground floor with a common wall with the school (on the eastern side of the school).

1.c. A prefabricated nursery building, that covers a single ground floor and is located beside the school.

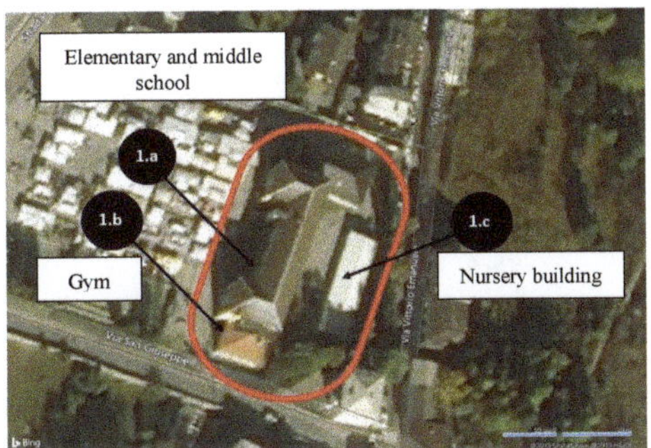

Figure 4. Buildings involved, source: www.bing.com/maps.

The buildings are equipped by two oil boilers characterized by different circuits and by different kinds of heaters (radiators, fan heaters, and air nozzles) for the schools, the gym, and the nursery; consequently, the absence of integration between each building is one of the critical issues from the energy point of view. The thermal efficiency values of the two traditional oil boilers with blast burners are 81.5% (generator of 300 kW) and 78.9% (generator of 130 kW). Regarding the domestic hot water (DHW) production there is a centralized generation combined with the heating generation. Other critical issues of the building are as follows:

- Significant energy leakage through the opaque casing (as shown by the values of thermic transmittance in Table 3);
- Obsolete regulation and balance systems (simple regulation on-off with no internal temperature compensation);
- Obsolete heat generation technology (oil boilers over 10 years old);
- Not clean energy sources (diesel fuel) and consequent high emission levels of CO_2.

Table 3. Oulx envelope system characteristics (before retrofitting).

	Before	
Element	**Thickness (mm)**	**Thermic Transmittance (W/m^{2}K)**
School external wall	400	0.847
Gym external wall	290	1.020
Nursery external wall	70	0.332
School upper-attic slab	200	2.401
Gym upper-attic slab	60	1.429
Nursery upper-attic slab	50	0.438

In addition, data in terms of energy costs and tariffs were collected for the actual situation, using non-renewable energy sources. Specifically, the information presented in Table 4 was provided by administrative municipal accounting, and current values are assumed to be the same due to the impossibility of accessing more recent information. Then, the litres of consumed diesel fuel in one year were calculated and a consumption of 57,746 litres/year was established to meet the needs of the three buildings.

Table 4. Energy costs for buildings involved.

Client	Cost	Years
Middle school	€46,857	2012
Elementary school	€17,620 (average)	2003–2012
Nursery	€5,050	2013

The mathematic model that shows the performances of the building and plants that are the object of this study was created with software certified by the Comitato Termotecnico Italiano (CTI). The resulting values have been validated taking into account the trends of utilization of the buildings, as shown in Table 5 below:

Table 5. Trend of building utilization.

Zone	Day of Utilization	Hours per Day	Internal Temperature Point Set When Used/Not Used
School	5	12	20 °C/16 °C
Gym	7	12	20 °C/16 °C
Nursery	5	12	22 °C/19 °C

The primary energy indicator (total (Qp), and that normalized with respect to the floor area (EP)) for the two services of space heating and domestic hot water are shown in Table 6 below. Specifically, the non-renewable, renewable, and total values of consumption are calculated.

Table 6. Oulx energy indicators.

Service	Qp,nren (kWh)	Qp,ren (kWh)	Qp,tot (kWh)	EP,nren (kWh/m^2)	EP,ren (kWh/m^2)	EP,tot (kWh/m^2)
Heating	491,432	0	491,432	172.98	0.0	172.98
Domestic hot water	37,919	0	37,919	13.35	0.0	13.35
TOTAL	529,350	0	529,350	186.32	0.0	186.32

After an energy analysis and identification of weaknesses and critical issues of the actual situation of the buildings pilot, different retrofit alternatives (Table 7) were studied in order to improve the current energy situation and minimize the environmental impact. Since the main purpose of the project was to facilitate consumers to become prosumers of RE and to become owners of RE energy plants (through the CSOP financing model), the first alternative concerns solely the replacement of the boilers with a unique biomass-fired one and regulation retrofitting. On the other hand, the subsequent alternatives intervene on the envelope of the buildings, insulating the external walls and roof with a growing thickness as the alternatives increase. Intervening only on the energy system is not enough; for a good result of the project it is, therefore, necessary to intervene on the envelope system, increasing its efficiency in order to reduce heat losses for transmission and ventilation. In this way the required winter load for the heating system will be lower. In addition, as mentioned previously, it is considered useful to reach nearly Zero Energy Building (nZEB) conditions and to obtain the incentives offered by the "Conto Termico". Table 8 shows the Oulx envelope system characteristics after intervention A4, where the results start to reach nZEB conditions. Consequently, through the energy simulation, Table 9 shows the results obtained in comparison to the current situation, considering the energy consumption of the building system from non-renewable sources (Qp,nren), the energy consumption of the building system from renewable sources (Qp,ren), and the CO_2 emissions consequent to fuel consumption. In addition, as shown in Table 9, the percentage of energy from renewable sources compared to the total energy used by the building is 80%. This value is defined as a "minimum requirement" derived from the Ministerial Decree of 26 June 2015 (https://www.mise.gov.it/images/stories/normativa/DM_requisiti_minimi_allegato1.pdf).

Table 7. Retrofit alternatives for Oulx pilot case study. nZEB: nearly Zero Energy Building.

Code of Simulation	Interventions
0.0	As-built simulation model.
0.1	As-built simulation model from real consumption (benchmark).
A1	Simulation 0 and replacement of the boilers with a unique biomass-fired one and regulation retrofitting.
A2	Simulation 1 and the upper-attic slab insulation (18 cm).
A3	Simulation 2 and external walls insulation for the school and the gym (18 cm).
A4	Simulation 1 and nZEB conditions obtained with the upper-attic slab insulation (40 cm), external walls insulation for the school and the gym (30 cm), and the nursery's external walls (25 cm).
The A5	Simulation 1 and nZEB conditions obtained with the upper-attic slabs insulation (50 cm for the school and the gym, 40 cm for the nursery), external walls insulation for the school and the gym (40 cm), and the nursery's external walls as built.
A6	Simulation 1 and nZEB conditions obtained with the replacement of the windows with more efficient components (transmittance: <1.0 W/m^{2}K), upper-attic slab insulation (15 cm for the school and the gym, 12 cm for the nursery), external wall insulation for the school and the gym (15 cm), and the nursery's external walls as built.

Table 8. Oulx envelope system characteristics (after intervention A4).

Element	After *	
	Thickness (mm)	Thermic Transmittance (W/m^{2}K)
School external wall	720	0.110
Gym external wall	610	0.112
Nursery external wall	320	0.103
School upper-attic slab	600	0.084
Gym upper-attic slab	460	0.082
Nursery upper-attic slab	450	0.073

* A4 is the first alternative reaching nearly Zero Energy Building (nZEB) conditions.

Table 9. Oulx energy simulation results.

Code of Simulation	Qp,tot	Qp,ren		Qp,nren		CO2 Emissions	
	(kWh/y)	(kWh/y)	% (ren/tot)	(kWh/a)	% (nren/tot)	(kgCO2eq/a)	% (VS 0.1)
0.0	529,350	-	-	529,350	100%	137,551	85.84%
0.1	616,697	-	-	549,061	100%	160,248	100%
A1	457,140	365,712	80%	91,428	20%	22,857	14.26%
A2	335,284	268,228	80%	67,057	20%	16,764	10.46%
A3	197,529	158,023	80%	39,506	20%	9876	6.16%
A4	177,276	141,821	80%	35,455	20%	8864	5.53%
A5	177,213	141,771	80%	35,443	20%	8861	5.53%
A6	177,638	142,110	80%	35,528	20%	8882	5.54%

4.2.2. Feasibility Analysis (KPI Selection and Evaluation)

After defining the appropriate retrofit alternatives (Table 7), and consequently simulating their energy performance results (Table 9), the definition of the different indicators has been made. These indicators assess an impact of defined alternatives not just regarding the energy aspects but considering all the sustainable aspects (i.e., environmental, economic, technical, and social). Based on indicator impact assessment, it is possible to identify the most feasible and sustainable project that will fit in each pilot study case. The criteria were primarily developed based on a review of existing literature and verified in a workshop in which the "Playing card" [15] method was employed, involving different parties as was detailed in [7].

Afterward, new modifications were introduced to select the final set of key performance indicators (KPIs) (Table 10). These last changes emerged as the project progressed, during different meetings and workshops, and they were explicitly detailed and accepted by the partners. The goal of selection process was to reduce the criteria to obtain a practical but still significant number that is sufficient for conducting a sustainability assessment.

Hereafter, the impact assessment methodology for each selected indicator, with respect to the different retrofitting measures developed previously, will be illustrated. The evaluation process provides quantitative and qualitative information giving a support for each retrofitting measurement. They can be classified into four main categories: environmental, economic, technical, and social. Table 10 shows the selected KPIs with which the different refurbishment alternatives are evaluated alongside environmental, economic, technical, and social aspects. Each detailed KPI matrix addresses—subject to availability of data and depending on the RES—some or all of the following KPIs.

Table 10. Key performance indicator matrix.

Category	Code	Indicator	Type	Data Source	Unit
Environmental	ENV1	Primary energy saving	Quantitative	Estimated or metered data	(kWh$_{primary\ energy}$/year)
	ENV2	Global CO_2 emission reduction	Quantitative	Estimated or metered data	(kg/year)
	ENV3	Local NO_X emission reduction	Quantitative	Estimated or metered data	(kg/year)
	ENV4	Local PM_{10} emission reduction	Quantitative	Estimated or metered data	(kg/year)
Economic	EC1	Payback period (PBP)	Quantitative	Calculation	(Years)
	EC2	Investment cost	Quantitative	Calculation	(Euro)
	EC3	Public incentives	Quantitative	Process documentation	(%)
	EC4	Savings on energy expenditure	Quantitative	Calculation	(Euro/year)
	EC5	Labour cost	Quantitative	Estimated or metered data	(Euro/year)
	EC6	Labour cost by a social cooperative	Quantitative	Estimated or metered data	(Euro/year)
	EC7	Material cost	Quantitative	Estimated or metered data	(Euro)
	EC8	Material cost purchased on the territory	Quantitative	Estimated or metered data	(Euro)
	EC9	Running cost	Quantitative	Calculation	(Euro/year)
	EC10	Type thermal account access (TAA) vs. Energy efficiency certificates (EEC)	Qualitative	Process documentation	(TAA/EEC)
Technical	T1	Increase of plant system efficiency	Quantitative	Estimated or metered data	(%)
	T2	Installed power reduction	Quantitative	Estimated or metered data	(kW)
Social	S1	Architectural impact	Qualitative	Process documentation	(Ordinal)

Environmental Indicators.

- ENV1—Primary energy saving. Primary energy that would be saved if the new plant was built. It is linked to the renewable nature of the investment and to the interventions on the building envelope. It was calculated with a specific software in which the material, thickness, thermic transmittance, and internal surface resistance were some of the inputs needed [9].
- ENV2—Global CO_2 emission reduction. The building's energy system CO_2 emissions are undoubtedly a criterion that should be assessed for the sustainable development of cities [16,17]. It is calculated comparing the current situation with the different alternatives proposed.
- ENV3—Local NOx emission reduction. NO_x produces toxic pollution that affects the health of individuals, also harming the environment, climate, and vegetation [18]. This also implies that there is an indirect impact on the social health of communities [19].
- ENV4—Local PM_{10} emission reduction. PM_{10} emissions are caused by fuel burning and heavy industrial processes and are very harmful to human health [18]. These emissions cause lung diseases, heart attacks and arrhythmias, cancer, atherosclerosis, childhood respiratory disease, and premature death.

Economic Indicators.

- EC1—Payback period (PBP). PBP, simple or discounted, is a popular criterion that represents the time in which negative and positive cash flows are equal. It represents the moment after which the expenses are amortized and there is the actual gain. This criterion gives immediate insight to investors in the event that there is a preference to shorten the PBP [20]. The payback period is assessed as shown in Equation (1):
- EC2—Investment cost. Many studies consider investment costs as the most important criterion to evaluate energy savings interventions. The investment cost involves all the costs related to refurbishment of the building and/or new heating system; it includes the purchase of building

material, technological installations, manpower, and set up of the cost for each individual element of the renovation project (building envelope and energy systems), as demonstrated in Table 11 [21,22].

- EC3—Public incentives. This is the percentage of savings linked to the share of investment cost covered by administrative incentives. The Stability Law confirmed the extension of 65% tax reductions for energy efficiency measures and 50% for restructuring buildings completed by the end of 2017 [23]. "Conto Termico" involves the following incentives:

  ○ Up to 65% of the expenditure incurred for nearly Zero Energy Buildings (nZEBs);
  ○ Up to 40% for wall and ceiling insulation, replacement of windows, solar shading, indoor lighting, building automation technologies, boilers;
  ○ Up to 50% for thermal insulation work in climate zones E/F and up to 55% in case of thermal insulation and replacement of window seals when combined with other interventions (heat pumps, solar thermal, etc.).

$$PBP = \frac{investment\ costs}{annual\ savings\ on\ energy\ expenditure} \tag{1}$$

Table 11. Oulx investment costs.

Materials/Service	Price (€/Unit)	Quantity (Unit)	Amount (€)
Wall insulation	100	2,000	200,000
Upper-attic insulation	100	1,200	120,000
audits	6250	1	6250
Building site	20,000	1	20,000
Lean concrete	2500	1	2500
Foundation	15,000	1	15,000
Walls	3750	4	15,000
Slab	12,500	1	12.500
Waterproofing	1000	1	1000
Passages	5000	1	5000
District pipes	20,000	1	20,000
Biomass boiler	70,000	1	70,000
Plant modifications	10,000	4	40,000
Control	20,000	1	20,000
Mounting	30,000	1	30,000
Project	20,000	1	20,000
Tele management	20,000	1	20,000
TOTAL			617,250

To have access to these incentives, there are some aspects to take into account such as certification by an accredited body that certifies compliance with the UNI EN 303-5 standard, useful thermal efficiency not lower than 87% + log (Pn) where Pn is the nominal power of the device, atmospheric emissions not above a certain value verified by an accredited body based on the relevant measurement method, and certification of the pellets used by an accredited certification body that certifies compliance with the UNI EN ISO 17225-2 standard, etc.

- EC4—Savings on energy expenditure. The savings on annual expenditure taking into account the primary energy savings calculated previously.
- EC5—Labour cost. It includes the salary of employees who are directly involved in production activities, services (such as general repairs and maintenance performance), and supervision. It is assumed to be 40% of investment costs, as an expert in the field suggested during an internal meeting [24,25].

- EC6—Labour cost by a social cooperative. The part of labour cost which will be covered by the social cooperative.
- EC7—Material cost. The costs of raw materials or parts that go directly into producing products or providing services. This cost was assumed to be only at the beginning of the project (one off) including aspects like the boiler, insulation, and concrete.
- EC8—Material cost purchased on the territory. This criterion evaluates the portion of material cost that remains in the territory. The territory is intended to be the Susa Valley.
- EC9—Running cost. This involves the energy costs plus maintenance costs. The maintenance costs are assumed to be 2% of investment cost according to [26].
- EC10—Type Thermal Account Access (TAA) vs. Energy Efficiency Certificates (EEC). This represents the access to the thermal account and energy efficiency certificates, Italian public incentives carried out by energy services management.

Technical Indicators.

- T1—Increase of plant system efficiency. This is the increase in the efficiency of the new system plant compared to the existing one [9].
- T2—Installed power reduction. This represents the reduction of installed power; it is always an aspect that contributes directly in energy reduction.

Social Indicators.

- S1—Architectural impact. This indicator evaluates the visual outcome that may be created by the application of retrofitting measurements for a city. When retrofit measures lead to aesthetic improvement of the city, this criterion has a higher value. Five scores of impact are presented in Table 12 according to the study conducted by Dall'O' et al. [27], with reference to specific measures. This criterion adopts an ordinal scale to rank the strategies, from the best to the worst.

Table 12. Architectural impact criterion.

Typology of Criterion	Description of Criterion	Numerical Value of Criterion	Description of Intervention
Positive	Great positive impact	1	External Thermal Insulation Composite Systems
	Positive impact	2	Windows replacement
Neutral	No impact	3	Roof insulation – Boiler replacement – Lightning replacement
Negative	Little negative impact	4	Photovoltaic panels
	Negative impact	5	Solar thermal collector

After the appropriate selection and the definition of the performance indicators, the next step consists in assessing each KPI and establishing the evaluation matrix shown in Table 13, which is fundamental to reach the final selection. It allows the comparison of each refurbishment alternative proposed in the preliminary analysis with the current situation, taking into account the selected KPIs. To complete it, the collaboration of different parties is necessary, since the indicators cover the different areas of the project, from the economic to the social, through the technical and environmental. EDILCLIMA software was employed to simulate the energy alternatives and to obtain the data, while for assessing each KPIs the specific method is used as explained above (Section 4.2.2).

Table 13. Evaluation matrix.

Category	Indicator	A1	A2	A3	A4	A5	A6
Environmental	ENV1 (kWh$_{primary\ energy}$/year)	525,269	549,640	577,191	581,242	581,254	581,169
	ENV2 (kg/year)	137,427	143,520	150,408	151,420	151,423	151,402
	ENV3 (kg/year)	94.55	98.94	103.89	104.62	104.63	104.61
	ENV4 (kg/year)	6.83	7.15	7.50	7.56	7.56	7.56
Economic	EC1 (PBP) (years)	8.3	11.7	16.5	16.3	16.3	16.3
	EC2 (Euro)	284,750	417,250	617,250	617,250	617,250	617,250
	EC3 (%)	40%	40%	40%	65%	65%	65%
	EC4 (Euro/year)	34,142	35,727	37,517	37,781	37,782	37,776
	EC5 (Euro/year)	136,250	136,250	136,250	136,250	136,250	136,250
	EC6 (Euro/year)	34,063	34,063	34,063	34,063	34,063	34,063
	EC7 (Euro)	148,500	281,000	481,000	481,000	481,000	481,000
	EC8 (Euro)	51,975	98,350	168,350	168,350	168,350	168,350
	EC9 (Euro/year)	39,523	33,156	26,962	25,630	25,459	25,490
	EC10 (TAA/EEC)	TAA	TAA	TAA	TAA	TAA	TAA
Technical	T1 (%)	9.80%	9.80%	9.80%	9.80%	9.80%	9.80%
	T2 (kW)	175	175	175	175	175	175
Social	S1 (-)	3	1	1	1	1	2

4.2.3. Feasibility Analysis (Best Scenario Selection)

To proceed with this step, an outranking MCA, called PROMETHEE (preference ranking organization method for enrichment evaluation), was chosen in order to outrank the different energy retrofit interventions proposed previously in each case study [28]. The target was to provide a comprehensive overview of the best alternative. Moreover, a sensitivity analysis was carried out, modifying the weights and preferences of each alternative, in order to observe how their ranking varies.

The PROMETHEE method belongs to the outranking category, which has been developed by Brans et al. [28]. The PROMETHEE method uses the partial aggregation and it is very useful in ranking a limited number of alternatives, considering conflicting criteria [29]. It is based on the pair-wise comparison, checking if one of two alternatives outranks the other or not [30]. Two specific types of information are necessary in order to implement this method, the criteria weights and the decision-maker's preference function for comparing the contribution of the alternatives in terms of each separate criterion [31].

In order to apply the chosen model, the "Visual PROMETHEE" is employed. First of all, the set of criteria were defined and added, which in the MCA must generally be in finite number. Therefore, the six different retrofit alternatives for Oulx case study had to be reconstructed within the program. It is necessary to give every criterion a direction of preference. Specifically, it must be decided whether the criterion must be minimized or maximized. With the maximization is given a greater preference to higher values; instead, with minimization, it is established that a greater value indicates a worse response than the alternative. Finally, for each criterion inserted, the measurement scale of the criterion that can be qualitative or quantitative must be established [32]. In the present case there were two qualitative criteria. For the "Architectural impact" indicator it was decided to use the "5-points" ordinal scale.

The other indicator that is considered as qualitative is access to "Conto Termico", and the corresponding scale was yes/no.

For all other indicators, quantitative criteria were set. The criteria for which the maximization choice was made are: primary energy saving, global CO_2 emission reduction, local NO_x emission reduction, local PM_{10} emission reduction, public incentives, savings on energy expenditure, increase of plant system efficiency, and installed power reduction. On the contrary, the criteria to which the minimization function was associated are: PBP, investment cost, labour cost, labour cost by a social cooperative, material cost, material cost purchased on the territory, and running cost. It was decided to classify all the criteria of the same type within the same cluster. Later, all previously processed data were added and the matrix was composed. Visual PROMETHEE allows us to quantify the degree of preference, indicated as $\pi(a, b)$, of a generic alternative "*a*" compared to "*b*", calculated as in Equation (2).

$$\pi\,(a,\,b) = \sum_{j=1}^{n} W_j P_j(a,\,b) \tag{2}$$

where W_j is the weight assigned to each *j*-th criterion and $P_j\,(a,\,b)$ is the preference function. For each criterion, a $P_j\,(a,\,b)$ represents a function of the difference between the two alternatives. Preference function is applied to decide how much the alternative *a* is preferred over the alternative *b*:

$$P_j\,(a,\,b) = F_j\left[d_j\,(a,\,b)\right],\quad 0 \le P_j\,(a,\,b) \le 1 \tag{3}$$

The value of preference function varies between 0 and 1 (0 for no preference or indifference, 1 for strict preference), meaning that the larger the deviations, larger the preferences. The preference function could be of different types: usual, U-shape, V-shape, level, linear, and Gaussian [33]. In this study, the V-shape (i.e., criterion with linear preference) preference function is considered for all the indicators, with the preference value calculated as the standard deviation of each indicator and without indifference value. PROMETHEE allows to calculate the outgoing and incoming flows for each alternative. The outgoing flow is indicated with $\varphi+$ and represents the measure of the robustness of the analysed alternative. The outgoing flow calculated as in Equation (4), varies between 0 and 1. The more $\varphi+$ approaches 1, the more preferable is the alternative considered in comparison to the others, on the other side, if equal to 0, the action in question does not has no advantage over the others.

$$\phi + (a) = \frac{1}{n-1}\sum_{b \neq a} \pi\,(a,\,b)\,\phi + (a)\,\in\,[0,\,1] \tag{4}$$

As far as the incoming flow is concerned, the notation $\varphi-$ represents the measure of the weakness of the action in analysis with respect to the other alternatives. Also this parameter varies between 0 and 1, but on the contrary, where $\varphi- = 0$ means that the selected alternative has a degree of weakness equal to zero, and therefore represents the best alternative; on the contrary $\varphi- = 1$ represents the worst one. Equation (5) is used for the calculation:

$$\phi - (a) = \frac{1}{n-1} \tag{5}$$

At this point it is possible to calculate the net flow simply as the difference of the outgoing one and the incoming one. The net flow allows us to directly compare the proposed alternatives and provide the ranking of alternatives as shown in Equation (6).

$$\phi-(a) = \phi+(a) - \phi-(a) \tag{6}$$

The result of the best alternative is presented after implementing the sensitivity analysis. A sensitivity analysis is proposed by changing different weights with respect to the Baseline alternative, according to stakeholders' interests and opinions (Table 14). This last part is useful to test the robustness of the model.

Table 14. Sensitivity analysis. w: weight; p: preference.

		ENV1	ENV2	ENV3	ENV4	T1	T2	S1			
Baseline	w	0.0625	0.0625	0.0625	0.0625	0.125	0.125	0.25			
	p	21,397	5349	3.85	0.28	9.8	175	0.76			
Change 1	w	0.059	0.059	0.059	0.059	0.059	0.059	0.059			
	p	21,397	5349	3.85	0.28	9.8	175	0.76			
Change 2	w	0.075	0.075	0.075	0.075	0.1	0.1	0.2			
	p	21,397	5349	3.85	0.28	9.8	175	0.76			
		EC1	EC2	EC3	EC4	EC5	EC6	EC7	EC8	EC9	EC10
Baseline	w	0.025	0.025	0.025	0.025	0.025	0.025	0.025	0.025	0.025	0.025
Change 1	p	1.66	179,387	13	3637	41,397	10,349	137,990	48,296	8627	1
Change 1	w	0.059	0.059	0.059	0.059	0.059	0.059	0.059	0.059	0.059	0.059

The baseline model assigns same weight for each category (i.e., environmental, economic, technical, and social), 25% each one, divided equally to the indicators. This means that the weight of each particular indicator will depend of the number of KPIs included in that category (Table 15).

- Each environmental indicator will get a weight of 0.0625 percent, obtained through the division of 25 percent by four indicators.
- Each economic indicator will get a weight of 0.025 percent, obtained through the division of 25 percent by 10 indicators.
- Each technical indicator will get a weight of 0.125 percent, obtained through the division of 25 percent by two indicators.
- Each social indicator will get a weight of 0.25 percent, obtained through the division of 25 percent by one indicator.

Change 1 proposes the same weight for each indicator (e.g., ENV1, EC1, T2), of 5.9 percent each (Table 16). This leads to different weights for each category of indicators:

- 23.5 percent for environmental indicators;
- 11.8 percent for technical indicators;
- 5.9 percent for social indicators; and
- 58.8 percent for economic indicators

Change 2 focuses on the two aspects that have more impact in the project, the environmental and economic categories. Taking into account the relevance of these two, a higher weight was assigned (30 percent each one), leaving the rest to social and technical aspects, divided equally (Table 17).

- 30 percent for the environmental category and 0.075 percent for each environmental indicator;
- 30 percent for the economic category and 0.03 percent for each economic indicator;
- 20 percent for the technical category and 0.1 percent for each technical indicator; and
- 20 percent for the social category and 0.2 percent for each social indicator

Table 15. Baseline results.

Alternative	A1	A2	A3	A4	A5	A6
Net phi	−0.3156	0.0020	0.0514	0.1042	0.1043	0.0538
Rank	6	5	4	2	1	3

Table 16. Change 1 results.

Alternative	A1	A2	A3	A4	A5	A6
Net phi	−0.0516	0.0208	−0.0629	0.0353	0.0354	0.0230
Rank	5	4	6	2	1	3

Table 17. Change 2 results.

Alternative	A1	A2	A3	A4	A5	A6
Net phi	−0.2923	−0.0158	0.0384	0.1032	0.1034	0.0631
Rank	6	5	4	2	1	3

From the model runs, by changing the weights, the best alternative is always A5 followed by A4 and A6, as shown in Tables 15–17. The main reason is because they reach nZEB conditions, obtaining a great public incentives. Simulation 5 and 4 only differ in the thickness of insulation, which is the reason why they obtain similar values of net phi. The lowest values are associated to Alternative 1 (just adding a biomass boiler).

4.3. Phase III: Target Group Involvement

4.3.1. Info-Events

During the months of November and December, 12 events were organized involving local institutions (e.g., mayors) and organizations that work in the area of Susa Valley. The purpose of these events was (1) to inform and share the research activities and the project results (mainly related to the technical analysis) with the Susa Valley community; (2) raise awareness among stakeholders about the energy community benefits and, finally, (3) to co-create an action plan, to be implemented in the following months, shared by all stakeholders for the definition of an energy community in the Susa Valley.

4.3.2. Workshops

On 7 February 2020, the first workshop was organized in Almese with the collaboration of Deutscher Caritas Verband and Cooperativa Sociale Amico. Indeed, thanks to their contribution (these groups have close ties to local people), 20 citizens were invited (through personal communications) to attend this event. The duration was about half of day and the educational approach was alternated with three periods of debate and activities, in which the participants were called to express their thoughts and opinions.

After a first section in which the aim of the project and the meaning of "energy transition", "renewable energy sources", "energy community", and "CSOP (financing model)" were described and explained, the first discussion was introduced related the characteristics of the heating system in their home. Pre-defined questions were asked as following:

"What type of heating system do you have in your home? What are the costs? Are you satisfied with your system? Do you think your heating bill is too high? Are you having problems keeping your home adequately heated?"

This process involved a free discussion, and the answers obtained showed that most of citizens were not satisfied with the energy expenditure since the heating bills were too high. The energy expenditure depends on several factors: the cost of the energy established by the supplier, the volume of the apartment to be heated, and the house typology, etc., but it has emerged that the level of efficiency of the envelope (windows, presence of wall insulation, etc.) is the factor that has the greatest impact.

Subsequently, the CSOP model was explained in more detail and the five key points of this model were highlighted. On the basis of this, each participant was asked to express their preference regarding only three elements of the CSOP through the question: *"What are the CSOP benefits you are most interested in?"*. Specifically, the key elements were written on sheets (one per sheet). Participants were given a maximum of three dots, one red dot to stick on the sheet with the most important benefit for them, and two green dots to put on the sheets with some benefits for them. In this way, two elements remained unchosen, that is, those that were not important to them. The results showed that the "small source of income" benefit was not successful because on the one hand, the income was small and, on the other hand individuals were slightly sceptical about obtaining money. Indeed, this option was only chosen by 11.4% of individuals (75% red and 25% green). For this reason, "environmental issues" and "low investment", both with a preference of 28.5%, were the more successful options. Specifically, it has been said that if a low investment was required, they would agree to contribute as it was an interesting project from which the whole community could benefit. In addition, advantages with respect to the statements: "the trusted administrator helps and represents consumers" (8.5%) and "independence from the national energy supply" (22.8%) were benefits that the participants were not interested in.

Finally, a debate was opened with the participants through the following question: *"In your opinion, what are the obstacles/problems in participating in a CSOP based on renewable energy?"* The obtained answers can be summarized in the three following points:

- Distrust because, being an innovative project, there is no one who can guarantee that the project will be successful. In this case no one can give feedback on the success of this type of project;
- Control and verification due to the disparity of investment of the various actors. This is to avoid situations of greater representation with greater investment;
- Bureaucracy: navigating the process and the necessary documentation could be complicated for simple citizens not working in the legal field.

Therefore, the workshop allowed us to understand the citizens' energy habits and to understand any problems attributable to the low efficiency of the building envelope and, consequently, to a high energy expenditure (for heating). Following the subsequent activity, the participants showed interest in the topic of energy communities and explained which benefits of the CSOP they were interested in and the barriers they could encounter. The collection of these data was fundamental to refine the survey questions.

5. Conclusions and Future Developments

In conclusion, the present study illustrated and described the three different phases underlying the creation of EC through a legal framework (CSOP): (1) the identification and description of selected buildings (preparation phase), (2) preliminary and feasibility analysis, and finally (3) target group involvement in implementing the CSOP model. Specifically, Phase I and Phase II were described in detail. The first action was data collection and the proposal of different retrofit measures in order to avoid (as much as possible) the use of fossil fuels in favour of renewable ones, to increase the efficiency of the building energy plant system and of the building envelope system, and finally, to reduce the energy consumption also through a change in behaviour. Once the different proposals were defined, the best solution was chosen through an MC analysis based on key performance indicators (KPIs) considering different stakeholders' opinions. The procedure was applied to a real case study (Oulx, Susa Valley), showing the different phases aimed at creating an energy community. Specifically, the actual situation of the involved buildings in Oulx was described and then several (six) appropriate retrofit alternatives were defined and simulated in order to obtain the future designed energy consumption and environmental emission values. In addition, indicators related to different sustainable aspects (not only energy-related or environmental, but also economic, technical, and social) were assessed in order to identify the most feasible and sustainable project to be carried out for the actual realization. Finally, first through completion an evaluation matrix and then a PROMETHEE application, it was possible to order and to rank the six proposed retrofit interventions. Considering also a sensitivity analysis in which a change of the weights and preferences of each indicators was carried out to highlight and observe a ranking variation, the final results showed that the best alternative was always A5 followed by A4 and A6. The main reason underlying this result was because the achievement of nZEB conditions is linked to numerous public incentives. In addition, simulations 5 and 4 were very similar and differed in the thickness of insulation; for this reason similar values of net phi were obtained. The lowest values were associated with alternative 1 (just adding a biomass boiler).

A future step is to encourage the active role of consumers (private or public users) since they undertake a crucial role in the EC, e.g., the participation in the new financial model based on co-ownership (CSOP). The main future task is to involve several target groups (women, those of low-income, and people affected by poverty) through a social analysis using the information collected with the questionnaire, events, and working group in order to identify the main drivers favouring/hindering participation in energy communities.

Author Contributions: S.T.M. formal analysis, investigation, methodology, writing—original draft preparation, supervision; M.V.D.N. formal analysis, investigation, methodology, writing—original draft preparation writing—original draft preparation, supervision; S.M. formal analysis, investigation, methodology, writing—original draft; P.L. funding acquisition, supervision. All authors have read and agreed to the published version of the manuscript.

Funding: This research was funded by the H2020 project, entitled Supporting Consumer Co-Ownership in Renewable Energies—"SCORE", grant number 784960- SCORE-H2020-EE-2016-2017/H2020-EE-2017-CSA-PPI.

Acknowledgments: The authors of this study wish to acknowledge the contributions of a number of colleagues, partners, and institutions involved in the SCORE project, including: Eng. Luca de Giorgis, Cooperativa La Foresta, Consorzio Forestale Alta Valle Susa, Cooperativa Sociale Amico, Prof. Jen Lowitzsch, Politecnico di Torino, Centre for the Study of Democracy, Město Litoměřice, Miasto Słupsk, Climate Alliance, co2online, Deutscher Caritas Verband, Europa Universität Viadrina Frankfurt (Oder), Federacja Konsumentów, Instytut Energetyki Odnawialnej, Porsenna.

References

1. IEA. *The Critical Role of Buildings*; IEA: Paris, France, 2019; Available online: https://www.iea.org/reports/the-critical-role-of-buildings (accessed on 3 April 2019).
2. Gielen, D.; Boshell, F.; Saygin, D.; Bazilian, M.D.; Wagner, N.; Gorini, R. The role of renewable energy in the global energy transformation. *Energy Strateg. Rev.* **2019**, *24*, 38–50. [CrossRef]
3. Brummer, V. Community energy–benefits and barriers: A comparative literature review of Community Energy in the UK, Germany and the USA, the benefits it provides for society and the barriers it faces. *Renew. Sustain. Energy Rev.* **2018**, *94*, 187–196. [CrossRef]
4. Romero-Rubio, C.; de Andrés Díaz, J.R. Sustainable energy communities: A study contrasting Spain and Germany. *Energy Policy* **2015**, *85*, 397–409. [CrossRef]
5. Lowitzsch, J. Investing in a Renewable Future–Renewable Energy Communities, Consumer (Co-) Ownership and Energy Sharing in the Clean Energy Package. *Renew. Energy Law Policy Rev.* **2019**, *9*, 14–36.
6. Lowitzsch, J. Consumer Stock Ownership Plans (CSOPs)—The Prototype Business Model for Renewable Energy Communities. *Energies* **2019**, *13*, 1–24. [CrossRef]
7. Moghadam, S.T.; Valentina, M.; Nicoli, D.; Giacomini, A.; Lombardi, P.; Toniolo, J. The role of prosumers in supporting renewable energies sources. In *IOP Conference Series: Earth and Environmental Science*; IOP Publishing: Bristol, UK, 2019; Volume 297. [CrossRef]

8. Di Nicoli, M.V.; Moghadam, S.T.; Lombardi, P. A Framework for Selecting the Best Refurbishment Alternative in Renewable Energies Towards. In Proceedings of the Energy for Sustainability Conference (EfS), Torino, Italy, 24–26 July 2019; pp. 24–26.

9. Wang, J.J.; Jing, Y.Y.; Zhang, C.F.; Zhao, J.H. Review on multi-criteria decision analysis aid in sustainable energy decision-making. *Renew. Sustain. Energy Rev.* **2009**, *13*, 2263–2278. [CrossRef]

10. Strantzali, E.; Aravossis, K. Decision making in renewable energy investments: A review. *Renew. Sustain. Energy Rev.* **2016**, *55*, 885–898. [CrossRef]

11. Simos, J. *Evaluer L'impact sur L'environnement: Une Approche Originale par L'analyse Multicritère et la Nègociation*; Polytechnique University Rom: Lausanne, Switzerland, 1990.

12. Lowitzsch, J.; Hanke, F. Consumer (Co-)ownership in RE, EE & the fight against energy poverty—A dilemma of energy transitions. *Renew. Energy Law Policy* **2019**, *9*, 5–22.

13. Lowitzsch, J. Consumer Stock Ownership Plans (CSOPs)—The Energy Communities. *Energies* **2020**, *13*, 118. [CrossRef]

14. Ahmed, A.; Gasparatos, A. Multi-dimensional energy poverty patterns around industrial crop projects in Ghana: Enhancing the energy poverty alleviation potential of rural development strategies. *Energy Policy* **2020**, *137*, 111123. [CrossRef]

15. Simos, J.; Maystre, L.Y. *L'evaluation Environnementale: Un Processus Cognitif Negocie*; EPFL: Lausanne, Switzerland, 1989.

16. Marinakis, V.; Doukas, H.; Xidonas, P.; Zopounidis, C. Multicriteria decision support in local energy planning: An evaluation of alternative Scenarios for the Sustainable Energy Action Plan. *Omega* **2017**, *69*, 1–16. [CrossRef]

17. Giaccone, A.; Lascari, G.; Peri, G.; Rizzo, G. An ex-post criticism, based on stakeholders' preferences, of a residential sector's energy master plan: The case study of the Sicilian region. *Energy Effic.* **2017**, *10*, 129–149. [CrossRef]

18. EEA. *Air Quality in Europe—2014 Report, No. 5*; EEA: Copenhagen, Denmark, 2014; Available online: https://www.eea.europa.eu/publications/air-quality-in-europe-2014 (accessed on 17 November 2014).

19. Jovanović, M.; Afgan, N.; Radovanović, P.; Stevanović, V. Sustainable development of the Belgrade energy system. *Energy* **2009**, *34*, 532–539. [CrossRef]

20. Doukas, H.C.; Andreas, B.M.; Psarras, J.E. Multi-criteria decision aid for the formulation of sustainable technological energy priorities using linguistic variables. *Eur. J. Oper. Res.* **2007**, *182*, 844–855. [CrossRef]

21. Cavallaro, F.; Ciraolo, L. A multicriteria approach to evaluate wind energy plants on an Italian island. *Energy Policy* **2005**, *33*, 235–244. [CrossRef]

22. Becchio, C.; Corgnati, S.P.; Delmastro, C.; Fabi, V.; Lombardi, P. The role of nearly-zero energy buildings in the transition towards Post-Carbon Cities. *Sustain. Cities Soc.* **2016**, *27*, 324–337. [CrossRef]

23. G. S. E. (GSE). Conto Termico 2.0 Allegato I Criteri di Ammissibilità Degli Interventi. Available online: https://www.gse.it/documenti_site/DocumentiGSE/Serviziperte/CONTOTERMICO/NORMATIVA/Allegatodecretointerministeriale16febbraio2016.PDF (accessed on 20 September 2011).

24. Torabi Moghadam, S.; Lombardi, P. An interactive multi-criteria spatial decision support system for energy retrofitting of building stock using CommunityVIZ to support urban energy planning. *Build. Environ.* **2019**, *163*, 106233. [CrossRef]

25. Dall'O', G.; Norese, M.F.; Galante, A.; Novello, C. A multi-criteria methodology to support public administration decision making concerning sustainable energy action plans. *Energies* **2013**, *6*, 4308–4330. [CrossRef]

26. Becchio, C.; Ferrando, D.G.; Fregonara, E.; Milani, N.; Quercia, C.; Serra, V. The cost-optimal methodology for the energy retrofit of an ex-industrial building located in Northern Italy. *Energy Build.* **2016**, *127*, 590–602. [CrossRef]

27. Dall'O', G.; Galante, A.; Torri, M. A methodology for the energy performance classification of residential building stock on an urban scale. *Energy Build.* **2012**, *48*, 211–219. [CrossRef]

28. Brans, J.P.; Vincke, P.; Mareschal, B. How to select and how to rank projects: The Promethee method. *Eur. J. Oper. Res.* **1968**, *24*, 228–238. [CrossRef]

29. De Montis, A.; De Toro, P.; Droste-franke, B.; Omann, I.; Stagl, S. Assessing the quality of different MCDA methods (MAUT Explained well). *Altern. Environ. Valuat.* **2005**, 99–184.

30. Dirutigliano, D.; Delmastro, C.; Torabi Moghadam, S. Energy efficient urban districts: A multi-criteria application for selecting retrofit actions. *Int. J. Heat Technol.* **2017**, *35*, S49–S57. [CrossRef]

31. Albadvi, A.; Chaharsooghi, S.K.; Esfahanipour, A. Decision making in stock trading: An application of PROMETHEE. *Eur. J. Oper. Res.* **2006**, *177*, 673–683. [CrossRef]

32. Bufardi, A.; Sakara, D.; Gheorghe, R.; Kiritsis, D.; Xirouchakis, P. Multiple criteria decision aid for selecting the best product end of life scenario. *Int. J. Comput. Integr. Manuf.* **2003**, *16*, 5266–5534. [CrossRef]

33. Vulevi, T.; Dragovi, N. Multi-criteria decision analysis for sub-watersheds ranking via the PROMETHEE method. *Int. Soil Water Conserv. Res.* **2017**, *5*, 50–55. [CrossRef]

Article

Urban Sustainability Audits and Ratings of the Built Environment

Constantinos A. Balaras [1,*], Kalliopi G. Droutsa [1], Elena G. Dascalaki [1], Simon Kontoyiannidis [1], Andrea Moro [2,*] and Elena Bazzan [2]

[1] Group Energy Conservation, Institute for Environmental Research and Sustainable Development, National Observatory of Athens, GR-15236 Athens, Greece; pdroutsa@noa.gr (K.G.D.); edask@noa.gr (E.G.D.); skonto@noa.gr (S.K.)

[2] iiSBE Italia, International Initiative for a Sustainable Built Environment, I-10138 Torino, Italy; elena.bazzan@iisbeitalia.org

* Correspondence: costas@noa.gr (C.A.B.); andrea.moro@iisbeitalia.org (A.M.)

Received: 16 October 2019; Accepted: 4 November 2019; Published: 7 November 2019

Abstract: Buildings and the built environment in cities are seen as both a source of, and solution to, today's economic, environmental and social challenges. The audit process to collect data and rate their sustainability levels is a demanding process given the complexity of the issues involved. Stakeholders often lack advanced knowledge on the sustainability issues involved, access to practical tools that match the local priorities and the overall resources to diagnose and evaluate the current state, analyse, assess and rank different scenarios, and monitor implementation and progress towards meeting sustainable development goals and local priorities. A new multicriteria European built environment assessment method that is supported by practical tools was developed in a transnational collaborative effort to support the assessment, planning, monitoring and overall decision-making process for rating the sustainability at the building or neighbourhood scale. The assessment system addresses the main sustainability issues (e.g., site and infrastructure, urban systems, energy and natural resources, emissions and environment, service quality, social aspects, economy), which are described and quantified with an "exhaustive" list of ~180 sustainability criteria and indicators, and a manageable number of common mandatory key performance indicators. The assessment system can satisfy the public administrations' needs for being easy to use, open access, flexible and adaptable tools in order to facilitate their efforts for developing effective sustainability plans.

Keywords: sustainability; buildings; neighbourhoods; decision-making process; key performance indicators; KPIs; built environment; audit; assessment tools

---

1. Introduction

The European buildings sector represents 41.7% of the total annual final energy in the European Union Member States (EU-28) or 442 million tonnes of oil equivalent (Mtoe) in 2017 (Figure 1), and is responsible for ~30% of the total carbon dioxide emissions [1]. During their life cycle, buildings also use half of all raw material extraction and a third of all water consumption [2]. Furthermore, the waste stream from the construction of buildings and civil infrastructure, demolition, road planning and maintenance (i.e., construction and demolition waste—CDW) is one of the heaviest and most voluminous waste streams that accounts for 25% to 30% of all waste generated in the EU-28 [3].

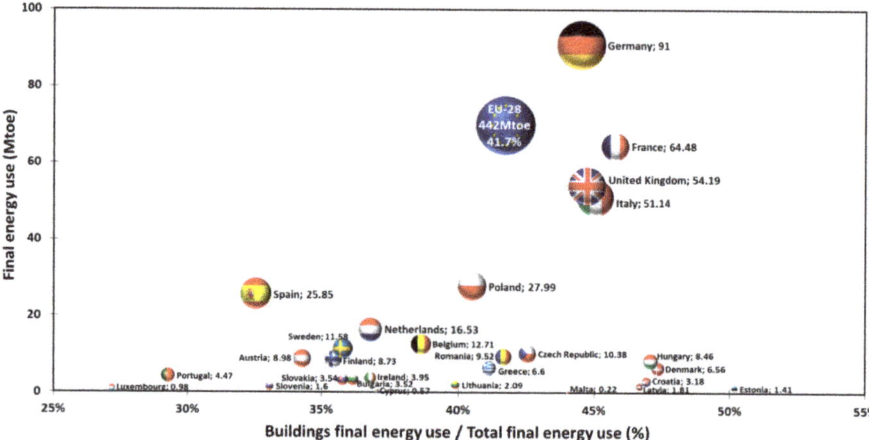

Figure 1. Final energy use (million tonnes of oil equivalent—Mtoe) in European buildings and ratio (%) of the buildings' final energy consumption to the total. The bubble size represents the total final energy use in each country; for EU-28, the value is not to scale. Data source: Eurostat.

According to the European Commission's urban development network, the European urban areas are home to over two-thirds of the EU's population and account for about 80% of the final energy use [4]. These urban areas are the engines of the European economy, but they are also places where persistent problems, such as unemployment, segregation and poverty, are most evident. Urban development is central to the EU's Regional Policy, which addresses the environmental, economic, social and cultural dimensions. An integrated approach is necessary in order to achieve sustainable urban renewals or new developments by incorporating environmental protection, education, economic development, social inclusion through strong partnerships between local citizens, civil society, industry and various levels of government.

Recognizing the importance of buildings and the built environment, the EU has initiated ambitious efforts to minimize the use of energy and natural resources in buildings, with radical resource efficiency and circular material flows in its Circular Economy Action Plan [5] to alleviate their environmental impacts. The 2030 EU climate and energy framework includes binding targets and policy objectives for reducing the greenhouse gas (GHG) emissions by at least 40% from 1990 levels, for increasing the share of renewables by at least 32% of final energy consumption, and for improving energy efficiency by at least 32.5% [6]. Member States are also obliged to adopt integrated National Climate and Energy Plans (NECPs) for the period 2021–2030 and develop national long-term strategies to ensure consistency with NECPs. One of the main instruments for addressing these challenges and the energy use in buildings is the Energy Performance of Buildings Directive (EPBD), recently amended by EU 2018/844 that entered into force on 9 July 2018, an integral part of the "Clean Energy for All Europeans" package [7]. EPBD encourages energy efficiency and promoting cost-effective building renovations, with the vision of a decarbonised building stock by 2050. As we move into the new era of nearly-zero-energy buildings (nZEB) as of January 2021, the next big challenge is the renovation of national building stocks. These large-scale efforts could best be served by addressing groups of buildings in urban neighbourhoods, considering synergies and energy interactions between individual buildings and the broader energy system at local level, towards the concept of zero-energy districts [8]. Although the evolution towards energy and spatial planning is challenging, good practices promoting bottom-up initiatives are emerging, focusing on neighbourhood scale oriented urban projects, using decentralised energy systems, local energy communities, energy districts, etc. [9].

The EU was also instrumental in shaping the global 2030 Agenda and the United Nations Sustainable Development Goals (SDGs) [10] and is a frontrunner for the long-term implementation of the SDGs that are further enhanced with EU's policies and integrated into all the Commission's priorities [11]. The 17 SDGs are the blueprint to achieve a better and more sustainable future for all, addressing the global challenges we face, including those related to energy, climate and environmental degradation in buildings and cities. The 2030 Agenda integrates in a balanced manner the three pillars of sustainable development—economic, social and environmental.

At the centre stage of the work related to the built environment is SDG-11 aiming to make cities inclusive, safe, resilient and sustainable, targeting sustainable urbanization and transport systems, resource efficiency, mitigation and adaptation to climate change, resilience to disasters, reducing adverse environmental impacts, safeguarding the cultural and natural heritage, and providing green and public spaces. In this context, the supporting goals in the areas of energy and climate include: SDG-7 to ensure access to affordable, reliable, sustainable and modern energy for all, by focusing on increased energy efficiency and the use of renewables for creating more sustainable and inclusive communities and resilience to environmental issues like climate change; SDG-13 to take urgent action to combat climate change and its impacts. Additional goals that are an integral part of sustainable development include: SDG-3 to ensure healthy lives and promote well-being for all by providing and facilitating access to health systems, reducing ambient pollution; SDG-6 to preserve clean water as a natural resource and combat chronic or recurring shortages of fresh water; SDG-8 to promote inclusive and sustainable economic growth, employment and decent work for all; SDG-9 to build resilient infrastructure, promote sustainable industrialization and foster innovation, including transportation, energy and information and communication technology; SDG-10 to reduce inequality by paying attention to the needs of disadvantaged and marginalized populations; SDG-12 to ensure sustainable consumption and production patterns, by promoting resource and energy efficiency, sustainable infrastructure, and providing access to basic services, green and decent jobs and a better quality of life; SDG-15 to combat desertification, halt and reverse land degradation, halt biodiversity loss in relation to urban growth; SDG-16 to promote just, peaceful and inclusive societies for sustainable development in an urban context; SDG-17 to facilitate inclusive partnerships between governments, the private sector and civil society, built upon principles and values, a shared vision, and shared goals that place people at the centre, at the global, regional, national and local level.

The Urban Agenda for the EU was launched in May 2016 with the Pact of Amsterdam as a new multi-level working method promoting cooperation between Member States, cities, the European Commission and other stakeholders in order to stimulate growth, liveability and innovation in the European cities and to identify and successfully tackle social challenges [12]. According to the first-ever SDG index and dashboards report for European cities that was recently released, no European capital city or large metropolitan area has yet fully achieved the SDGs [13]. As illustrated in Figure 2 major challenges lie ahead. The SDG agenda may not be fully achieved without the involvement of cities. Addressing unsustainable patterns of consumption and production, and climate change and environmental degradation, extreme poverty, unemployment and socio-economic disparities, mandates the engagement of regional and local authorities. Overall, cities in Europe perform best on SDG-3 (Health and Well-Being), SDG-6 (Clean Water and Sanitation), SDG-8 (Decent Work and Economic Growth) and SDG-9 (Industry, Innovation and Infrastructure). By contrast, performance is lowest on SDG-12 (Responsible Consumption and Production), SDG-13 (Climate Action) and SDG-15 (Life on Land). As expected, the definition of territorial levels and metropolitan areas and standardize subnational data and indicators, revealed major gaps in available information in order to monitor all the SDGs.

(a)   (b)

Figure 2. (**a**) European cities Sustainable Development Goals (SDGs) index for sustainable cities and communities (SDG-11); (**b**) overall SDG city scores. Source: https://euro-cities.sdgindex.org.

The various aspects of sustainable development in an urban context include energy, environment, transportation, infrastructure and services, land use, natural resources, and social well-being among others, as well as mandating specific actions and significant efforts. The SDGs are further enhanced through national action plans, with regional level and finally local level priorities and goals. More than ever, a local push is needed to improve sustainability efforts following a bottom-up approach of local actions that will effectively drive the processes to meet the SDGs in the spirit of the concept "Think Globally, Act Locally". However, developing, monitoring and assessing local, regional and national plans towards sustainable development at building and neighbourhood scale, considering the plethora of SDGs and sustainability issues, are complex undertakings. These efforts can be overwhelming for local and regional authorities that may not have the expertise and personnel. Accordingly, there is a need to facilitate local authorities and municipalities to act quickly and accelerate progress.

Existing Systems for Rating and Labelling

Energy and environmental audits in industry, tourism, commerce and the buildings sector have been used to collect the appropriate data that is essential for a systematic analysis in order to identify, quantify and report on the opportunities for improved performance. There are several available schemes for building energy audits that depend on the project intent and procedure (e.g., energy performance assessment, rating, certification or labelling), the specific operating conditions, the building type, among other factors [14]. The use of the term "energy audit" can be subjective and can vary from country to country since they are conducted in varying degrees or levels of technical detail, accuracy and complexity based on the purpose they serve. In some cases, this is done intentionally to reflect certain attributes, levels of complexity or stand-out in the market as a tailored process to a specific scheme and thus differentiate from other competing processes. Sometimes, it may also be an unintentional result in an effort to directly link required processes to different legal acts and relevant regulations that may apply. Some examples include survey, screening, diagnosis, inspection, review, preliminary (detailed) audit or preliminary (detailed) assessment, or as it relates to financial assessments like an investment-grade audit or feasibility study.

Practically all schemes include some common stages: preliminary contacts (e.g., client interview to define project intent, collect preliminary information), intake (e.g., collect available data like drawings, energy bills or metered data, perform an on-site visit, collect field data, complete checklists, audit forms and protocols, verify estimates and default values, perform in-situ measurements), analysis (e.g., rating, benchmarking, perform calculations or simulations, define a baseline to investigate energy conservation measures and assess scenarios, determine a list of cost-effective recommendations with quantified savings), and results (e.g., meet and present results to the client, generate reports and other deliverables). Some schemes may have distinct characteristics (e.g., use specific calculation tools that will determine the input data, or deliver distinct results like an energy performance certificate or prepare documents and specifications for tenders).

Sustainability audits in an urban context are more elaborate since they involve various issues and themes that need to be addressed [15]. Sustainability is also being adopted into building codes at different levels of government and with varying motivations. The approach taken reflects local societal perceptions, political priorities, national policies and economic factors [16]. The creation of standards or codes that define a level of performance for sustainable buildings has emerged as a need within the industry. However, there are different approaches due to wide variations in economic, social, political and technological conditions and priorities in different countries and jurisdictions around the world. Rating systems provide a method that one can voluntarily adopt and comply with various sustainability measures that meet a pre-defined set of requirements. Standards are also being developed as a collection of criteria for meeting the acceptable requirements at a high level of performance. They may be adopted in building codes or simply used as a level of performance that a project may comply by. For example, the ASHRAE Standard 189.1 that is recognized as a leading green standard around the world and forms the technical basis for the International Green Construction Code (IgCC), includes mandatory criteria in several sustainability issues and themes, site, construction, materials, energy, indoor environmental quality, water, etc. [17].

At building scale, various voluntary sustainability rating systems and labelling schemes have been developed, e.g., BREEAM (https://www.breeam.com/), CASBEE (http://cabee.eu/), Green Star (https://new.gbca.org.au/green-star/), LEED (https://new.usgbc.org/leed) and Protocollo ITACA (http://itaca.org/), to facilitate the process for reducing energy use and environmental impacts during construction, management and operational phases [18]. The systems include different performance indicators that are used as metrics with fixed weighting and scoring systems to determine how well the sustainability objectives are achieved, facilitate the decision-making process, assess specific project requirements or ensure compliance with regulations and norms [19–21]. The indicators quantify what one is trying to achieve, and depending on specific project needs and priorities one may need to use several of them at different stages of the work or process. The indicators can be expressed as numerical values (e.g., building's energy use intensity in order to assess different performances or compare against other benchmarks; water consumption per building occupant, etc.), or ratios and percentages (e.g., percent of renewables that cover power or heat demand; percent of recycled waste, etc.).

A voluntary reference framework known as LEVEL(s) is also being developed for the European Commission [22] providing a common European framework of common indicators to measure the sustainability performance of buildings across their whole life cycle, focusing on GHG emissions, resource efficiency, water use, health and comfort, resilience and adaptation to climate change, cost and value. Each indicator links the building's individual characteristics (currently referring to only residential and office buildings) and impacts to sustainability priorities, facilitating users to consider key concepts and building-scale indicators, following specific guidelines and standardized calculations for each indicator.

Several systems have also been extended to urban scale, e.g., BREEAM Communities, CASBEE for Urban Development, LEED for Neighbourhoods and Protocollo ITACA Urban Scale. The main aspects for sustainable cities address similar performance indicators like the ones for building scale, and include more categories, for example, urban transport, supply and distribution networks, social factors, etc. [23].

A new European multicriteria assessment method has been developed that enhances existing knowhow in a holistic system for accessing urban sustainability of the built environment at neighbourhood scale. This complements the existing public approaches at building and city scales, so that it is more suitable and manageable to handle by municipalities. The following sections outline the main structure of the method and tools for addressing the sustainability issues for buildings and neighbourhoods, the generic framework with an emphasis on the energy and environmental indicators, the key performance indicators, the results from nine European pilots, providing details for the application in Greece, and the training system that includes educational material developed for decision-makers and technical professionals.

2. The Common European Sustainable Built Environment Assessment for Mediterranean Cities (CESBA MED) Method

The Common European Sustainable Built Environment Assessment for Mediterranean cities (CESBA MED) was a collaborative effort of several European organizations from seven countries. The work is structured around the UN 17 SDGs, aiming to support users and their efforts towards a sustainable future. The initial concept of the assessment method and tool was a reference decision-making process that was originally developed for the building scale [24] and then extended at neighbourhood scale. The following sections outline and briefly discuss the process for converging on the number and type of sustainability indicators that are considered in the method, the normalization and scoring process, the development of the generic framework, and the national tools.

2.1. Sustainability Indicators and Key Performance Indicators (KPIs)

The approach taken in this work was to first develop a generic framework that includes an "exhaustive" list of sustainability indicators that cover all relevant themes, given that there is still no consensus on a specific number or types of indicators. This way one can have access to a comprehensive database that includes different performance indicators from which to select the ones that meet local priorities and needs, or best fit the project intent. A minimum number of key performance indicators are defined and used in order to ensure that the core sustainability issues can be addressed in a satisfactory manner.

Accordingly, the first step was to critically review 14 transnational European projects and public assessment systems, in order to derive a representative list of indicators at building and neighbourhood scales that address the main sustainability pillars [25]. A total of 216 indicators were identified, critically reviewed and finally grouped under the main sustainability issues.

The structure of the method organizes the information in Issues, Categories and Criteria-Indicators [26]. The "Issues" identify the general themes that are essential for assessing the sustainability at building and neighbourhood (urban) scales. The sustainability Issues for the building scale include: A-Site and infrastructures, B-Energy and resources, C-Environment, D-Indoor Environmental Quality (IEQ), E-Service quality, F-Social, cultural and perceptual aspects and G-Economy. The seven sustainability Issues for the neighbourhood scale include: A-Urban systems, B-Economy, C-Energy, D-Emissions, E-Natural resources, F-Environment and G-Social aspects.

The "Categories" under each Issue describe its specific aspects that group relevant Criteria and Indicators. Each Issue includes a different number of Categories. The building scale includes 25 Categories. For example, under the issue "IEQ" there are four categories: Indoor air quality and ventilation, Air temperature and relevant humidity, Daylight and illumination and Noise and acoustics. The neighbourhood scale includes 23 Categories. For example, the issue "Energy" includes two categories: Non-renewable energy sources, and Renewable and clean energy sources.

The "Criteria" detail the specific aspects of a Category and represent the main assessment entries used to characterize a building or an urban area. The "Indicators" quantify the performance with respect to each criterion. In principle, several indicators can be associated with the same criterion, since one can define multiple strategies to quantify the building or urban area performance with regard to a specific criterion. For example, building energy use intensity (EUI) can be expressed as kWh/m^2 or kWh/m^3 and in some cases energy use per employee (e.g., for an office building) or energy per bed (for hotels), depending on the characteristic functions of a building. For simplicity in this work, only one indicator is associated with each criterion. The metrics are used to quantify the performance and determine how well the sustainability objectives are achieved.

Tables A1 and A2 summarize the various sustainability Issues, Categories and Indicators for building scale [27] and neighbourhood scale [28]. Different numbers of criteria-indicators are included under a given category, each one of them describing a particular aspect of the corresponding category. For example, at the neighbourhood scale, Category 'C.2 Renewable and clean energy' includes fourteen Indicators, e.g., share of on-site renewables on total final or primary energy consumption for residential or non-residential buildings, share of electricity production from renewables on public or private property, total electricity from renewables that is exported from the area, total electricity from renewables used in or exported from the area, share of thermal energy from renewables on public or private property, etc.

Some indicators may appear under both scales (e.g., energy use at the building scale and for all buildings in the area at the neighbourhood scale). For example, at the building scale under the Issue 'B. Energy and Resources', the Category 'B.1 Life Cycle Non-Renewable Energy' includes the Criterion 'B.1.2 Final Thermal Energy Use' and 'B.1.3 Final Electrical Energy Use'. Aggregating the relevant information for all the buildings in the area, one can derive the equivalent indicators at the neighbourhood scale (i.e., B.1.1 for each building and C.1.1 for all buildings in the area). Sometimes qualitative criteria are used instead of quantitative ones. In this case, the expert's assessment is based on the prescribed reference descriptions in order to assess and score the specific performance. For example, at the building scale under Issue 'F. Social, Cultural, Perceptual', Category 'F.2 Culture and Heritage' that includes Criterion 'F.2.1 Compatibility of urban design with local cultural values' is qualitatively assessed with an indicator of whether the architectural design features related to the urban design are incompatible, marginally- or fully-compatible. Similarly, at the neighbourhood scale, under Issue 'G. Social Aspects', Category 'G.6 Management and Community Involvement' the Criterion 'G.6.3 Community involvement in urban planning activities' is qualitatively assessed with an indicator that reflects different levels of citizens' engagement in the planning process, from a non-participatory process (to reflect performance below standard) to full co-decision with delegated citizen power (to reflect an ideal performance).

A limited number of key performance indicators (KPIs) were selected from the various indicators as mandatory minimum requirements in order to be able to address the main sustainability issues, which are also identified in Tables A1 and A2. For example, one commonly accepted metric to measure a building's energy use performance is the energy use intensity (EUI in kWh/m^2), which can be used to benchmark against similar buildings or with best-practices and assess energy efficiency measures within buildings. The KPIs are defined and calculated following common standardized procedures. This work considered most of the LEVEL(s) indicators in the process of selecting the KPIs for the building scale. The results from the normative KPI calculations can then be used as a passport for comparing different buildings, areas, regions or countries, on a common basis.

The organization of the sustainability issues, the selection of the most applicable criteria, performance indicators and KPIs followed an iterative process at various stages of the work. The first step was to review, analyse and organize the knowhow generated from 14 EU projects and systems [25] and the work in the new LEVEL(s) indicators [29]. Each national team of the CESBA MED partnership engaged and collaborated with local committee experts in six EU countries to elaborate the issues and indicators, in order to ensure that they are representative and cover national needs and priorities in local context.

For reaching a wider consensus, the work progress on the performance indicators and the proposed KPIs were also reviewed and elaborated with other European experts and project representatives during two sprint workshops. The final list of the KPIs for building and neighbourhood scale was fixed following the nine national pilot tests performed by the partners in six EU countries. As a result, some KPIs were excluded due to the limited availability of the input data, e.g., quantities of building construction materials and recyclable content that have been used for existing buildings or other public works in the area, thus ensuring the applicability of the approach and the use of the indicators in the field.

2.2. Normalization and Scoring

All sustainability assessment and rating systems use a normalization process in order to convert the indicator values into a common basis (scale). The various indicators are diverse in nature, have numerical values with different orders of magnitude and correspond to physical quantities with different units or in some cases include qualitative scores. The normalized scores of the individual indicator values are then aggregated using different weights to calculate a score for the corresponding categories and issues, and finally a total sustainability score for a building or a neighbourhood.

2.2.1. Indicator Scores

Each indicator value is a dimensionalized and rescaled value (Figure 3) in an interval from −1 (performance below standard) to +5 (advanced performance) [26], following a similar concept with Protocollo ITACA [30]. For example, the score value at "0" corresponds to the minimum acceptable performance of an indicator in compliance with minimum standard regulation mandates defined by law (e.g., an EUI for new buildings or the percentage use of renewables), or the value of current practice in case of no regulations (e.g., percentage of employment, length of pedestrian and bicycle paths). The score value at "+5" corresponds to excellence or ideal performance (e.g., an EUI for a nearly zero energy building, or very-high employment rate for the residents in a neighbourhood). Values of indicators below minimum standards or current practice are assigned to a score of "−1".

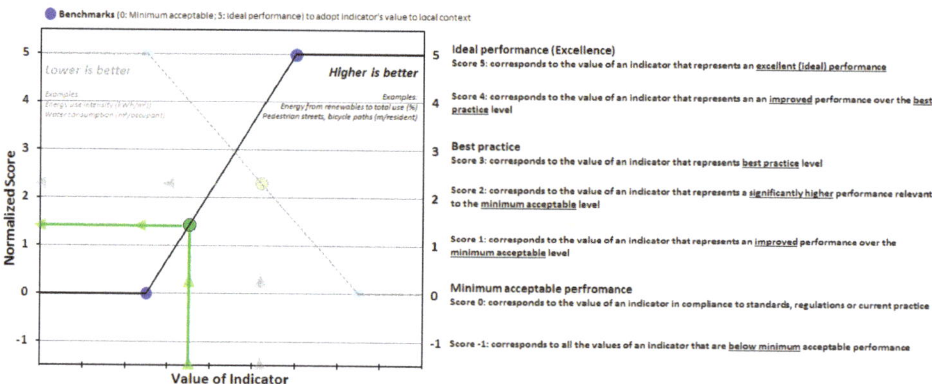

Figure 3. Normalizing and scoring process. For an indicator following the principle "higher is better," the linear correlation is illustrated with the black solid line, while the principle "lower is better" is illustrated with the grey dashed line.

For simplicity, individual scores are defined by linear interpolation between the two limits (i.e., "0" and "+5"). For each indicator, the numerical values at the two limits are adapted to the local context by using appropriate national, regional or local benchmarks. For some indicators, higher performance corresponds to a higher normalized score, following the principle that "higher is better", thus the slope of the linear correlation (from 0 to +5) is positive (e.g., the percentage use of renewables, the length of pedestrian and bicycle paths in a neighbourhood). In this case, a higher value of the indicator corresponds to higher performance and thus it receives a higher normalized score. For others, the normalized score follows the principle that "lower is better" (e.g., a low EUI for buildings or low water consumption), thus the linear correlation (from 0 to +5) has a negative slope. In this case, a lower value of the indicator corresponds to higher performance and thus it receives a higher normalized score.

The national and local benchmarks for each indicator have been predefined at the appropriate values for (ideal) excellent practice (corresponding to "+5" in the normalized score), the minimum acceptable performance (corresponding to "0") and below standard (corresponding to "−1"). These values are already included in the national and local versions of the method (see Section 3.1). If necessary, the user can adjust them according to the local characteristics (e.g., energy use intensities for the local buildings, water consumption in the area, etc.).

2.2.2. Sustainability Score

The calculations for the sustainability score are weighted in terms of the regional, local or project priorities. The weighting factors are properly estimated values that reflect the relative importance of characteristics compared to others. This way, the user has an opportunity to place the desirable emphasis on specific sustainability issues and performance indicators, to reflect regional variations and add local context. The weighted score of each Indicator is calculated by using different multiplicative factors to adapt its normalized score (see Section 2.2.1) as illustrated in Figure 4a. The following discussion reviews the various weighting factors that are taken into account at different stages of the calculations, starting at the overarching level of the sustainability Issues and then at the more detailed level for addressing the characteristics of each indicator.

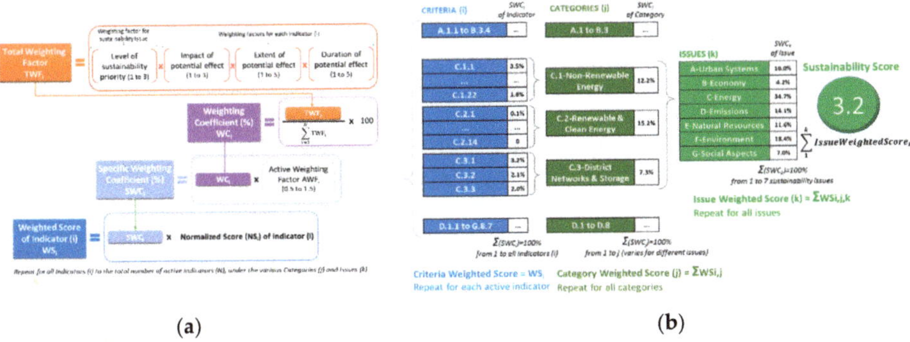

(a)   (b)

Figure 4. Overview of the calculation processes at the neighbourhood scale for the: (**a**) Indicator's (i) weighted score; (**b**) weighting allocation and aggregation of scores for Criteria, Categories, Issues and sustainability score. For illustration purposes, the active indicators are identified with a numerical value, while for the inactive indicators (not selected), the specific weighting coefficients are shown as 0. The graphic does not include all the criteria and categories that are listed in Table A2.

For each one of the seven sustainability Issues, it is possible to define its level of priority on a scale from 1 (less important) to 3 (most important or more relevant). For example, the level of sustainability priority for the energy issue may be set at 2 (i.e., considered of average importance if the buildings have an average energy performance and good exploitation of renewables), while the issue of emissions in an area with major environmental problems the assigned priority may be set at level 3 (i.e., considered a major issue).

Beyond the mandatory KPIs, one can select an appropriate number of active indicators that best fit the local needs and project-specific priorities under each Category and Issue. As a default, the weighting factors are equally distributed among the active indicators so that the weightings equal 100%. These coefficients may then be adjusted to place more emphasis on a specific indicator.

The total weighting factor (TWF$_i$) for each indicator (Figure 4a) is calculated as the product of the following weighting factors that account for the:

- Level of sustainability priority for the Issue that includes the specific indicator, which is rated using a 1 to 3 points scale described above, and for each Indicator the:

- Impact of potential effect, rated using a 1 to 3 points scale, i.e., 1-minor, 2-moderate, 3-major),
- Extent of potential effect, rated using a 1 to 5 points scale depending on the spatial coverage, i.e., 1-block, 2-neighborhood, 3-district, 4-urban/region, 5-global),
- Duration of potential effect, rated using a 1 to 5 points scale, i.e., 1 for 1 to 3 years, 2 for 3 to 10 years, 3 for 10 to 30 years, 4 for 30 to 75 years, 5 for greater than 75 years).

For example, based on the above, the global warming potential indicator (C.1.3 at building scale, Table A1) the Issue (C-Environment) can be weighted with 3 (i.e., the environment is considered a major issue), and assigning for the indicator a weighting factor of 3 (major impact), 5 (global potential effect) and 5 (duration >75 years). The on-site use of renewables in buildings (C.2.1 at neighbourhood scale, Table A2), the Issue (C-Energy) can be weighted with 2 (i.e., energy is considered an average issue in an area where all buildings use solar collectors), and assigning for the indicator a weighting factor of 3 (major impact), 2 (neighbourhood potential effect) and 3 (duration 10 to 30 years). Since the specification of these weighting factors is not a trivial process, the national versions of the method (see Section 3.1) include national default values, although a user can always adjust them.

The weighting coefficient (WC_i) that accounts for the relative importance of an indicator among the selected ones is calculated as a percentage of the ratio of the individual TWF_i to the total for all active indicators (Figure 4a). To further fine-tune the weighting coefficients, the values may be adjusted using another multiplicative factor to account for the possible importance of an indicator in the context of a specific project or its potential impact on more than one criterion, categories or even under different issues. The active weighting factor (AWF) is set at 0.5 (i.e., lowering the Indicator's weight by half) or 1.5 (i.e., increasing its weight by half). For example, the AWF for C.1.20 Energy use for public lighting (Table A2) can be set to 1.5 (i.e., 50% more important) because of its importance in a project not only in terms of energy savings, but also in relation to the perceived safety of public areas (G.8.3) and even aesthetics (G.8.7). Finally, the normalized score of each indicator is multiplied by the specific weighting coefficient (SWC) to obtain the weighted score (WS) of each indicator.

Overall, the process provides the ability to use different weights for adjusting each indicator (criterion), category and issue, according to local environmental, social and economic priorities and scenarios under assessment. Although altering the weighting system may be perceived as a manipulation of the results in order to improve the overall scores, the intent is to allow sufficient user flexibility for adapting the method to the local and project-specific priorities. Alternatively, to safeguard the process, the default weights can be reviewed, agreed upon and then locked by the decision-makers, before allowing third party interaction.

The normalized scores associated with all active indicators (criteria) in the same category, e.g., are aggregated to produce a single weighted score for each category. For example, the criteria weighted scores for C.1.1 up to potentially C.1.22, C.2.1 up to C.2.14 and C.3.1 to C.3.3 at neighbourhood scale (Table A2) are used to obtain the category weighted scores for C.1, C.2 and C.3, shown in Figure 4b. Then, the scores for all categories in the same issue are further aggregated to produce a single weighted score for each issue (e.g., C-Energy in Figure 4b). Finally, the results from all seven issues are aggregated to produce a concise total sustainability score for the project.

2.2.3. The Generic Framework

The CESBA MED Generic Framework (GF) is the general, all-inclusive starting version of the tool that supports the assessment method with all seven issues, categories and indicators available for the building and neighbourhood scales (Tables A1 and A2). The total number of indicators in the GF that one can potentially select from and use is 153 for the building scale [27] and 178 for the neighbourhood scale [28]. This "exhaustive" list of performance indicators is an excellent starting point for developing national and local tools by selecting and using only the ones that are relevant according to national, local sustainability priorities and project intent. For practical purposes, one should select a manageable number of indicators from the complete list under the various issues and categories that for a given project best match the local sustainability issues, priorities and strategic policies. During the development of the national tools (see Section 3.1) this exercise was elaborated for adapting the GF Tools in six national versions and then to local context during the specific pilot projects. Always, the minimum number of indicators are the key performance indicators that were determined as a result of the iterative process for developing the CESBA MED GF that finally reached 13 KPIs for the building scale [31], including most of the LEVEL(s) indicators, and 16 KPIs for the neighbourhood scale [32]. The KPIs are collected and stored in a "Passport" that constitutes a depository of common and comparable data. The two-page CESBA Passport provides some general information on the project and details the KPI values. These results enable a consistent comparison of the key information on the sustainability performance of buildings and neighbourhoods for exchanging and sharing information and good practices between different areas, cities, regions and countries. A single page CESBA Certificate is a concise single-page information sheet that captures the scores for each of the seven sustainability issues and can be used to display and communicate the overall performance.

2.2.4. The GF Tools

All the indicators are analytically presented in the building and neighbourhood scale GF tools. The presentation of each indicator (Figure 5a) includes background information, an overview of relevant calculation steps that one must follow for KPIs (according to standards) and for others based on recommended good practices that may be adapted according to national or local practices. Supporting references and other resources are also included and the user may also write-in other relevant notes. The input is the calculated value for the corresponding indicator (e.g., the energy use intensity in kWh/m^2) for the specific project. Under the assessment criteria, the tool automatically transfers the default benchmark values that correspond to the scale ($-1, 0, 3, 5$). These values may be further adjusted, if necessary, in order to accommodate for some specific characteristics for a given project (e.g., adjust the energy use intensity benchmarks for historic buildings that may not strictly comply with conventional high-performance standards). Entering the numerical value of the indicator, the tool estimates a weighted score. The user may also include a target performance value for reference and comparative assessment. As an option, there is also a place holder for a third-party score that can be used during verification.

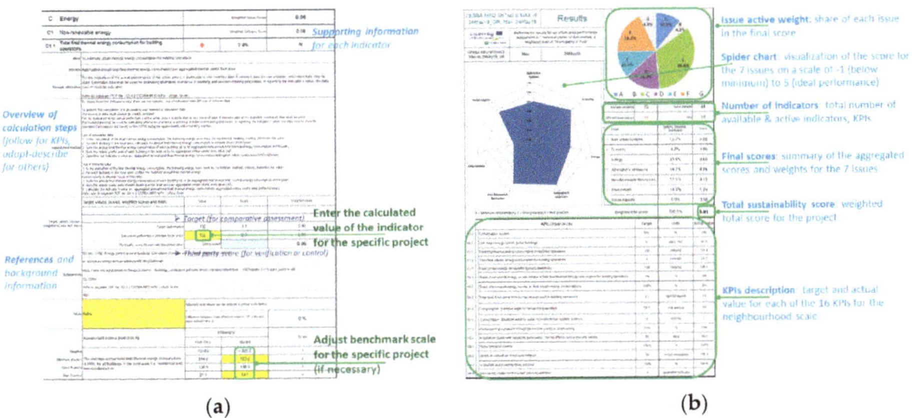

(a)   (b)

Figure 5. Excerpts from the Generic Framework (GF) tool: (**a**) Presentation an indicator and interface for benchmarking and scoring; (**b**) overview of the main final results.

The final results (Figure 5b) summarize the performance assessment for the building or the urban area, along with insights on the importance of the different issues in the final score, number of active indicators and detailed overview of the KPIs. The specific scores for each one of the seven sustainability issues are listed and are also illustrated in a spider chart to easily understand and communicate results, by identifying the sustainability issues with strong performance (scores close to 5) or the weaker ones (scores close to 0) and the total sustainability score for the project. A detailed presentation of the results for each one of the KPIs includes the corresponding target and actual value.

2.3. The Decision Making Process

The CESBA MED Tools are intended to support decision-makers and managers of public and municipal building stocks in the implementation of more sustainable renovation plans or the new developments, combining the building and the neighbourhood scales [33]. The process should consider the buildings in their urban environment and look for synergies between groups of buildings in the area in order to optimize energy planning in the context of a sustainability performance assessment.

The CESBA MED method and tools (Figure 6) can support all project phases. Instrumental in the whole process is the engagement of the people. Urban developments affect a wider community of citizens, workers, commuters, visitors, etc. Therefore, it is essential that all affected parties, including residents and businesses, are actively involved in all stages of the process, from the early diagnosis, for shaping the developments that affect them. Empowering local communities through regular and meaningful consultations and engagement, improves transparency through more open governance and greater public participation of citizens and other local stakeholders and helps reach greater public acceptance through a sense of ownership.

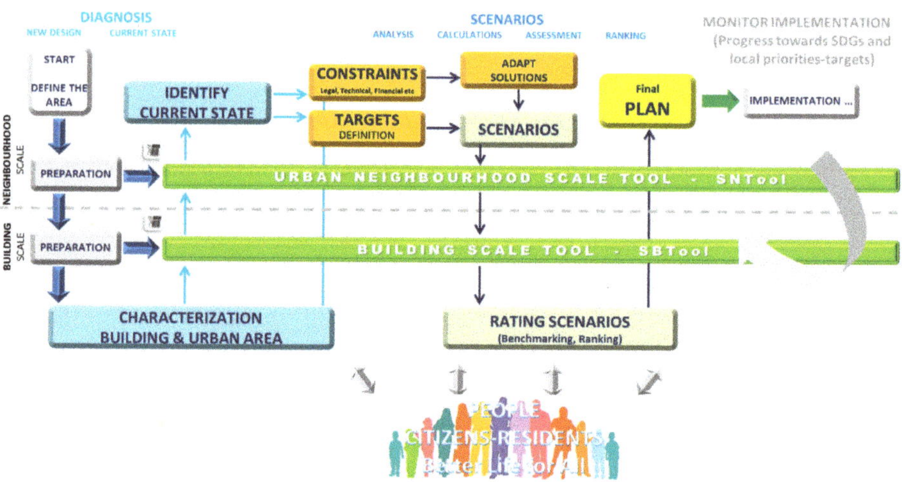

Figure 6. Schematic of the main stages and overall decision-making process.

Assessing the actual performance at the current state takes a snapshot of the existing condition and characteristics, and identifies the critical sustainability issues. Accordingly, one can assess the potential performance resulting from the implementation of different renovation scenarios in order to identify the most cost-effective and sustainable one. Similarly, for new developments, it is possible to assess the potential performance of alternative planning options in order to identify the most cost-effective sustainable development scenario. In both cases, after implementation, it is possible to monitor and evaluate the progress at different stages, the effectiveness of implemented actions and the achievement of the sustainability performance targets.

2.3.1. Diagnosis

The first step is to define the physical boundaries and decide which of the surrounding infrastructures are relevant. The physical boundaries of the urban area may be derived considering the spatial coverage along with the legal and administrative lines, the property ownership status and land use, the social and economic characteristics of the area, the period of building construction and the energy supply infrastructure, etc. The neighbourhood of a small urban scale area (e.g., block/cluster of buildings) may include 5–15 buildings with a traditional composition extending over 200–400 m in size that can be crossed in 10–15 min walk, with 200–1500 inhabitants.

During the preparation phase, the appropriate input data is collected in order to create a sufficient knowledge basis. Like in every audit process [14] it is essential to collect good quality data is essential, since this will have a direct impact on the overall quality of the process and it is critical for securing accurate and realistic final results for quantifying the respective indicators. Sometimes specific characteristics cannot be determined or measured in a practical way with an acceptable cost, at least for routine audits, while even the perception of building or neighbourhood characteristics can deviate significantly from one assessor to the other. Accordingly, it is crucial to find the right level of simplification so that the audit and data processing is time-efficient while obtaining results that are close to the most detailed analysis as possible.

Accessibility to reliable data and information is indispensable to adequately assess the sustainability performance of the urban environment. This will allow the adoption of good monitoring practices, resulting in better policy formulation and implementation. In general, data acquisition may be time-consuming. Information may be scattered among different administrative bodies of the municipalities and other organizations (e.g., building authorities, cadastral office, land surveying office) and other resources like census data, municipality and regional reports (e.g., operational programs), existing energy performance certificates, energy supply companies, along with publicly accessible resources (e.g., Google Earth, Open Street Map), etc. In all cases, a site visit will be necessary in order to perform field inspections of the buildings and the neighbourhood and to collect missing data or verify and extend available information.

Educated assumptions or use of default values may be needed for quantifying some indicators. However, one must consider the trade-offs between the effort involved to measure specific data; what accuracy can be reached, how much effort will be involved, how much time will be required and what is the relevance or impact on the results. With the exception of the calculations for the KPIs that follow specific normative procedures according to standards, the final decision depends on the relevant expertise and past experience of the user/assessor. In any case, one needs to be aware of the uncertainties or inaccuracies involved in a given process as a result of the assumptions that will be made or imposed by specific calculation procedures and the ways to interpret and use the results.

2.3.2. Scenarios

As a first step, the information collected during the diagnosis is used for a SWOT analysis. This way one can prepare more applicable scenarios that will exploit the area's main strengths and opportunities in terms of sustainability and take corrective measures that are responsive to its weaknesses, while accounting for legal–technical–financial–environmental constraints that may limit the range of possible retrofit strategies. Legal constraints may result from building codes, mandates for improving the energy performance of buildings, and cultural heritage protection regulations. Technical constraints may limit the use of some technologies in building renovations, e.g., space availability for on-site installation of renewables on building rooftops or facades or near-by areas. Financial constraints are often the largest obstacles in renovation projects. Available funding sources must be secured early in the planning phase, taking advantage of different financing instruments. For building renovations, one needs to consider the financial status of the building owners, as well as the tenants, in order to avoid negative social impacts like gentrification. Environmental constraints are usually related to the local climatic conditions which may not favour some technologies or the exposure of building roofs and facades to solar radiation for the proper exploitation of thermal solar or photovoltaics.

Early in the process, one must define clear and measurable targets that should be achieved by the project, covering all main aspects of sustainability, e.g., environment-energy, economy and social. The targets must be SMART, i.e., Specific (clearly defined), Measurable (quantifiable), Attainable (realistic and achievable), Relevant (for energy retrofitting of urban districts) and Time-bound (with a specific time plan of when they can be achieved). Environmental targets should consider the means to improve energy performance, reduce GHG emissions, increase the share of renewables, prioritize the use of sustainable materials, reduce soil sealing and increase open natural areas. Targets related to the economy should consider means to improve return on investment, exploit the use of different instruments for financing, maintain affordable property and value of land, secure resources to strengthen economic feasibility and secure sustainable growth and enhance local labour force participation. Social targets should avoid gentrification that may result from energy renovations, improve district surroundings (e.g., open spaces, accessibility, heat island), improve transport infrastructure and mobility, encourage community involvement and citizen's engagement in near- and long-term planning, strengthen public services and improve safety and security.

The scenarios for improving the performance of a neighbourhood should consider all the buildings in the area and seek synergies and opportunities to increase energy performance by prioritizing central energy supplies and district energy systems versus individual solutions, use environmentally friendly materials, enhance open green public spaces, improve public transport and mobility and improve public infrastructures. At the building scale, the priority is to improve energy performance of public buildings, reduce energy use and emissions from non-renewables, integrate renewables (e.g., consider thermal solar for domestic hot water or combi systems, use photovoltaics with appropriate energy storage and/or smart grids), expand central energy supply (e.g., natural gas network) and increase energy efficiency by prioritizing central energy networks over individual solutions. Engaging the citizens and building occupants in the process can provide valuable input to the experts and technical teams.

Different scenarios are evaluated along the following lines: (a) Selecting and optimizing energy renovations at a building and neighbourhood scale (i.e., reducing energy demand, increasing energy performance by prioritizing central energy supplies versus individual solutions, integrating renewables with appropriate energy storage and/or smart grids); (b) considering other interventions for improving public transportation and mobility, enhancing green spaces, and other public infrastructures; (c) exploiting different business models and financing instruments; and finally (d) identifying the desirable scenario that will address the municipality's objectives and priorities.

Considering the final sustainability score for each scenario, one can select the best one that meets expectations and plans of the municipality or the public authority having jurisdiction, in-line with the local sustainability objectives and priorities, or the owner's intent (e.g., an authority that manages public buildings). The results can be easily communicated to the decision-makers to document the current state, summarise the proposed strategies of the final plan and illustrate the anticipated improved sustainability performance. Once a decision is taken to proceed with implementation, the concept will have to be elaborated in more detail including a cost–benefit analysis, exploit different business models and financing instruments, issue tenders and conclude with a contract. Following implementation, the local tools are ready to be used to assess whether the goals and objectives have been met, document actual progress for improving sustainability and monitor progress towards the performance targets. The results should be properly publicized and communicated so that they gain visibility and acceptance.

2.4. Training System

To further facilitate the process, users are also supported by a comprehensive electronic CESBA MED Training System for self and in-house education, training and professional development on sustainability. The developed material facilitates the proper use of the method and tools, improves the knowledge base and enhances the understanding of the various sustainability issues by different target groups and stakeholders (e.g., engineers, technical staff, decision and policymakers) to set up and implement high quality and sustainable urban plans, and supports continuous learning [34]. The training material is organized in different modules [35], including the GF concept and the multicriteria assessment methodology, the decision-making process, case studies, the assessment criteria of the contextualized tool at building and neighbourhood scale with a detailed presentation of the KPIs along with calculation examples. The electronic training material is accessible through an open e-learning platform (https://cesba-med.research.um.edu.mt/moodle/course/index.php) and is available in different languages, e.g., Catalan, Croatian, English, French, Greek, Italian and Spanish. The educational material was successfully used during 17 national pilot training courses that were held in the participating countries with about 275 participants.

3. Results and Discussion

The CESBA MED system was used in the field during nine pilot studies in six countries (Table A3) to demonstrate its applicability in diverse applications at different building uses and urban areas (e.g., social housing, 19th century historic buildings), scales (e.g., a building block, a university campus and different size urban neighbourhoods) and project intents (e.g., renovations or new developments). In some cases, different renovation scenarios were also assessed and the results are also summarized in Table A3. The pilots served two main purposes. First, working together with local experts and municipalities, the goal was to develop the national versions of the tools, by selecting a suitable number of indicators, translating the tools and incorporating representative national weights for the different sustainability issues and benchmarks for normalizing the indicator values. This way, the existing default values in the national versions of the tools are ready to be fine-tuned, to better match the local characteristics (e.g., energy use intensities for the local buildings, water consumption in the area, etc.). Second, they were used as a final test phase for verifying that the selected KPIs can be realistically used in the field. The goal was to ensure that the input data are commonly available during the building and urban audits, so that the KPIs can be consistently calculated. The national pilots also revealed some interesting information on the most popular sustainability indicators that were selected by each national team, illustrating the emphasis and the priorities given by the participating municipalities.

3.1. National Tools

The GF Tools are available in English, while the nationally contextualized assessment tools are available in different languages (i.e., Catalan, Croatian, French, Greek, Italian and Spanish). The national tools include the same KPIs, but use a different number of categories, criteria and indicators (Table 1) that best fit in the national and local context and their sustainability priorities.

Table 1. Overview of the sustainability issues, categories and criteria-indicators used in the generic framework (GF) and the national framework tools.

	GF	Italy (A)	Italy (B)	France (A)	France (B)	Spain (A)	Spain (B)	Malta	Greece	Croatia	Average
Building Scale											
Issues	7	7	7	7	7	7	7	7	7	7	7
Categories	25	15	18	8	8	19	21	11	16	15	15
Indicators	153	16	31	16	19	38	40	36	31	27	28
KPIs	13	13	13	13	13	13	13	13	13	13	13
Neighbourhood Scale											
Issues	7	7	7	7	7	7	7	7	7	7	7
Categories	23	14	20	11	13	16	20	20	16	20	17
Indicators	178	34	46	16	19	33	59	66	44	38	39
KPIs	16	16	16	16	16	16	16	16	16	16	16

Each team selected from the pool of indicators included in the Generic Framework the ones that are most relevant according to their national sustainability priorities and are commonly encountered at regional-local issues. For example, the generic framework for the neighbourhood scale includes a total of 23 categories and 178 criteria-indicators, while the national tool in Greece uses a total of 16 categories and 44 criteria-indicators. The only core set of criteria that is mandatory and included in all national tools, are the KPIs that represent internationally recognized priorities for sustainability assessment.

According to the pilot test results, the selected number of sustainability criteria averaged 28 indicators at building scale and 39 indicators at neighbourhood scale (Table 1). The sustainability issue that has attracted more emphasis based on the number of selected indicators (Figure 7) was "B-Energy and Resources" with 32% of the total number indicators used at building scale and "G-Social Aspects" with 26% of the total at neighbourhood scale.

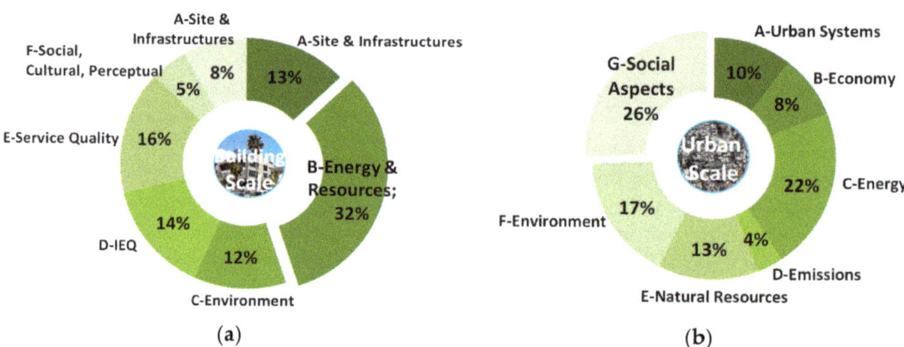

Figure 7. Overview of the average breakdown of the selected criteria-indicators used in the national tools for the seven sustainability issues for (**a**) the building scale; (**b**) the neighbourhood scale.

For each one of the selected indicators, the national teams in collaboration with local committee experts specified the local benchmarks, i.e., the values that correspond to the appropriate local excellent practice (corresponding to "+5" in the normalized score), the minimum acceptable performance (corresponding to "0" in the normalized score) and below minimum standard (corresponding to "−1" in the normalized score). This information was used to benchmark the values for each indicator and normalize them on the -1 to 5 points scale (see Section 2.2.1).

The benchmarks for the KPIs from the different regions are summarized in Table 2. The values can provide initial guidance during future developments and adaptations of similar tools in other regions. The empty cells in Table 2 (i.e., B.1.10 for embodied energy, C.3.2 for recycled solid waste, D.1.9 for ventilation rates) refer to cases with missing information.

Table 2. Summary of the key performance indicators (KPIs) benchmarks (minimum and best values) used in the national framework tools for the building and neighbourhood scales.

Code	Key Performance Indicators (KPIs)	Units	Value	Italy (A)	Italy (B)	France (A)	France (B)	Spain (A)	Spain (B)	Malta	Greece	Croatia	Average
				BUILDING SCALE									
B.1.1	Primary energy use	kWh/m^2/y	Minimum	80	140	48	140	225	292		310.6	90	165.7
			Best	30	23	0	0	70	58		87.6	55	40.5
B.1.2	Final thermal energy use	kWh/m^2/y	Minimum	70	80	40	130	22	75		69.1	50	67.0
			Best	20	10	0	30	12	20		11.5	10	14.2
B.1.3	Final electrical energy use	kWh/m^2/y	Minimum	30	23	40	140	75	70		99.4	30	63.4
			Best	20	5	0	0	20	30		29.1	0	13.0
B.1.5	Renewables in final thermal energy use	%	Minimum	30	25	25	10	30	30		16	20	23.3
			Best	100	50	100	100	100	100		80	90	90.0
B.1.6	Renewables in final electrical energy use	%	Minimum	40	35	10	10	40	40		20	5	25.0
			Best	100	75	200	100	100	100		100	90	108.1
B.1.10	Embodied non-renewable primary energy	MJ/m^2	Minimum	2500		180	900				6230	14	1964.8
			Best	1000		90	504				3000	3	919.4
B.4.5	Water consumption for indoor uses	m^3/occupant/y	Minimum	40	47	40	90	100	11		6	5.5	34.2
			Best	25	23	20	20	20	5		1.5	2	13.8
C.1.3	Global warming potential	kg CO$_2$ eq./m^2/y	Minimum	30	28	20	80	30	96.31		7.5	40	43.1
			Best	0	5	5	5	10	19.26		2	5	5.9
C.3.2	Solid waste categories recycled	%	Minimum	50	14	0.4	0.4	15			57	28	23.5
			Best	80	100	1	1	100			100	100	68.9
D.1.9	Ventilation rate	Lt/s/m^2	Minimum	10	0.35	0.5		6			0.29	2.77	3.3
			Best	20	0.49	0.9		12			0.83	6	6.7
D.2.2	Thermal comfort index	%	Minimum	10	10	10	10	25	10		25	25	15.6
			Best	0	6	5	0	5	0		5	5	3.3
G.1.4	Operational energy cost	€/m^2/y	Minimum	20	10.7	15	15	60	35		18.9	7.5	22.8
			Best	10	1.75	5	5	40	10		4.7	1.5	9.7
G.1.5	Operational water cost	€/m^2/y	Minimum	5	1.55	10	13		5		0.59	0.5	5.1
			Best	1	0.7	3	2.3		1		0.15	0.2	1.2
				NEIGHBOURHOOD SCALE									
A.1.7	Land conservation	%	Minimum	0.5	7	15	10	4	10	10	10	2	7.6
			Best	2	42	30	20	15	20	28	20	10	20.8
B.3.3	Operational energy cost for public buildings	€/m^2/yr	Minimum	7.4	10	14	14	20	13.56	100	17.7	100	33.0
			Best	4	3	3.5	3.5	10	3.33	0	4.1	0	3.5
C.1.1	Total final thermal energy consumption for buildings	kWh/m^2/yr	Minimum	70	80	40	50	75	76.23	50	314	100	95.0
			Best	30	10	0	0	20	33.8	0	21.1	50	18.3
C.1.4	Total final electric energy consumption for buildings	kWh/m^2/yr	Minimum	50	23	12	55	70	29.85	25	64.2	75	44.9
			Best	20	5	0	5	20	10.88	5	7.9	50	13.8
C.1.7	Total primary energy consumption for buildings	kWh/m^2/yr	Minimum	322	72	40	140	225	152	50	461.9	100	173.7
			Best	242	50	0	0	70	15	15	38.2	70	55.6

Table 2. *Cont.*

C.2.1	On-site renewables in total final thermal energy consumption	%	Minimum	20	25	25	30	25	25	25	4	5	20.4
			Best	100	50	100	100	90	90	90	14	30	73.8
C.2.7	On-site renewables in total final electrical energy consumption	%	Minimum	20	35	25	35	15	15	35	1	20	22.3
			Best	100	75	200	75	75	75	75	47	35	84.1
D.1.2	Total GHG Emissions from energy use in buildings	kg CO_2 eq./m^2/yr	Minimum	22.5	13	20	30	30	30	80	46	22	32.6
			Best	0	11	5	10	10	10	30	5	15	10.7
E.1.6	Water consumption in residential buildings	m^3/occupant/yr	Minimum	65	47.45	40	68	150	150	15	62.1	250	94.2
			Best	61	23.7	20	30	40	60	5	18.6	100	39.8
E.1.7	Water consumption in public buildings	m^3/m^2	Minimum	1	1.3	5	1.1	15	15		0.65	5	5.5
			Best	0.5	0.6	2	0.4	5	5		0.33	3	2.1
F.1.3	Recharge of groundwater through permeable paving/landscaping	%	Minimum	20	40	20	20	20	20	20	15	20	21.7
			Best	40	60	70	100	70	70	100	80	80	74.4
F.2.3	Ambient air quality (PM10) above acceptable limits	days/yr	Minimum	35	35	30	30	15	15		35	20	26.9
			Best	25	0	11	11	11	11		0	15	10.5
G.2.1	Proximity of residents to public transport	%	Minimum	70	60	50	0	30	30	30	50	5	36.1
			Best	100	100	100	100	100	100	100	100	40	93.3
G.2.4	Pedestrian & bicycle network	m/100 inhabitants	Minimum	14	43	15	200	20	5	5	2	0	33.8
			Best	42	129	40	50	80	40	40	20	500	104.6
G.4.2	Proximity of residents to key services	%	Minimum	80	30	30	30	30	50	50	50	20	41.1
			Best	100	80	100	100	80	100	100	90	70	91.1
G.6.3	Community involvement in urban planning (qualitative score)	level (score)	Minimum	0	3	0	0	0	0	0	0	0	0.3
			Best	5	5	5	5	5	5	5	5	3	4.8

The weights used for each one of the seven sustainability issues from 1 (less important) to 3 (most important or more relevant) defined in the national versions of the tools reflect the local priorities, policies or project intent. As summarized in Table 3, for the building scale, the sustainability issue "B-Energy and Resources" stands out as the strongest priority. Along with "C-Environment" are the two most prominent issues, averaging together ~80%. For the neighbourhood scale, lower weights were consistently used for "B-Economy" illustrating that once there is a commitment for sustainable development, the economic criteria have a lower priority. The sustainability issue related to "C-Energy" stands out by averaging 26.6% among all the pilots, although different regions have other priorities in terms of where they focus their efforts by allocating higher weights.

Table 3. Summary of the weights on different sustainability issues used in the national framework tools for the building and the neighbourhood scales.

Sustainability Issues	Italy (A)	Italy (B)	France (A)	France (B)	Spain (A)	Spain (B)	Malta	Greece	Croatia	Average
Building Scale										
A-Site and Infrastructures	0.0%	0.0%	0.0%	0.0%	4.9%	11.6%	7.0%	6.5%	7.6%	4.2%
B-Energy and Resources	58.0%	69.8%	72.0%	72.0%	62.9%	54.9%	31.6%	28.5%	51.2%	55.7%
C-Environment	23.0%	24.3%	25.0%	25.0%	19.5%	20.4%	23.6%	36.6%	19.5%	24.1%
D-IEQ	11.0%	4.2%	2.0%	2.0%	2.1%	1.5%	2.0%	0.5%	8.0%	3.7%
E-Service Quality	0.0%	0.0%	0.0%	0.0%	7.9%	8.1%	20.7%	12.6%	3.2%	5.8%
F-Social, Cultural, Perceptual	0.0%	0.0%	0.0%	0.0%	1.9%	2.7%	12.0%	4.3%	5.1%	2.9%
G-Economy	8.0%	1.8%	2.0%	2.0%	0.8%	0.8%	3.1%	11.0%	5.5%	3.9%
Neighbourhood Scale										
A-Urban Systems	11.6%	10.4%	18.9%	0.0%	6.5%	10.2%	13.5%	10.8%	12.2%	10.4%
B-Economy	1.7%	6.6%	5.0%	1.8%	9.1%	3.6%	1.8%	4.2%	4.6%	4.3%
C-Energy	41.1%	18.4%	30.5%	28.2%	26.7%	25.9%	16.2%	33.7%	21.5%	26.9%
D-Emissions	6.9%	14.3%	23.6%	33.9%	7.3%	12.7%	5.8%	14.7%	13.3%	14.7%
E-Natural resources	6.9%	14.1%	3.4%	8.7%	7.3%	10.1%	11.7%	11.5%	14.3%	9.8%
F-Environment	18.3%	15.7%	9.4%	9.9%	31.3%	23.8%	28.7%	18.2%	9.0%	18.3%
G-Social Aspects	13.4%	20.5%	9.1%	17.4%	15.4%	13.6%	22.3%	6.9%	25.0%	15.9%

Note: Cells a green highlight identify the Issues with the Highest Weight in each national tool

3.2. Results from the Hellenic Pilot

Greece has introduced a regulatory framework to support urban policy based on the law on spatial planning and sustainable development (L.4447/2016) and has drafted the national spatial and development strategy that includes medium- and long-term targets. Sustainable urban development strategies are implemented within 13 regional operational programmes, giving priority to regional capital cities or to functional areas with a certain size of population [36]. Urban and territorial strategies address 10 thematic objectives, including environmental protection and resource efficiency, network infrastructures in transport and energy, low-carbon economy, climate change adaptation and risk prevention, social inclusion, etc. The sustainable urban development strategies cover practically all thematic areas, while cities have launched broad and substantial public consultations in defining their strategies. In most cases, the municipalities have commissioned external experts to draft their plans due to limited human resources and expertise.

The CESBA MED pilot in Greece was performed at the Municipality of Fylis, north-west of Athens. Over the years, the municipality has exhibited genuine efforts, activities and communicated plans for improving its energy and carbon footprint, by implementing various initiatives on energy conservation and sustainable development aiming to regenerate into a thriving area. The pilot focused at a central area of the Municipal Unit of Ano Liosia, including the town hall and a school complex, covering 27.1 ha, half of which is built, and 1330 residents. There are about 360 buildings (including seven public buildings, the town hall, an indoor sports hall and five school buildings) in the selected area, 55% of which residential, 23% mixed-use and 22% non-residential buildings, dominated by low rise buildings (75% are one to two-storey high).

Starting with the initial 178 indicators in the GF for the neighbourhood scale, the first screening rule was their applicability to the national conditions, resulting in 118 indicators. In consultation with the national local committee members this number of indicators was judged not to be practical for an assessment tool, so the second screening rule was to exclude those that were very difficult to define or assess during an audit, resulting in 77 indicators. Consulting with members from several municipalities trying to focus on the indicators that would be most relevant from a local authorities' point of view, along with their perception with regard to data availability, the number was reduced to 57 indicators. Finally, in close collaboration with the Municipality of Fylis, the final number of indicators was limited to 44 that have been documented with their selection rationale [37]. The sustainability issues with the highest importance (weighted at 3) were C- Energy, D- Atmospheric Emissions and G-Social Aspects, followed by B-Economy, E-Non-renewable Resources and F-Environment (weighted at 2) and relatively lower for A-Built Urban Systems (weighted at 1). The specific indicator weights, the corresponding benchmarks, the input data and information collected during the audits, along with the calculation processes and results are also documented in the national pilot reports.

The municipality's sustainability priorities address most of the issues identified in the SWOT analysis (Figure 8). They include efforts to reduce energy use and cost (e.g., renovate municipal buildings for improved energy performance, use energy-efficient public lighting), to exploit renewables for electricity generation (in part through energy communities) to cover the needs of public buildings and low-income residents. Energy communities are a partnership or legal entity between different local energy users, citizens, social organizations, local or city authorities, small and medium-sized local businesses, which enable energy consumers to produce, consume and share a part or all of its needed energy from renewables on a collective basis. Greece was the first EU-28 country that introduced comprehensive legislation on the establishment and operation of the energy communities (Law 4513/23.1.2018). Since then, about 60 energy communities have been formed around the country and several energy projects are under development.

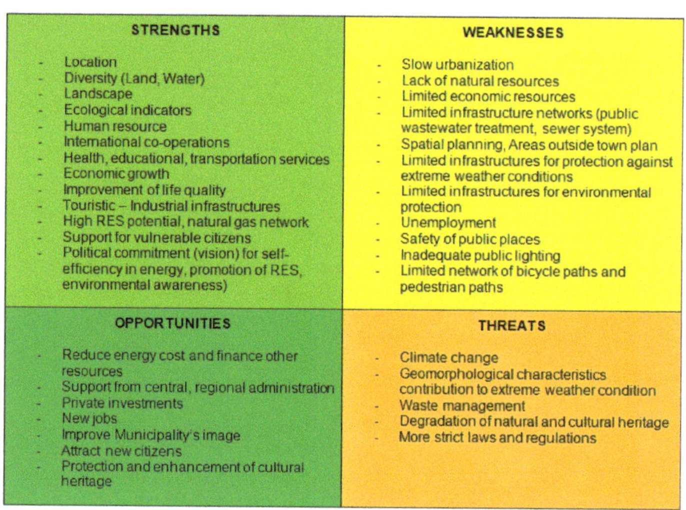

Figure 8. SWOT analysis for the Hellenic pilot.

Other integral elements of the municipality's plan is to utilize the new main natural gas supply network, to install PVs and solar thermal in schools and sports facilities, to reduce CO_2 emissions in the area (e.g., expand the pedestrian and low traffic roads, along with bicycle lanes, reintroduce municipal local transportation and public parking), and to engage citizens in the decision making process for the development of green areas for the urban regeneration of neighbourhoods. Self-efficiency in water, recycling (e.g., strengthening the collection and recycling of waste) and anti-flooding protection are secondary targets.

For sharing the detailed results, the area's characteristics and KPI values are summarized in the two-page CESBA Passport (Figure 9). For a higher-level communication of the overall sustainability performance, the results are summarized in the CESBA Certificate (Figure 9) that graphically illustrates the scores for each one of the seven sustainability issues and the resulting total sustainability score. The Passport can be used to exchange information on the common KPI performance, while the Certificate communicates how a project contributes value to the various sustainability issues.

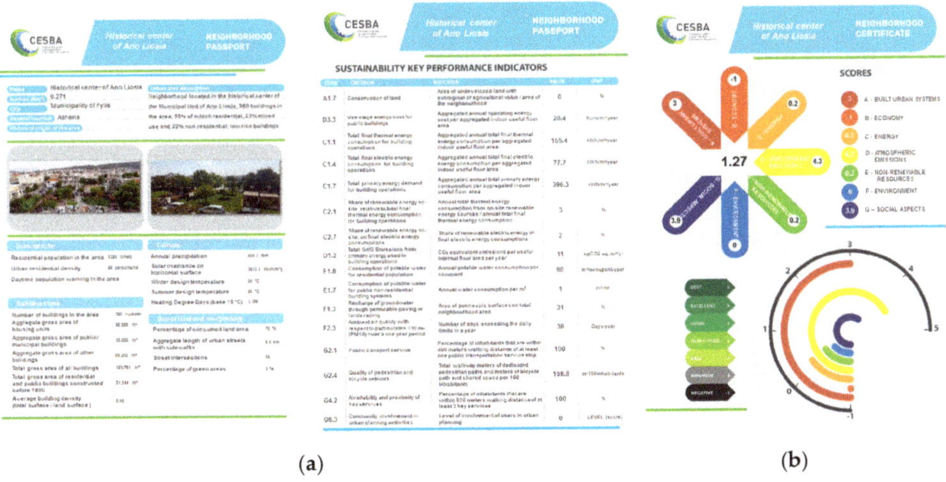

(a)   (b)

Figure 9. Representative results for the urban area assessed in the Hellenic pilot: (**a**) the two-page Common European Sustainable Built Environment Assessment (CESBA) Passport; (**b**) the CESBA Certificate.

Overall, the data collection during the audit process mandated a very close collaboration with the various municipality directorates, divisions and departments (e.g., technical services, building infrastructures, city plan and urban environment, roads and public spaces, financial, municipal assets, cleaning and recycling, green areas, etc.), access to previous studies and municipality's assessment plans, data from the national statistical authority, information from the depository of energy performance certificates, and other sources. Retrieving the input data was a time-consuming process, while sometimes provisionally available data proved difficult or even impossible to retrieve (e.g., plans or reports were not electronically archived, while hard copies were lost or misplaced). This underlines the fact that data availability is as important as data accessibility. Among the direct benefits resulting from the early stages of implementation of the CESBA MED system was the motivation and support it provided to get organized and initiate a housekeeping process within the municipality. There is value even when relevant efforts reach the point of simply identifying missing information. Following the principle that one cannot assess or manage what cannot be measured or quantified through the appropriate indicators, one can initiate practical actions for monitoring the necessary data that can prove beneficial in the future.

4. Conclusions

In their efforts to achieve local-regional-national and international Sustainable Development Goals (SDGs), municipalities and public authorities need flexible and easy to use methods and tools to facilitate their efforts and overcome the burdens of addressing the complexities of the issues involved. CESBA MED is a new open and flexible multicriteria assessment system structured around the UN and EU SDGs that can be used to quantify and include sustainability issues in the decision-making process.

It supports users throughout the process in order to initiate, organize, adapt, evaluate and identify the best sustainable renovation strategies for buildings or neighbourhoods, and monitor progress towards achieving sustainability targets.

Compared to other sustainability audit and rating systems, CESBA MED offers an open-source assessment system for measuring the sustainability at building and neighbourhood scale in a harmonized approach. Cities can easily adapt it to local context by selecting and using the most suitable indicators, incorporating weighting factors that reflect local targets, priorities and policies, and have their own tailored system, which strengthens a sense of local ownership. The assessment results are comparable among cities at national and transnational levels.

At building scale, CESBA MED addresses seven sustainability issues, including: A-Site and infrastructures, B-Energy and resources, C-Environment, D-Indoor environmental quality, E-Service quality, F-Social, cultural and perceptual aspects and G-Economy, which are described and quantified with 153 sustainability criteria/indicators. Among them, 13 KPIs have been selected as mandatory indicators, which represent the priority sustainability transnational issues. At the neighbourhood scale, seven sustainability issues are addressed, including: A-Urban systems, B-Economy, C-Energy, D-Emissions, E-Natural resources, F-Environment and G-Social aspects, which are described and quantified with 178 sustainability criteria/indicators, including 16 KPIs.

The generic framework and common tools are available in English and different languages, while the assessment and rating approach have been contextualized to national (local) context for Croatia, France, Greece, Italy, Malta and Spain. Nine pilots performed in six Mediterranean countries demonstrated the applicability and adaptability of the CESBA MED system in practice for diverse applications at different scales, and verified the practical use of the KPIs in the field.

The assessment system can be used to carry out a sustainability diagnosis of buildings and neighbourhoods, to set up performance targets and to assess suitable retrofit or new development scenarios, in order to integrate sustainability in urban planning efforts. At this stage, CESBA MED does not include specific cost-related information for the various scenarios. Future work will consider the integration of relevant information since this will add practical value and facilitate the cost/benefit analysis for implementation. Furthermore, although the pilots provided sufficient confidence in the use of the overall method in the field, additional work will be necessary in order to test all the indicators included in the generic framework. In some cases, it may be necessary to reconsider some indicators. For example, A.2.4 Extent and connectivity of bicycle paths are expressed in km/1000 residents. Apparently, an area with a very low number of inhabitants will result in very high value for A.2.4, even for a small bicycle route. Although this will not be an issue in a densely populated urban area, it may be more appropriate to consider an indicator expressed as km/resident. With the exception of KPIs, in cases that a specific indicator may not be appropriate in local context, one can adapt the existing indicator to a more suitable one, provided that the benchmarks are also adjusted accordingly, along with the weighting factors, if necessary. In other cases, one may wish to use alternative indicators to quantify a criterion. For example, there are several indicators to evaluate environmental impacts (e.g., using the quantities of GHG emissions or the global warming potential), energy consumption (e.g., expressing the energy use intensity per unit area or per unit volume at building scale or per capita at neighbourhood scale), or thermal comfort conditions (e.g., using the standard effective temperature—SET or the predicted mean vote—PMV), etc. In principle, a method that includes several alternative indicators for some or all of the criteria may appear more flexible and advantageous. However, this is not the case for local authorities targeted by CESBA MED that need a straight forward and easy to implement tool, taking into account their limited human resources and expertise to fully understand the pros and cons of selecting and using different indicators and alternative paths.

In this direction and to facilitate implementation, the CESBA MED system is also supported by an electronic training system that provides open access to educational material in different languages for different target groups (e.g., engineers, technical staff, decision and policymakers). The material can be used for self and in-house education, training and professional development to improve the knowledge base and understanding of the various sustainability issues and indicators, strengthen the capacity of local stakeholders to develop efficient policies and implement integrated local action plans for sustainable urban development.

The results from this work motivated the development of eight clear, actionable recommendations targeted to policymakers for promoting a new culture of the built environment in Europe [38]. Some notable good practices are already in place, illustrating the potential applicability of CESBA MED. For example, Protocollo ITACA [30] that is an environmental label promoted by the Italian regions for the evaluation and classification of buildings, is based on the transnational building scale tool [24], the reference assessment methodology adopted by CESBA MED. Since 2004, it was accepted by the Conference of Presidents of the Italian Regions and has been contextualised and used at local level by several Italian regions. Since 2015, Protocollo ITACA is the Italian national standard for the assessment of the sustainability of buildings and it is legally binding. Similar statutory audit obligations and regulatory actions for buildings may be adopted in other countries to help implement the European initiative level(s), and extended for neighbourhoods, cities and regions. Along these lines, the City Council of Sant Cugat del Vallès in Spain is using the CESBA MED method in the sustainable development of new buildings and urban areas. For example, during the design phase of new urban areas, developers are required to provide the appropriate data to calculate the CESBA MED indicators, in order to assess their proposals.

Future work will focus on extending the CESBA MED approach to regions and possibly national scales. The ambition is to facilitate and improve the effectiveness and impact of action plans and policies, towards a sustainable future for all.

Author Contributions: Conceptualization, A.M. and CESBA MED partners; methodology, A.M. and CESBA MED partners; software (Hellenic), K.G.D., S.K., C.A.B. and E.G.D.; Software, A.M., S.K., iiSBE and CESBA MED partners; formal analysis, E.B., C.A.B., K.G.D., E.G.D., S.K. and A.M.; investigation (Hellenic), C.A.B., K.G.D., E.G.D. and S.K.; data curation, C.A.B., E.B., K.G.D., E.G.D., S.K., A.M. and CESBA MED partners; writing—original draft, C.A.B.; writing—review and editing, C.A.B., K.G.D., E.G.D., S.K., A.M. and E.B.; visualization, C.A.B.; supervision (Hellenic), C.A.B.; project technical coordinator, A.M.; funding acquisition, A.M. and CESBA MED partners.

Funding: This research was co-funded by the European Regional Development Fund and national authorities, grant number INTERREG MED 2014-20 990; The APC was also funded by the same sources.

Acknowledgments: The work was performed in the frame of the European project CESBA MED (https://cesba-med.interregmed.eu) in a collaborative effort of 12 European organizations from 7 countries, coordinated by the City of Torino, Italy and A. Moro, iiSBE Italia as the technical coordinator of the work. The project was part of the Interreg Mediterranean programme (INTERREG MED 2014-20 990) co-financed by the European Regional Development Fund and national authorities. This paper reflects the views only of the authors who have made every effort to prepare this material for the benefit of the public in light of current and available information. It does not represent the opinion of the European Union. Neither the European Union institutions and bodies nor the authors may be held responsible for the use which may be made of the information contained therein.

Appendix A

Table A1. Sustainability issues, categories and indicators in the Common European Sustainable Built Environment Assessment for Mediterranean Cities (CESBA MED) generic frameworks for building scale [27].

Building Scale	
Issues	**Categories**
	A.1—Site Regeneration and Development
A—Site and Infrastructures	**A.1.1** Protection and restoration of wetlands (expert assessment, score); **A.1.2** Protection and restoration of coastal environments (expert assessment, score); **A.1.3** Reforestation for carbon sequestration, soil stability and biodiversity (expert assessment, score); **A.1.4** Development or maintenance of wildlife corridors (expert assessment, score); **A.1.5** Remediation of contaminated soil, groundwater or surface water (expert assessment, score); **A.1.6** Shading of building(s) by deciduous trees (%); **A.1.7** Use of vegetation to provide ambient outdoor cooling (ratio of total vegetated surface area by total site area); **A.1.8** Use of native plant types (%); **A.1.9** Provision of public open spaces (% of land within the site); **A.1.10** Provision and quality of children's play areas (expert assessment, score); **A.1.11** Facilities for small-scale food production for residential occupants (expert assessment, score); **A.1.12** Provision and quality of bicycle pathways and parking (expert assessment, score); **A.1.13** Provision and quality of walkways for pedestrian use (expert assessment, score)
	A.2—Urban Design
	A.2.1 Efficiency of land use through development density (%); **A.2.2** Need for commuting transportation through provision of mixed uses at site (expert assessment, score); **A.2.3** Impact of orientation on the passive solar potential of building (expert assessment, score); **A.2.4** Impact of site and building orientation on natural ventilation during warm seasons (wind pressure differential in Pa); **A.2.5** Impact of site and building orientation on natural ventilation during cold seasons (wind pressure differential in Pa)
	A.3—Project Infrastructures and Services
	A.3.1 Supply, storage and distribution of surplus thermal energy amongst buildings (%); **A.3.2** Supply, storage and distribution of surplus PV electrical energy amongst buildings (%); **A.3.3** Supply, storage and distribution of surplus hot water amongst buildings (%); **A.3.4** Supply, storage and distribution of surplus rainwater and greywater amongst buildings (%); **A.3.5** Provision of facility to produce energy from solid waste (expert assessment, score); **A.3.6** Provision of solid waste collection and sorting services (expert assessment, score); **A.3.7** Composting and re-use of organic sludge (expert assessment, score); **A.3.8** Provision of split grey/potable water services (%); **A.3.9** Provision of surface water management system (expert assessment, score); **A.3.10** On-site treatment of rainwater, storm water and greywater (expert assessment, score); **A.3.11** On-site treatment of liquid sanitary waste (%); **A.3.12** Provision of on-site communal transportation systems (expert assessment, score); **A.3.13** Provision of on-site parking facilities for private vehicles (%); **A.3.14** Connectivity of roadways (Mean distance between intersections, m); **A.3.15** Provision of access roads and facilities for freight or delivery (expert assessment, score); **A.3.16** Provision and quality of exterior lighting (expert assessment, score)

Table A1. *Cont.*

	B.1—Life Cycle Non-Renewable Energy
B—Energy and Resources	**B.1.1 *** Primary energy use (kWh/m^2/y); **B.1.2 *** Final thermal energy use (kWh/m^2/y); **B.1.3 *** Final electrical energy use (kWh/m^2/y); **B.1.4** Energy from renewables to total primary energy (%); **B.1.5 *** Renewables in final thermal energy use (%); **B.1.6 *** Renewables in final electrical energy use (%); **B.1.7** Use of renewable energy for all building operations (kWh/m^2/y); **B.1.8** Use of non-renewable energy for all building operations (kWh/m^2/y); **B.1.9** Consumption of non-renewable energy for project-related transportation-commuting (kWh/m^2/y); **B.1.10 *** Embodied non-renewable primary energy (MJ/m^2/y)
	B.2—Electricity Peak Demand
	B.2.1 Electrical peak demand for building operations (W/m^2); **B.2.2** Scheduling of building operations to reduce peak loads on generating facilities (W/m^2)
	B.3—Materials
	B.3.1 Re-use of suitable existing structures (%); **B.3.2** Protection of materials during construction phase (expert assessment, score); **B.3.3** Efficient use of material quantities for structural and building envelope components (kg/m^3); **B.3.4** Use of virgin non-renewable materials (%); **B.3.5** Use of recycled materials (%); **B.3.6** Efficient use of finishing materials (%); **B.3.7** Ease of disassembly, re-use or recycling (expert assessment, score)
	B.4—Potable-, Rain-, Grey-Water
	B.4.1 Embodied water in original construction materials (Lt/m^3); **B.4.2** Water consumption for indoor uses (m^3/m^2/y); **B.4.3** Use of water for irrigation purposes (m^3/m^2/y); B.4.4 Use of water for building systems (m^3/m^2/y); **B.4.5 *** Potable water consumption for indoor uses (m^3/occupant/y)
	C.1—Greenhouse Gas Emissions
	C.1.1 Emissions from embodied energy in original construction materials (kgCO$_{2\text{-eq}}$/m^2); **C.1.2** Emissions from embodied energy in construction materials used for maintenance or replacement (kgCO$_{2\text{-eq}}$/m^2); **C.1.3 *** Global Warming Potential (kgCO$_{2\text{-eq}}$/m^2/y)
	C.2—Other Atmospheric Emissions
C—Environment	**C.2.1** Emissions of ozone-depleting substances during facility operations (leaks CFC$_{eq}$ g/m^2/y); **C.2.2** Acidifying emissions during facility operations (SO$_2$eq kg/m^2/y); **C.2.3** Emissions leading to photo-oxidants during facility operations (Ethene eq. kg/m^2/y)
	C.3—Solid and Liquid Wastes
	C.3.1* Construction and demolition waste during life cycle (kg/m^2); **C.3.2** Solid waste categories recycled (%); **C.3.3** Liquid effluents from building operations transported off the site (m^3/y)
	C.4—Impacts on Project Site
	C.4.1 Recharge of groundwater through permeable paving or landscaping (%); **C.4.2** Changes in biodiversity on the site (expert assessment, score); **C.4.3** Adverse wind conditions at grade around tall buildings (expert assessment, score)

Table A1. *Cont.*

	C.5—Other Local and Regional Impacts
	C.5.1 Impact of building on access to daylight or solar energy potential of adjacent property (%); **C.5.2** Impact of construction process on local residents and commercial facility users (expert assessment, score); **C.5.3** Impact of building occupancy on peak load capacity of public transportation (%); **C.5.4** Impact of private vehicles used by building occupants on peak load capacity of local road system (%); **C.5.5** Potential for project operations to contaminate nearby water aquifer or wetland (m); **C.5.6** Annual thermal changes to lake water or sub-surface aquifers ($^\circ$C); **C.5.7** Contribution to Heat Island Effect from roofing, landscaping and paved areas (variance $^\circ$C); **C.5.8** Atmospheric light pollution caused by project exterior lighting (%)
D—Indoor Environmental Quality (IEQ)	**D.1—Indoor Air Quality and Ventilation**
	D.1.1 Pollutant migration between occupancies (expert assessment, score); **D.1.2** Formaldehyde concentration (μg/m^3); **D.1.3 *** TVOC concentration in indoor air (μg/m^3); **D.1.4** CO$_2$ concentrations in indoor air (ppm); **D.1.5** Effectiveness of natural ventilation during cooling season (ach); **D.1.6** Effectiveness of natural ventilation during intermediate seasons (ach); **D.1.7** Effectiveness of natural ventilation during heating season (ach); **D.1.8** Air movement in mechanically ventilated occupancies (m/s); **D.1.9 *** Ventilation rate (Lt/s/m^2)
	D.2—Air Temperature and Humidity
	D.2.1 Time outside of thermal comfort range (%); **D.2.2 *** Thermal comfort index (PPD %); **D.2.3** Appropriate air temperature and relative humidity in mechanically cooled occupancies (expert assessment, score); **D.2.4** Appropriate air temperature in naturally ventilated occupancies (expert assessment, score)
	D.3—Daylight and Illumination
	D.3.1 Appropriate daylighting in primary occupancy areas (DF %); **D.3.2** Control of glare from daylighting (max ratio of contrast)
	D.4—Noise and Acoustics
	D.4.1 Noise attenuation through the exterior envelope (STC); **D.4.2** Transmission of facility equipment noise to primary occupancies (NRC); **D.4.3** Noise attenuation between primary occupancy areas (STC)
E—Service Quality	**E.1—Safety and Security**
	E.1.1 Risk to occupants and facilities from fire (expert assessment, score); **E.1.2** Risk to occupants and facilities from flooding (expert assessment, score); **E.1.3** Risk to occupants and facilities from earthquake (expert assessment, score); **E.1.4** Risk to occupants from incidents involving biological or chemical substances (expert assessment, score); **E.1.5** Maintenance of core building functions during power outages (days); **E.1.6** Personal security for building users during normal operations (expert assessment, score)
	E.2—Functionality and Efficiency
	E.2.1 Appropriateness of type of facilities for occupant needs (expert assessment, score); **E.2.2** Suitability of interior layout(s) for required functions (expert assessment, score); **E.2.3** Appropriateness of space for required functions (expert assessment, score); **E.2.4** Exterior access and unloading facilities for freight or delivery (expert assessment, score); **E.2.5** Service quality and efficiency of vertical or horizontal transportation systems (expert assessment, score); **E.2.6** Spatial efficiency (%); **E.2.7** Volumetric efficiency (%)

Table A1. *Cont.*

	E.3—Controllability
	E.3.1 Effectiveness of facility management control system (expert assessment, score); **E.3.2** Capability for partial operation of facility technical systems (expert assessment, score); **E.3.3** Degree of local control of lighting systems (m^2); **E.3.4** Degree of personal control of technical systems by occupants (expert assessment, score)
	E.4—Flexibility and Adaptability
	E.4.1 Ability for building operator or tenant to modify facility technical systems (expert assessment, score); **E.4.2** Potential for horizontal or vertical extension of structure (expert assessment, score); **E.4.3** Adaptability constraints imposed by structure or floor-to-floor heights (expert assessment, score); **E.4.4** Adaptability constraints imposed by building envelope and technical systems (expert assessment, score); **E.4.5** Adaptability to future changes in type of energy supply (expert assessment, score)
	E.5—Operation and Maintenance
	E.5.1 Operating functionality and efficiency of key facility systems (expert assessment, score); **E.5.2** Adequacy of the building envelope for maintenance of long-term performance (expert assessment, score); **E.5.3** Durability of key materials (expert assessment, score); **E.5.4** Maintenance management plan (expert assessment, score); **E.5.5** On-going monitoring and verification of performance (expert assessment, score); **E.5.6** Retention of as-built documents (expert assessment, score)
	F.1—Social Aspects
F—Social, Cultural, Perceptual	**F.1.1** Universal access on-site and within the building (expert assessment, score); **F.1.2** Access to direct sunlight from living areas of dwelling units (%); **F.1.3** Visual privacy in principal areas of dwelling units (%); **F.1.4** Access to private open space from dwelling units (%)
	F.2—Culture and Heritage
	F.2.1 Compatibility of urban design with local cultural values (expert assessment, score); **F.2.2** Provision of public open space compatible with local cultural values (expert assessment, score); **F.2.3** Impact of the design on existing streetscapes (expert assessment, score); **F.2.4** Use of traditional local materials and techniques (%); **F.2.5** Maintenance of the heritage value – external (expert assessment, score); **F.2.6** Maintenance of the heritage value—internal (expert assessment, score)
	F.3. Perceptual Aspects
	F.3.1 Impact of tall structure(s) on existing view corridors (expert assessment, score); **F.3.2** Quality of views from tall structures (expert assessment, score); **F.3.3** Sway of tall buildings in high wind conditions (m); **F.3.4** Perceptual quality of site development (expert assessment, score); **F.3.5** Aesthetic quality of facility exterior (expert assessment, score); **F.3.6** Aesthetic quality of facility interior (expert assessment, score); **F.3.7** Access to exterior views from interior (expert assessment, score)
	G.1—Operational Cost
G—Economy	**G.1.1** Construction cost (€/m^2); **G.1.2** Operating and maintenance cost (€/m^2); **G.1.3** Life-cycle cost (€/m^2); <u>**G.1.4** *</u> Operational energy cost (€/m^2/y); <u>**G.1.5** *</u> Operational water cost (€/m^2/y); **G.1.6** Investment risk (expert assessment, score); **G.1.7** Affordability of residential rental or cost levels (%)

* Key Performance Indicator (KPI).

Table A2. Sustainability issues, categories and indicators in the CESBA MED generic frameworks for neighbourhood scale [28].

NEIGHBOURHOOD SCALE	
Issues	**Categories**
	A.1—Urban Structure and Form
A—Urban Systems	**A.1.1** Concentration of lots to the total area (%); **A.1.2** Urban compactness (m^3/m^2); **A.1.3** Building plot ratios (%); **A.1.4** Residential density (inhabitants/hectare); **A.1.5** Urban street canyons H/W aspect ratio (%); **A.1.6** Homogeneity of the urban fabric (%); <u>**A.1.7** *</u> Conservation of land (%)
	A.2—Transportation Infrastructure
	A.2.1 Walking distance to public transport for area residents (% of residential buildings located within 500m); **A.2.2** Walking distance to public transport for area workers and students (%); **A.2.3** Extent and connectivity of pedestrian streets and walkways (%); **A.2.4** Extent and connectivity of bicycle paths separated from vehicular traffic (km/1000 residents); **A.2.5** Cyclomatic complexity of the street network (-); **A.2.6** Connectivity of the street network (number/km^2); **A.2.7** Street network connection and accessibility (%); **A.2.8** Scale of the street network (m); **A.2.9** On-street and indoor parking spaces relative to local population (%); **A.2.10** Intermodality facilities (expert assessment, score)
	B.1—Economic Structure and Value
B—Economy	**B.1.1** Affordability of building property (%); **B.1.2** Affordability of building rental (%); **A.1.3** Long-term risk for capital investments (%); **B.1.4** Impact of land values on adjacent areas (%); **B.1.5** Impact of construction and operations on the local economy (%); **B.1.6** Percent of building vacancies (%)
	B.2—Economic Activity
	B.2.1 Income equity for households (GINI index 0-1); **A.2.2** Percent of average annual per-capita income of residents to region's total (%); **A.2.3** Employment rate (%); **A.2.4** Economic viability of local commercial businesses (%); A.2.5 Economic contribution from tourism activity (€/resident)
	B.3—Cost and Investment
	B.3.1 Adequacy of social housing (expert assessment, score); **B.3.2** Public contribution in residential retrofitting investments (%); <u>**B.3.3** *</u> Operational energy costs for public buildings (€/m^2/y); **B.3.4** Levels of total public and private investment (€/resident)
	C.1—Non-Renewable Energy
C—Energy	<u>**C.1.1** *</u> Total final thermal energy consumption for all buildings (kWh/m^2/y); **C.1.2** Total final thermal energy consumption for residential buildings (kWh/m^2/y); **C.1.3** Total final thermal energy consumption for non-residential buildings (kWh/m^2/y); <u>**C.1.4** *</u> Total final electrical energy consumption for all buildings (kWh/m^2/y); **C.1.5** Total final electrical energy consumption for residential buildings (kWh/m^2/y); **C.1.6** Total final electrical energy consumption for non-residential buildings (kWh/m^2/y); <u>**C.1.7** *</u> Total primary energy consumption for all buildings (kWh/m^2/y); **C.1.8** Average total primary energy use of residential buildings to the local minimum value (%); **C.1.9** Average total primary energy use of non-residential buildings to the local minimum value (%); **C.1.10** Average total primary energy use for heating of residential buildings to the local minimum value (%); **C.1.11** Average total primary energy use for heating of non-residential buildings to the local minimum value (%); **C.1.12** Average total primary energy use for cooling of residential buildings to the local minimum value (%); **C.1.13** Average total primary energy use for cooling of non-residential buildings to the local minimum value (%); **C.1.14** Average total primary energy use for DHW of residential buildings to the local minimum value (%); **C.1.15** Average total primary energy use for DHW of non-residential buildings to the local minimum value (%); **C.1.16** Average total primary energy use for indoor lighting of residential buildings to the local minimum value (%); **C.1.17** Average total primary energy use for indoor lighting of non-residential buildings to the local minimum value (%); **C.1.18** Electrical peak demand for non-residential buildings (MW); **C.1.19** Time periods of electrical peak loads (h/y); **C.1.20** Energy use for public lighting (kWh/m^2/y); **C.1.21** Energy use for local public transport (km/MJ); **C.1.22** Final energy use for building demolition or dismantling (kWh/m^2/y)

Table A2. *Cont.*

	C.2—Renewable and Clean Energy
	C.2.1 * On-site renewables in total final thermal energy consumption for all buildings (%); **C.2.2** On-site renewables in total final energy consumption for residential buildings (%); **C.2.3** On-site renewables in total final energy consumption for non-residential buildings (%); **C.2.4** On-site renewables in total primary energy consumption for all buildings (%); **C.2.5** On-site renewables in total primary energy consumption for residential buildings (%); **C.2.6** On-site renewables in total primary energy consumption for non-residential buildings (%); **C.2.7** * On-site renewables in total final electrical energy consumption for all buildings (%); **C.2.8** Aggregated electrical energy generation from renewables located on public properties (MWh/y); **C.2.9** Aggregated electrical energy generation from renewables located on private properties (MWh/y); **C.2.10** Aggregated electrical energy generated from renewables that is exported from the local area (MWh/y); **C.2.11** Percentage of electrical energy use from renewables (%); **C.2.12** Aggregated thermal energy generation from renewables located on public properties (MWh/y); **C.2.13** Aggregated thermal energy generation from renewables located on private properties (MWh/y); **C.2.14** Aggregated thermal energy generated from renewables that is exported from the local area (MWh/y)
	C.3—District Networks and Storage
	C.3.1 Waste heat recovery from building operations (%); **C.3.2** Mid- and long-term thermal storage capacity of geothermal energy sinks (%); **C.3.3** Mid-term storage of electrical energy (GWh)
D—Emissions	D.1—Atmospheric Pollution
	D.1.1 GHG emissions from embodied energy in materials used for construction, maintenance or replacements (tons CO_2eq./1000 m^2); **D.1.2** * Total GHG emissions from energy use in buildings (kgCO_2eq./m^2/y); **D.1.3** Aggregate ozone-depleting emissions from 3-year building operations (tons CO_2eq./1000 m^2); **D.1.4** Aggregate of acidifying-emissions from 5-year building operations (kg CO_2eq./1000 m^2); **D.1.5** Aggregate GHG emissions from the use of private vehicles (tons CO_2eq./y); **D.1.6** Aggregate GHG emissions from the use of public transport (tons CO_2eq./10,000 passengers); **D.1.7** Total GHG Emissions from buildings, private and public mobility (tons CO_2eq./1000 residents)
E—Natural Resources	E.1—Potable-, Rain-, Grey-Water
	E.1.1 Availability of a public municipal water supply (%); **E.1.2** Provision of split grey/potable water services in buildings (%); **E.1.3** Re-use of rainwater in residential buildings (%); **E.1.4** Re-use of rainwater in non-residential buildings (%); **E.1.5** Re-use of storm water (%); **E.1.6** * Water consumption in residential buildings (m^3/occupant/y); **E.1.7** *Water consumption in non-residential buildings (m^3/m^2/y); **E.1.8** Consumption of potable water for irrigation purposes (m^3/1000 m^2/y); E.1.9 Intensity of water purification treatment (kWh/m^3/y)
	E.2—Solid and Liquid Wastes
	E.2.1 Proximity of residents to solid waste and recycling collection points (%); **E.2.2** Separate collection and disposal of solid waste and recycling (%); **E.2.3** Solid waste from construction and demolition projects retained in the area for re-use or recycling (%); **E.2.4** Solid waste from residents' activities and facility operations sent out of the area for re-use, recycling or disposal (%); **E.2.5** Composting and re-use of organic sludge (%); **E.2.6** Public wastewater that is disposed or treated (%); **E.2.7** Liquid effluents from building operations that are exported (%); **E.2.8** Potential for building operations to contaminate nearby bodies of water (expert assessment, score); **E.2.9** Cumulative annual thermal changes to lake water or sub-surface aquifers (expert assessment, score)
	E.3—Usage, Retention and Maintenance
	E.3.1 Consumption of non-renewable material resources for building construction or renovation (Tons/1000 m^2); **E.3.2** Efficient use of materials for construction of infrastructures (Tons/1000 m^2); **E.3.3** Reused or recycled materials for building construction or renovation (%); **E.3.4** Adaptive re-use of existing buildings and structures (%); **E.3.5** Preservation and maintenance of existing buildings and structures (expert assessment, score); **E.3.6** Heritage preservation of existing buildings (expert assessment, score)

Table A2. *Cont.*

	F.1—Environmental Impacts
F—Environment	**F.1.1** Impact of construction activities on natural environment and land preservation (expert assessment, score); **F.1.2** Impact of construction activities or landscaping on soil stability or erosion (expert assessment, score); **F.1.3** * Recharge of groundwater through permeable paving/landscaping (%); **F.1.4** Impacts on local biodiversity (expert assessment, score); **F.1.5** Urban heat island effects (°C); **F.1.6** Impact on access to daylight or solar energy potential of contiguous buildings (%); **F.1.7** Impact of local building user population on peak load capacity of public transport system (%); **F.1.8** Impact of private vehicles used by local population on peak load capacity of the local road network (expert assessment, score); **F.1.9** Atmospheric light pollution from exterior lighting systems of building (Night sky brightness cd/m^2); **F.1.10** Atmospheric light pollution from public lighting (Night sky brightness cd/m^2); **F.1.11** Albedo of building and paving surfaces (number);
	F.2—Outdoor Environmental Quality
	F.2.1 Ambient air quality (PM2.5) above acceptable limits (days/y); **F.2.2** Ambient air quality (PM2.5) above acceptable limits over a week (days/week); **F.2.3** * Ambient air quality (PM10) above acceptable limits (days/y); **F.2.4** Ambient air quality (PM10) above acceptable limits over a week (days/week); **F.2.5** Ambient air quality (CO) above acceptable limits (days/y); **F.2.6** Ambient air quality (O$_3$) above acceptable limits (days/y); **F.2.7** Olfactory air quality (odours) in the area (number of anecdotal reports/y); **F.2.8** Adverse wind conditions at grade around low-rise buildings (expert assessment, score); **F.2.9** Adverse wind conditions at grade around high-rise buildings (expert assessment, score); **F.2.10** Area above acceptable ambient daytime noise limits (%); **F.2.11** Area above acceptable ambient nigh time noise limits (%); **F.2.12** Summer thermal comfort conditions (SET); **F.2.13** Winter thermal comfort conditions (SET)
	F.3—Ecosystems and Landscape
	F.3.1 Green zones and recreation areas availability (%); **F.3.2** Proximity to green zones and recreation areas (m); **F.3.3** Green zones and recreation areas density (%); **F.3.4** Contamination of soil and water resources in undeveloped land (%); **F.3.4** Surface water management to prevent flooding (expert assessment, score); **F.3.6** Tree shade open area coverage (%); **F.3.7** Green roofs (%); **F.3.8** Vegetated walls and other building surfaces (m^2); **F.3.9** Presence or potential for wildlife corridors (expert assessment, score); **F.3.10** Ecological diversity (expert assessment, score); **F.3.11** Ecological sensitivity (expert assessment, score); **F.3.12** Walking or bicycling nature trails (km/1000 residents); **F.3.13** Condition of surface freshwater systems (expert assessment, score); **F.3.13** Condition of groundwater and subsurface aquifers (expert assessment, score); **F.3.14** Viability of adjacent wetlands and urban marine environments (expert assessment, score)
	G.1—Accessibility and Safety
G—Social Aspects	**G.1.1** Building accessibility by physically disabled persons (%); **G.1.2** Sidewalks and other pedestrian paths accessibility by physically disabled persons (%); **G.1.3** Barrier-free accessibility in outdoor public areas (%); **G.1.4** Accessibility to and use of public transport for physically disabled persons (%); **G.1.5** Adequacy of signage and traffic safety measures (expert assessment, score)
	G.2—Traffic and Mobility Services
	G.2.1 * Proximity of residents to public transport (%); **G.2.2** Availability of car sharing services (%); **G.2.3** Measures to limit passing through car and truck traffic (expert assessment, score); **G.2.4** * Pedestrian and bicycle network (m/100 inhabitants); **G.2.5** Availability of sheltered bicycle parking facilities (%)
	G.3—Communication Services
	G.3.1 Coverage of broadband communication network (%); **G.3.2** Access to a broadband communication network (%)
	G.4—Facilities and Services
	G.4.1 Availability and proximity of key food and retail services (%); **G.4.2** * Proximity of residents to key services (%); **G.4.3** Proximity of residents to a primary school (%); **G.4.4** Proximity of residents to a secondary school (%); **G.4.5** Proximity of residents to children's' play facilities (%); **G.4.6** Proximity of residents to leisure facilities (%); **G.4.7** Proximity of residents to indoor sports facilities (%)

Table A2. *Cont.*

G.5—Local Food Production
G.5.1 Local food production (m²/inhabitant); **G.5.2** Residents' access to and use of urban agricultural plots (%)
G.6—Management and Community Involvement
G.6.1 Involvement of residents in community affairs (%); **G.6.2** Community management of urban facilities and urban spaces (%); **G.6.3** * Community involvement in urban planning activities (expert assessment, score); **G.6.4** Individual access to community facilities and key services during off-hours (%)
G.7—Society, Culture, Heritage
G.7.1 Compatibility of urban design with local cultural values (expert assessment, score); **G.7.2** Compatibility of public open space with local cultural values (expert assessment, score); **G.7.3** Compatibility of new building designs with existing streetscapes (expert assessment, score); **G.7.4** Use of traditional local materials and techniques (expert assessment, score); **G.7.5** Maintenance of UNESCO or other protected landscapes (expert assessment, score)
G.8—Acceptance and Perception of Conditions
G.8.1 Impact of tall structure(s) on existing view corridors (expert assessment, score); **G.8.2** Panoramic and scenic routes or viewpoints (expert assessment, score); **G.8.3** Perceived safety of public areas for pedestrians (expert assessment, score); **G.8.4** Impact of commercial signage on the visual environment (expert assessment, score); **G.8.5** Impact of overhead electric distribution system on the visual environment (expert assessment, score); **G.8.6** Perceptual quality of area development (expert assessment, score); **G.8.7** Aesthetic quality of new facility exteriors (expert assessment, score)

* Key Performance Indicator (KPI).

Appendix B

Table A3. Overview of national pilots [1], KPI values and neighbourhood scale sustainability issue scores, and scenario results (where available).

Neighbourhood pre1950 with mixed uses: 1.1 km², 613 buildings, social housing (101,109 m²), other (93,099 m²); 12,607 inhabitants; recently has undergone a major urban transformation (e.g., underground railway to reunite parts of divided urban areas, new low circulation roads). **KPIs** *: 0.5/8.2/235/78.2/403/0/1.23/86/63.5/0.8/17.2/118/100/12.1/100/0   **Sustainability scores** **: 0.4/0.1/1/−1/0.9/−1/2.6; Total: 0.62
Neighbourhood for social housing: 1 km², 333 buildings, residential (226,983 m²) and commercial (3960 m²) buildings; 4455 inhabitants.   **KPIs** *: 7.2/9.7/76.3/17.4/252/3/3/34/48.7/0.9/61/22/90.1/84.9/97.3/4   **Sustainability scores** **: 4.6/0.9/−1/0.3/1.2/2.3/2.3; Total: 1.42   **Scenario**: Renovate two building; Redevelop area used as barracks (construct some new buildings and recover some existing structures); Develop new sports areas, green spaces, access roads; Introduce new services for residents; Motivate local food production for personal use; Enable participative involvement of residents   **KPIs** *: 7.2/9.7/74.3/17.4/181.1/3/3.1/33.9/48.7/0.9/60.8/22/90.1/84.9/97.3/5   **Sustainability scores** **: 4.6/0.9/−1/1.1/1.2/2.4/3.4; Total: 1.75

Table A3. *Cont.*

France (A)	**Urban renewal area**: 1.2 km^2, 3600 residential buildings, 1372 social housing; 11,000 inhabitants. **KPIs ***: 7/NA/NA/NA/NA/NA/NA/NA/50/NA/NA/8/97/63.4/100/3.5 **Sustainability scores ****: −1/NA/NA/NA/1/NA/3.8; Total: 0.57
France (B)	**New development of an industrial and urban brownfield**: 0.08 km^2, 23 buildings, residential (95,000 m^2), office (25,000 m^2) and commercial (8600 m^2) buildings; 4700 inhabitants. **KPIs ***: 0/5.9/41/7/53/32/25/8/62/8/30/11.5/80/94/100/3 **Sustainability scores ****: NA/3.8/1/5/2.2/4.1/3.8; Total: 3.30
Spain (A)	**Building block with mixed uses**: 0.013 km^2, 17 buildings, 19th and early 20th century residential (56,750 m^2), office (2468 m^2) and retail (5736 m^2) buildings, with Art Nouveau architecture; 766 inhabitants, 1750 daytime working population. **KPIs ***: 0/16.6/54.8/53.3/172.2/1.2/0.73/26.4/35.8/0.58/0.12/5/100/115.5/100/2 **Sustainability scores ****: 1.3/1.4/−1/0.9/5/1.7/4.8; Total: 1.64 **Scenario**: Form a cooperative among building owners to finance renovations, Recover space for public use, Reduce loads (insulate building envelopes, green roofs) and energy costs, Exploit renewables (building-integrated PVs), Use efficient equipment (heat pumps for heating, cooling, DHW), Reduce emissions, Remove access barriers to buildings, Improve IEQ **KPIs ***: 4.0/10.0/20.1/75/75/40/40/20/25/0.10/10/0/100/250/100/3 **Sustainability scores ****: 1.9/4.4/2.8/2.5/5.0/1.7/4.8; Total: 3.08
Spain (B)	**Neighbourhood with mixed uses**: 0.44 km^2, 625 buildings, residential (560,673 m^2), office (554,869 m^2) and retail (537,203 m^2) buildings; 11,060 inhabitants. **KPIs ***: 2.7/7.1/47.5/33.3/124.6/1.6/0.03/31.2/49.1/4.9/15.4/6.7/100/16.2/100/2 **Sustainability scores ****: 1.9/2.2/−1/−1/4.4/1.8/3.8; Total: 1.02 **Scenario**: Expand pedestrian streets; Encourage light mobility (reduce emissions, noise); Expand green and leisure areas (reduce heat island effects) **KPIs ***: 2.7/7.1/47.5/33.3/1.6/0.9/0.03/31.2/49.1/4.9/15.4/6.7/100/30/100/4 **Sustainability scores ****: 2.1/2.2/−1/0.1/4.4/2.1/4.4; Total: 1.60
Malta	**University Campus**: 0.27 km^2, 30 buildings, office (194,674 m^2) and retail (1296 m^2) buildings; Daytime population 14,000 people. **KPIs ***: 27.9/NA/16.1/103.2/233.8/64/16.4/76.5/NA/7.9/18.3/12.6/60/26/100/3 **Sustainability scores ****: 2.7/2.7/1.8/0.3/3.1/0.5/2.8; Total: 1.85
Greece	**Neighbourhood historical centre with mixed uses**: 0.271 km^2, 360 buildings, residential (55,588 m^2), office (13,000 m^2) and retail (60,203 m^2) buildings; 1329 inhabitants. **KPIs ***: 0/20.4/155.4/77.7/396.3/3/2/11/60/1/31/38/100/188.8/100/0 **Sustainability scores ****: 3/−1/0.2/4.3/0.2/0/3.9; Total: 1.27 **Scenario**: Form an energy community and work towards power generation from renewables to cover needs for municipal buildings and low income households; Exploit renewables (roof PVs on public buildings); Use efficient equipment (high performance heat pumps for heating, cooling); Connect all buildings to new natural gas central network; Replace all lighting fixtures in public spaces with LED; Increase recycling collection bins; Reintroduce municipality transportation with two lines; Improve safety **KPIs ***: 0/1.9/132.0/72.1/341.4/4/6/6.5/59.5/0.6/31/38/100/188.8/100/1 **Sustainability scores ****: 3.1/−1/2.2/4.8/1.9/0/4.1; Total: 2.24

Table A3. *Cont.*

Croatia	**Neighbourhood with mixed uses**: 0.95 km^2, 354 buildings, residential (32,376 m^2) and commercial (4000 m^2) buildings; 1368 inhabitants. **KPIs** *: 2/8/64/194/147/3/0/22/49/3.8/79/−/40/138.9/0/0 **Sustainability scores** **: 1.5/3.2/0.2/1.0/2.1/1.0/0.6; Total: 1.03 **Scenario**: Encourage electromobility (install fast-charging stations, provide electric bicycles and scooters for public use); Refurbish family houses (target 40 households, renovate building envelopes to reduce loads, install high performance heat pumps or wood pellet boilers, and install roof PVs); Implement passive measures to reduce energy demand in residential buildings and reduce urban heat island effect; Promote the use of stone or recycled materials for paving pedestrian paths; Increase natural ground areas and green open spaces to reduce flood risk; Promote the use of rainwater for irrigation; Increase local food production and exchange of eco-food production to cover half of the total local demand; Collect local biomass for composting and/or heating/cooking; Enable participative involvement of residents. **KPIs** *: 2/8/53/60/139/11/20/22/100/3.8/79/20.8/70/138.9/0/3 **Sustainability scores** **: 1.5/4.4/1.9/2.5/3.9/1.0/3.2; Total: 2.58

[1] The national pilots were performed by the CESBA MED partners: in Croatia by Energy Institute Hrvoje Požar, in France by Auvergne-Rhône-Alpes Energy Environment Agency and EnvirobatBDM, in Greece by National Observatory of Athens, in Italy by iiSBE Italia, City of Torino, and City of Udine, in Malta by University of Malta and in Spain by Government of Catalonia and City of Sant Cugat del Vallès. * **Neighbourhood KPIs**; for descriptive names see Table A1: A.1.7 (%)/B.3.3 (€/m^2/y)/C.1.1 (kWh/m^2/y)/C.1.4 (kWh/m^2/y)/C.1.7 (kWh/m^2/y)/C.2.1 (%)/C.2.7 (%)/D.1.2 (kgCO$_2$eq/m^2/y)/E.1.6 (m^3/occupant/y)/E.1.7 (m^3/m^2)/F.1.3 (%)/F.2.3 (days/y)/G.2.1 (%)/G.2.4 (m/100 inhabitants)/G.4.2 (%)/G.6.3 level(score); ** **Neighbourhood sustainability issue scores**: A-Urban systems/B-Economy/C-Energy/D-Emissions/E-Natural resources/F-Environment/G-Social aspects; Scores: −1 (below minimum), 0 (minimum acceptable performance), 1 (improved performance over minimum), 2 (significantly higher over minimum), 3 (best practice), 4 (improved over best practice), 5 (ideal performance).

References

1. Eurostat. *Energy Statistical Country Datasheets*; European Commission, DG Energy, Unit A4: Brussels, Belgium, 2019; Available online: https://ec.europa.eu/eurostat/web/energy/data/energy-balances (accessed on 10 July 2019).

2. Dodd, N.; Donatello, S.; Garbarino, E.; Gama-Caldas, M. *Identifying Macro-Objectives for the Life Cycle Environmental Performance and Resource Efficiency of EU Buildings*; Joint Research Centre, European Commission: Seville, Spain, 2015; 116p, Available online: https://www.construction-products.eu/application/files/1615/2466/0561/151222_Resource_Efficient_Buildings_Macro_objectives_WP_Final_version.pdf (accessed on 10 July 2019).

3. *Construction and Demolition Waste (CDW)*; Waste streams, Environment, European Commission: Brussels, Belgium; Available online: https://ec.europa.eu/environment/waste/construction_demolition.htm (accessed on 10 July 2019).

4. *Regional Policy*; Urban Development; European Commission: Brussels, Belgium; Available online: https://ec.europa.eu/regional_policy/en/policy/themes/urban-development/ (accessed on 15 October 2019).

5. *Circular Economy Action Plan*; Environment, European Commission: Brussels, Belgium; Available online: https://ec.europa.eu/environment/circular-economy/index_en.htm (accessed on 7 October 2019).

6. *2030 Climate & Energy Framework*; Climate Strategies & Targets; European Commission: Brussels, Belgium; Available online: https://ec.europa.eu/clima/policies/strategies/2030_en (accessed on 8 October 2019).

7. *Clean Energy for all Europeans Package*; Energy; European Commission: Brussels, Belgium; Available online: https://ec.europa.eu/energy/en/topics/energy-strategy-and-energy-union/clean-energy-all-europeans (accessed on 7 October 2019).

8. Saheb, Y.; Shnapps, S.; Paci, D. *From Nearly-Zero Energy Buildings to Net-Zero Energy Districts*; EUR 29734 EN; Publications Office of the European Union: Luxembourg, Luxembourg, 2019.

9. De Pascali, P.; Bagaini, A. Energy Transition and Urban Planning for Local Development. A Critical Review of the Evolution of Integrated Spatial and Energy Planning. *Energies* **2019**, *12*, 35. [CrossRef]

10. *UN Sustainable Development Goals*; United Nations: New York, NY, USA, 2019; Available online: https://www.un.org/sustainabledevelopment/sustainable-development-goals/ (accessed on 10 July 2019).

11. COM. *739 Communication from the Commission to the European Parliament, the Council, the European Economic and Social Committee and the Committee of the Regions*; European Commission: Strasbourg, France, 22 November 2016; Available online: https://eur-lex.europa.eu/legal-content/EN/TXT/PDF/?uri=CELEX:52016DC0739&from=EN (accessed on 7 October 2019).

12. *Urban Agenda for the EU*; European Commission: Brussels, Belgium; Available online: https://ec.europa.eu/futurium/en/urban-agenda (accessed on 5 October 2019).

13. Lafortune, G.; Zoeteman, K.; Fuller, G.; Mulder, R.; Dagevos, J.; Schmidt-Traub, G. The 2019 SDG Index and Dashboards Report for European Cities (prototype version). Sustainable Development Solutions Network (SDSN) and the Brabant Center for Sustainable Development (Telos), 2019. Available online: https://www.sustainabledevelopment.report/news/first-ever-sdg-index-and-dashboards-report-for-european-cities/ (accessed on 16 October 2019).

14. Balaras, C.A.; Dascalaki, E.G. Chapter 9.1—Energy Audits of Existing Buildings. In *Handbook of Energy Efficiency in Buildings*, 1st ed.; Asdrubali, F., Desideri, U., Eds.; Butterworth-Heinemann Elsevier: Oxford, UK, 2018; pp. 677–713.

15. Barbano, G.; Eßig, N.; Mittermeier, P.; Orova, M.; Beagon, P.; Claudi, L.; Gómez-Salcedo, J.F.; Kiedaisch, F. *Definition of Sustainable Key Performance Indicators*; NEWTREND Project, H2020 Program; European Commission: Brussels, Belgium, 2016; Available online: http://newtrend-project.eu/wp-content/uploads/2015/11/NewTREND_WP2_D2.2_KPI_GB04_V5.2.pdf (accessed on 16 October 2019).

16. Lawrence, T.; Balaras, C.A.; Means, J.K. A Comparison of How Sustainability and Green Building Standards are Being Adopted into Building Construction Codes within the United States and the EU. In Proceedings of the International Conference on Sustainable Built Environment SBE 16—Strategies—Stakeholders—Success Factors, Hamburg, Germany, 8–11 March 2016; pp. 42–51. Available online: https://pure.tugraz.at/ws/portalfiles/portal/3050178/SBE16Hamburg_ConferenceProceedings.pdf (accessed on 8 October 2019).

17. *ANSI/ASHRAE/USGBC/IES Standard 189.1—Standard for the Design of High Performance Green Buildings Except Low-Rise Residential Buildings*; ASHRAE: Atlanta, GA, USA, 2017; Available online: https://www.ashrae.org/standards-research--technology/standards--guidelines (accessed on 8 October 2019).

18. Mattoni, B.; Guattari, C.; Evangelisti, L.; Bisegna, F.; Gori, P.; Asdrubali, F. Critical review and methodological approach to evaluate the differences among international green building rating tools. *Renew. Sustain. Energy Rev.* **2018**, *82*, 950–960. [CrossRef]

19. Chethana, I.M.; Illankoon, S.; Tam, V.W.Y.; Le, K.N.; Shen, L. Key credit criteria among international green building rating tools. *J. Clean. Prod.* **2017**, *164*, 209–220. [CrossRef]

20. He, Y.; Kvan, T.; Liu, M.; Li, B. How green building rating systems affect designing green. *Build. Environ.* **2018**, *133*, 19–31. [CrossRef]

21. Sicignano, E.; Di Ruocco, G.; Stabile, A. Quali—A Quantitative Environmental Assessment Method According to Italian CAM, for the Sustainable Design of Urban Neighbourhoods in Mediterranean Climatic Regions. *Sustainability* **2019**, *11*, 4603. [CrossRef]

22. *Level(s)—Taking Action on the Total Impact of the Construction Sector, Joint Research Centre (JRC)*; Publications Office of the European Union: Luxembourg, Luxembourg, 2019; Available online: https://ec.europa.eu/environment/eussd/pdf/LEVELS_REPORT_en.pdf (accessed on 7 October 2019).

23. Martos, A.; Pacheco-Torres, R.; Ordóñez, J.; Jadraque-Gago, E. Towards successful environmental performance of sustainable cities: Intervening sectors. A review. *Renew. Sustain. Energy Rev.* **2016**, *57*, 479–495. [CrossRef]

24. iiSBE. *International Initiative for a Sustainable Built Environment*; Ottawa, ON, Canada, 2007; Available online: www.iisbe.org (accessed on 8 October 2019).

25. Balaras, C.A.; Dascalaki, E.G.; Droutsa, K.G.; Moro, A.; Barbano, G.; Chanussot, L.; Cazas, J.; Zidar, M.; Bačan, I.; et al. *D3.1.1 Transnational Indicators and Assessment Methods for Buildings and Urban Areas*; CESBA MED Consortium: Marseille, France, 2017; 409p, Available online: https://cesba-med.interreg-med.eu/fileadmin/user_upload/Sites/Efficient_Buildings/Projects/CESBA_MED/D3.1.1_Indicators_CESBA_MED_Final-V1.5_June_2017.pdf (accessed on 8 October 2019).

26. Moro, A. *D3.3.1 Testing Protocol—Assessment Methodology*; CESBA MED Consortium: Marseille, France, 2017; 100p, Available online: https://cesba-med.interreg-med.eu/results/deliverables/ (accessed on 16 October 2019).

27. Moro, A. *D3.4.1—CESBA MED SBT Building Generic Framework*; CESBA MED Consortium: Marseille, France, 2019; 78p, Available online: https://cesba-med.interreg-med.eu/results/deliverables/ (accessed on 16 October 2019).

28. Moro, A. *D3.4.2—CESBA MED SNT Urban Generic Framework*; CESBA MED Consortium: Marseille, France, 2019; 83p, Available online: https://cesba-med.interreg-med.eu/results/deliverables/ (accessed on 16 October 2019).

29. Dodd, N.; Cordella, M.; Traverso, M.; Donatello, S. *Level(s)—A common EU Framework of Core Sustainability Indicators for Office and Residential Buildings, Part 3: How to Make Performance Assessments Using Level(s)*; Joint Research Centre, European Commission: Seville, Spain, 2017; 211p, Available online: https://susproc.jrc.ec.europa.eu/Efficient_Buildings/docs/170816_Levels_EU_framework_of_building_indicators.pdf (accessed on 7 October 2019).

30. ITACA. *Istituto per l' Innovazione e Trasparenza Degli Appalti e la Compatibilità Ambientale*; Institute for Transparency of Contracts and Environmental Compatibility: Rome, Italy; Available online: http://www.itaca.org (accessed on 7 October 2019).

31. Moro, A. *D3.4.3a CESBA MED KPIs—Building Scale*; CESBA MED Consortium: Marseille, France, 2019; 30p, Available online: https://cesba-med.interreg-med.eu/results/deliverables (accessed on 16 October 2019).

32. Moro, A. *D3.4.3b CESBA MED KPIs—Urban Scale*; CESBA MED Consortium: Marseille, France, 2019; 28p, Available online: https://cesba-med.interreg-med.eu/results/deliverables (accessed on 16 October 2019).

33. Moro, A. *D3.3.2 v.1.1 Model of Decision Making Process for Sustainable Urban Areas and Public Buildings*; CESBA MED Consortium: Marseille, France, 2018; 15p, Available online: https://cesba-med.interreg-med.eu/results/deliverables/ (accessed on 16 October 2019).

34. Borgaro, P. *D4.2.2 v.1.4 Training System Framework*; CESBA MED Consortium: Marseille, France, 2018; 26p, Available online: https://cesba-med.interreg-med.eu/results/deliverables/ (accessed on 16 October 2019).

35. Borg, R.P. *D5.2.1—elearning Courses*; CESBA MED Consortium: Marseille, France, 2019; 19p, Available online: https://cesba-med.interreg-med.eu/results/deliverables/ (accessed on 16 October 2019).

36. *STRAT-Board: Territorial and Urban Strategies Dashboard*; Greece Country Fact Sheet; European Commission: Brussels, Belgium; Available online: https://urban.jrc.ec.europa.eu/strat-board/#/factsheetcountry?id=EL&name=Greece&fullscreen=yes (accessed on 15 October 2019).
37. Droutsa, K.G.; Balaras, C.A.; Dascalaki, E.G.; Kontoyiannidis, S. *D3.3.2-GR Pilot Test in Greece, Assessment Report—Hellenic Pilot Test Results*; CESBA MED Consortium: Marseille, France, 2019; 15p, Available online: https://cesba-med.interreg-med.eu/results/deliverables (accessed on 16 October 2019).
38. Torrent, L.; Wadel, G.; Sagrera, A. *D5.3.1 CESBA MED Policy Paper*; CESBA MED Consortium: Marseille, France, 2019; 15p, Available online: https://cesba-med.interreg-med.eu/results/deliverables (accessed on 16 October 2019).

MDPI

St. Alban-Anlage 66

4052 Basel

Switzerland

Tel. +41 61 683 77 34

Fax +41 61 302 89 18

www.mdpi.com

Energies Editorial Office

E-mail: energies@mdpi.com

www.mdpi.com/journal/energies